Collins

NEW GCSE SCIENCE

Science B

For Specification Modules B1, B2, C1, C2, P1 and P2

OCR

Gateway Science

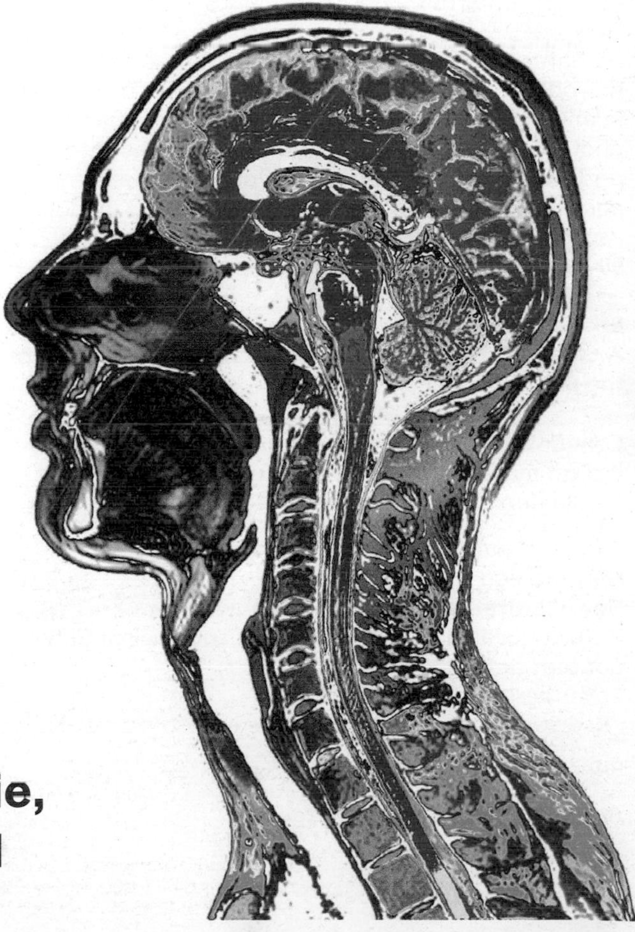

Series Editor: Chris Sherry

Authors: Colin Bell,
David Berrington, Brian Cowie,
Ann Daniels, Sandra Mitchell
and Louise Smiles

Student Book

William Collins' dream of knowledge for all began with the publication of his first book in 1819. A self-educated mill worker, he not only enriched millions of lives, but also founded a flourishing publishing house. Today, staying true to this spirit, Collins books are packed with inspiration, innovation and practical expertise. They place you at the centre of a world of possibility and give you exactly what you need to explore it.

Collins. Freedom to teach

Published by Collins
An imprint of HarperCollinsPublishers
77 – 85 Fulham Palace Road
Hammersmith
London
W6 8JB

10 9 8 7 6 5 4 3 2 1

ISBN-13 978-0-00-741537-3

British Library Cataloguing in Publication Data
A Catalogue record for this publication is available from the British Library

Commissioned by Letitia Luff
Project managed by Tammy Poggo and Hart McLeod
Production by Kerry Howie

Edited, proofread, indexed and designed by
Hart McLeod
New illlustrations by Simon Tegg
Picture research by Caroline Green and Thelma Gilbert
Concept design by Anna Plucinska
Cover design by Julie Martin
'Bad Science' pages based on the work of Ben Goldacre

Printed and bound by L.E.G.O. S.p.A. Italy

Acknowledgements – see page 320

Contents

Biology

B1 Understanding organisms

B1 Introduction	8
Fitness and health	10
Human health and diet	14
Staying healthy	18
The nervous system	22
Preparing for assessment: Applying your knowledge	26
Drugs and you	28
Staying in balance	32
Controlling plant growth	36
Variation and inheritance	40
Preparing for assessment: Research and collecting secondary data	44
Checklist	46
Exam-style questions: Foundation Tier	48
Exam-style questions: Higher Tier	50

B2 Understanding our environment

B2 Introduction	52
Classification	54
Energy flow	58
Recycling	62
Interdependence	66
Preparing for assessment: Applying your knowledge	70
Adaptations	72
Natural selection	76
Population and pollution	78
Sustainability	84
Preparing for assessment: Analysis and evaluation	88
Checklist	90
Exam-style questions: Foundation Tier	92
Exam-style questions: Higher Tier	94

Chemistry

C1 Carbon chemistry

C1 Introduction 96

Making crude oil useful 98
Using carbon fuels 102
Clean air 106
Making polymers 110

Preparing for assessment: Applying your knowledge 114

Designer polymers 116
Cooking and food additives 120
Smells 124
Paints and pigments 128

Preparing for assessment: Planning and collecting primary data 132

Checklist 134

Exam-style questions: Foundation Tier 136

Exam-style questions: Higher Tier 138

C2 Chemical resources

C2 Introduction 140

The structure of the Earth 142
Construction materials 146
Metals and alloys 150
Making cars 154

Preparing for assessment: Applying your knowledge 158

Manufacturing chemicals – making ammonia 160
Acids and bases 164
Fertilisers and crop yield 168
Chemicals from the sea: the chemistry of sodium chloride 172

Preparing for assessment: Research and collecting secondary data 176

Checklist 178

Exam-style questions: Foundation Tier 180

Exam-style questions: Higher Tier 182

Physics

P1 Energy for the Home

P1 Introduction 184
Heating houses 186
Keeping homes warm 190
A spectrum of waves 194
Light and lasers 198
Preparing for assessment: Applying your knowledge 202
Cooking and communicating using waves 204
Data transmission 208
Wireless signals 212
Stable Earth 216
Preparing for assessment: Analysis and evaluation 220
Checklist 222
Exam-style questions: Foundation Tier 224
Exam-style questions: Higher Tier 226

P2 Living for the future (energy resources)

P2 Introduction 228
Collecting energy from the Sun 230
Generating electricity 234
Global warming 238
Fuels for power 242
Preparing for assessment: Applying your knowledge 246
Nuclear radiations 248
Exploring our Solar System 252
Threats to Earth 256
The Big Bang 260
Preparing for assessment: Planning and collecting primary data 264
Checklist 266
Exam-style questions: Foundation Tier 268
Exam-style questions: Higher Tier 270

Bad Science for schools 272
Carrying out controlled assessments in GCSE science 278
How to be successful in your GCSE Science written examination 291
Glossary 300
Index 309
Internet search terms 318
Modern periodic table 319

How to use this book

Welcome to Collins New OCR Gateway Science B

The main content

Each two-page lesson has three sections corresponding to the Gateway specification:

> For Foundation tier you should understand the work in the first and second sections.

> For Higher tier you need to understand the first section and then concentrate on the second and third sections.

Each section contains a set of level-appropriate questions that allow you to check and apply your knowledge.

Look for:

> 'You will find out' boxes

> Internet search terms (at the bottom of every page)

> 'Did you know' and 'Remember' boxes

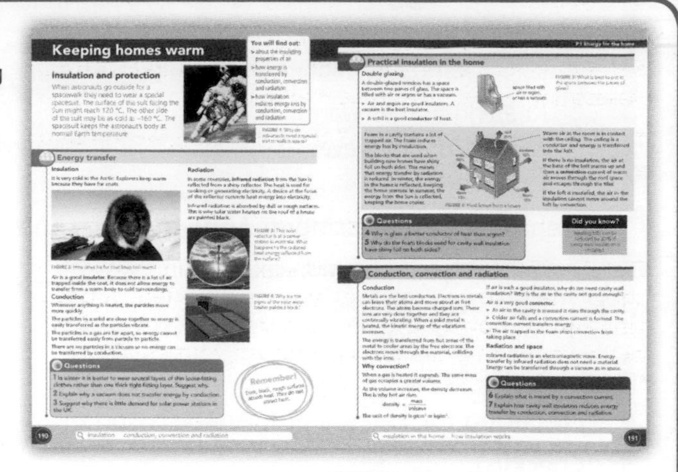
Keeping homes warm

Remember!

To cover all the content of the OCR Gateway Science specification you should study the text and attempt the exam-style questions.

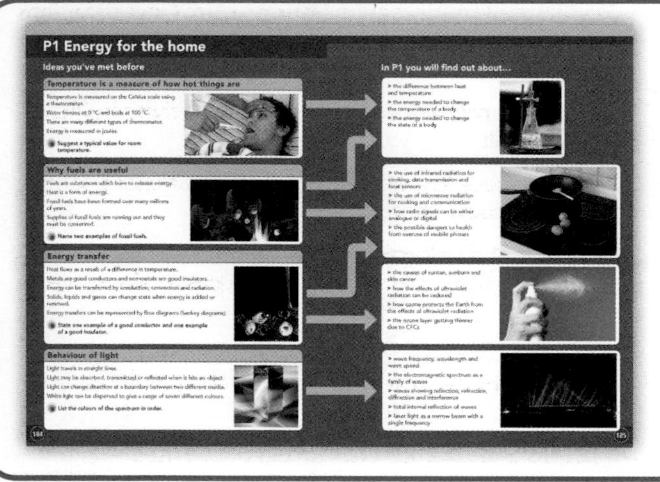
P1 Energy for the home

Module introductions

Each Module has a two-page introduction.

Link the science you will learn with your existing scientific knowledge to give you an overview before starting each Module.

Checklists

Each Module contains a checklist.

Summarise the key ideas that you have learned so far and look across the three columns to see how you can progress.

Refer back to the relevant pages in this book if you find any points you're not sure about.

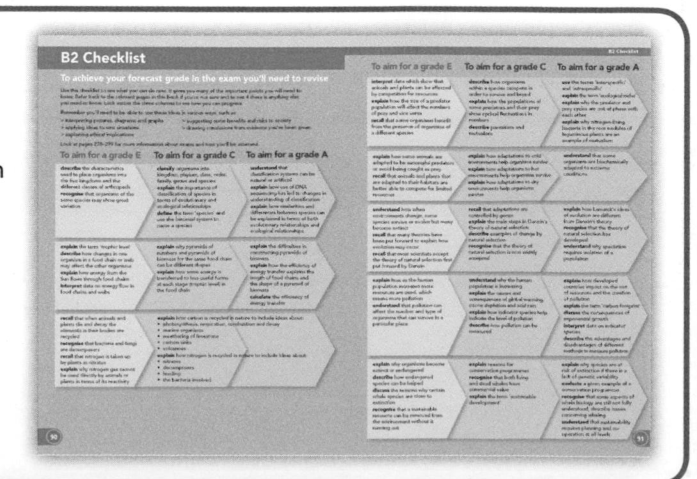
B2 Checklist

Exam-style questions

Every Module contains practice exam-style questions for both Foundation and Higher tier, labelled with the Assessment Objectives that it addresses.

Familiarise yourself with all the types of question that you might be asked.

Worked examples

Detailed worked examples with examiner comments show you how you can raise your grade. Here you will find tips on how to use accurate scientific vocabulary, avoid common exam errors, improve your Quality of Written Communication (QWC), and more.

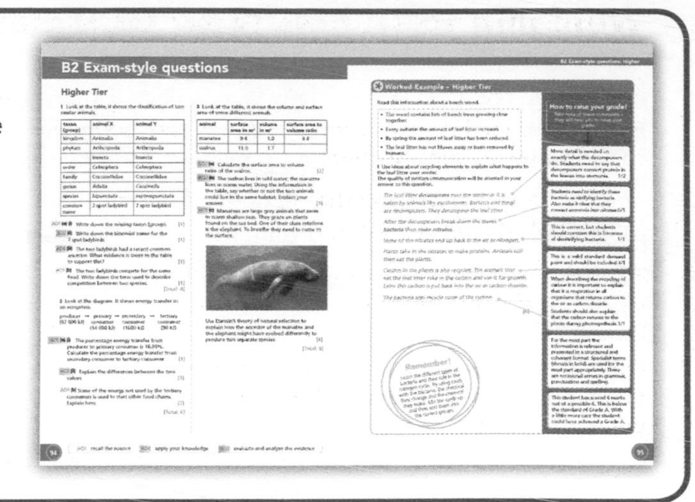
B2 Exam-style questions

Higher Tier

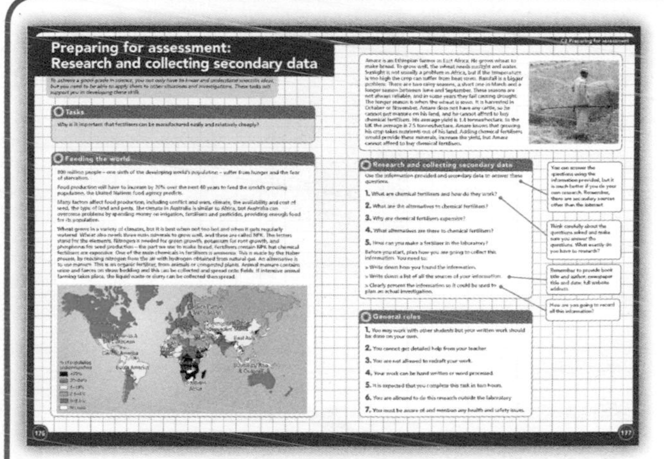
Preparing for assessment: Research and collecting secondary data

Preparing for assessment

Each Module contains preparing for assessment activities. These will help build the essential skills that you will need to succeed in your practical investigations and Controlled Assessment, and tackle the Assessment Objectives in your written exam.

Each type of preparing for assessment activity builds different skills.

> Applying your knowledge: Look at a familiar scientific concept in a new context.

> Research and collecting secondary data: Plan and carry out research using handy tips to guide you along the way.

> Planning and collecting primary data: Plan and carry out an investigation with support to keep you on the right track.

> Analysis and evaluation: Process data and draw conclusions from evidence. Use the hints to help you to achieve top marks.

Bad Science

Based on *Bad Science* by Ben Goldacre, these activities give you the chance to be a 'science detective' and evaluate the scientific claims that you hear everyday in the media.

Assessment skills

A section at the end of the book guides you through your practical work, your Controlled Assessment tasks and your exam, with advice on: planning, carrying out and evaluating an experiment; using maths to analyse data; the language used in exam questions; and how best to approach your written exam.

B1 Understanding organisms

Ideas you've met before

Organisation of cells into tissues, organs and body systems

Plant and animal cells make up tissues, for example muscular tissue, photosynthetic tissue.

Different tissues form an organ, for example heart.

Different organs work together in a body system, for example digestive system.

- Name two other examples of cells, tissues, organs and systems.

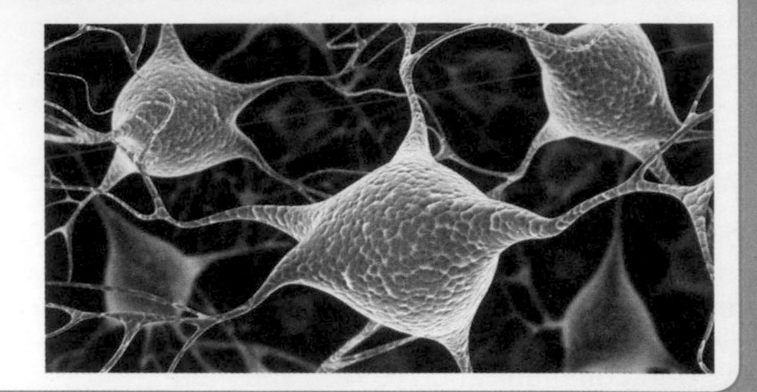

Behaviour and health can be affected by diet, drugs and disease

A healthy balanced diet contains different nutrients in the correct quantities.

Smoking is harmful to health since it contains many harmful chemicals.

Drugs can be medicines and help people suffering with pain or disease.

Some drugs are taken for other reasons and are illegal.

- Name two useful drugs (medicines) and two illegal drugs.

All living things show variation

Living organisms have similarities because they are alive.

Living organisms show differences even though they are the same species. This is variation.

Some differences are inherited and some are due to the environment.

- Name the characteristics of life shown by all organisms.

Behaviour is influenced by internal and external factors

All organisms respond to stimuli.

Some reactions are by instinct, for example reflexes.

Some actions can be learned; some actions can be conditioned.

- Describe a reflex action.

In B1 you will find out about...

- your sense organs such as your eyes
- your blood pressure
- human characteristics

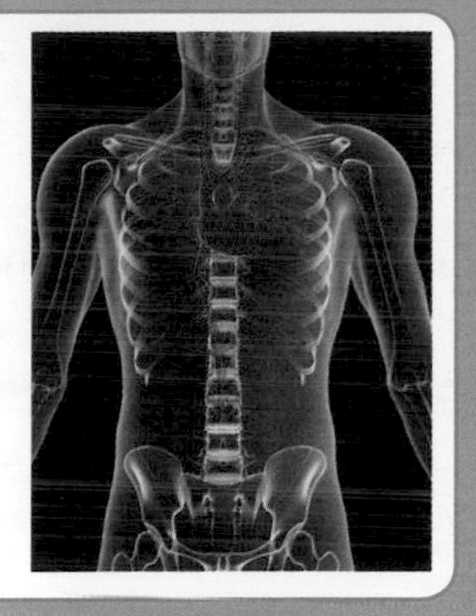

- the significance of drug classification
- consequences of protein deficiency
- transmission of infectious diseases
- genetic disorders
- alcohol content of drinks

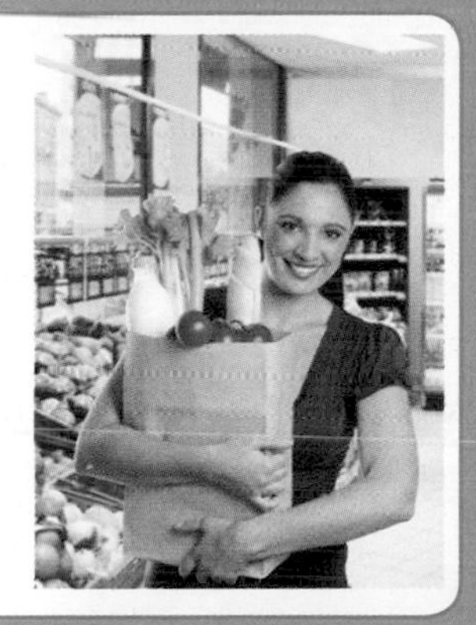

- human chromosomes
- consequences of faulty genes
- X and Y chromosomes and inheritance
- genetic diagrams

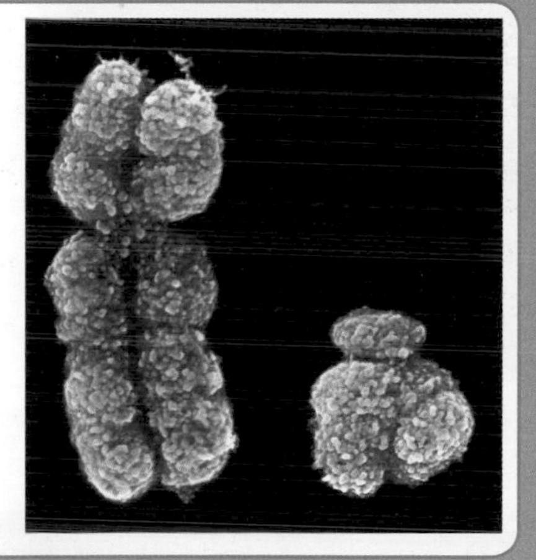

- the reflex arc
- the nerve impulse
- controlling your internal environment
- the importance of controlling blood sugar levels
- plant growth

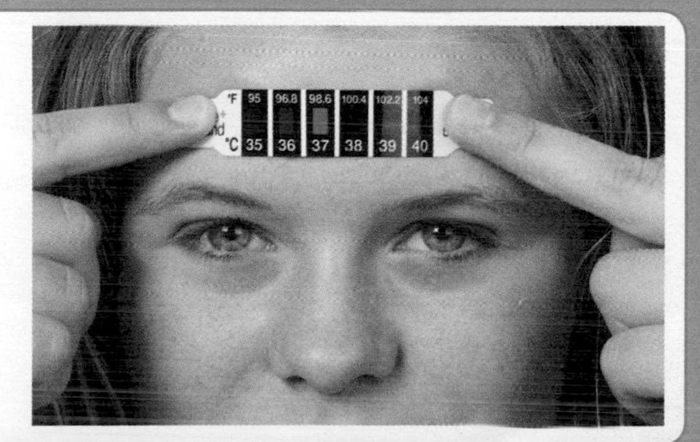

Fitness and health

You will find out:

- > why blood in arteries is under pressure
- > about measuring blood pressure
- > about the dangers of high and low blood pressure

Heart size

A human heart is about the size of an orange. It deals with 5 litres of blood. The heart of a blue whale is about the size of a small car. It deals with 10 000 litres of blood. One of the whale's arteries is so big you could crawl through it.

FIGURE 1: Whale watching. Why does a blue whale need a very big heart?

Blood pressure

Joe goes to a sports centre for football training. His **blood pressure** is checked.

> Heart muscle contracts to make a heart beat. This squeezes the blood through blood vessels called **arteries**.

> The blood in arteries is under pressure.

The blood needs to be under pressure so it can reach all parts of the body. This is important, because the blood carries food and oxygen to cells all around the body.

Remember!

Blood vessels taking blood away from the heart are called arteries. Blood vessels returning blood to the heart are called veins.

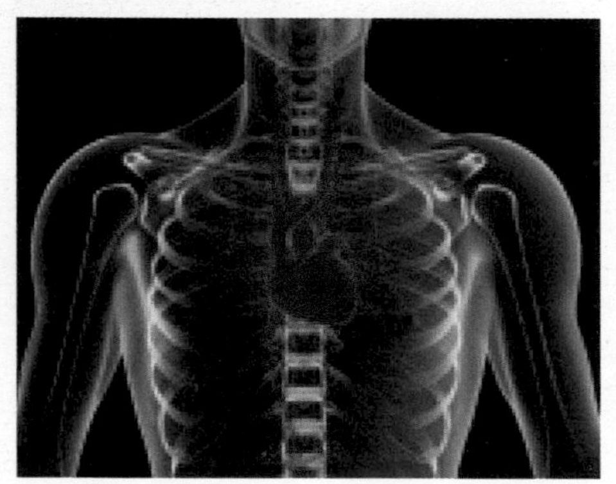

FIGURE 2: The human heart. What is it made of?

Questions

1 What type of blood vessel carries blood under pressure?

2 What puts blood under pressure?

3 Why is it important for blood to reach all parts of the body?

Fitness and health

After a lot of training sessions of physical activity, Joe thinks he is very fit. He is surprised when he catches a cold.

Joe's coach explains that being fit does not stop bacteria and viruses entering the body and causing infections.

Joe now knows the difference between fitness and being healthy:

> fitness is the ability to do physical activity

> healthy is being free from disease.

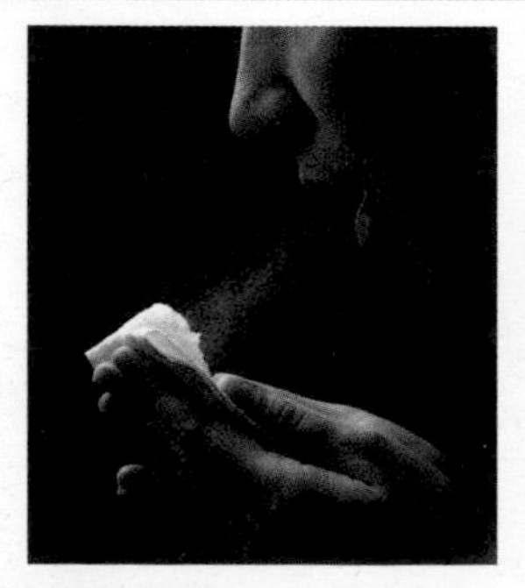

FIGURE 3: When someone coughs or sneezes, what are they spreading?

Did you know?

Many dinosaurs had problems with their blood pressure.

This dinosaur's head was 8 m above its heart. Enormous blood pressure was needed to pump blood around its body.

artery photographs   blood pressure   human heart

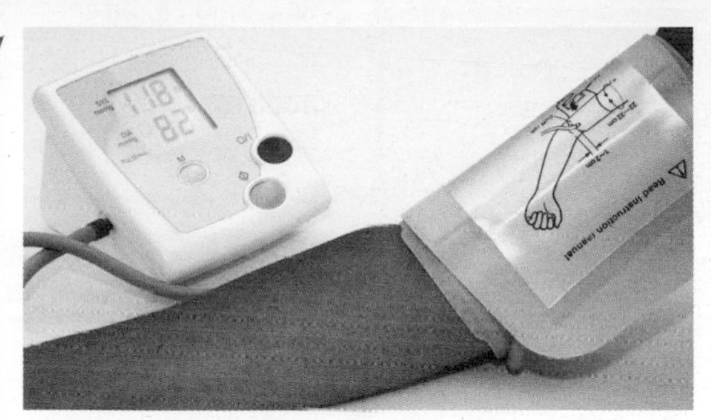

FIGURE 4: Why does Joe sit down to have his blood pressure taken?

Blood pressure questionnaire

Questions	Notes	Answers Yes	Answers No
1 Do you take regular exercise?	Strong heart muscles will lower blood pressure		✓
2 Do you eat a healthy balanced diet?	Reducing salt intake will lower blood pressure		✓
3 Are you overweight?	Being overweight by 5 kg raises blood pressure by 5 units	✓	
4 Do you regularly drink alcohol?	A high alcohol intake will increase blood pressure and damage both liver and kidneys	✓	
5 Are you under stress?	Relaxation will lower blood pressure	✓	

FIGURE 5: What changes could Joe's mum make to lead a healthier lifestyle?

Measuring blood pressure

Joe finds out that blood pressure is measured in millimetres of mercury. This is written mmHg.

Blood pressure has two measurements:

> **systolic pressure** is the maximum pressure the heart produces

> **diastolic pressure** is the blood pressure between heartbeats.

Joe's blood pressure is 95/60 mmHg, a normal result for a teenager.

Joe's mum visits her doctor every month. She is worried about high blood pressure. She fills in a questionnaire about her lifestyle.

The sports centre has measured Joe's mum's blood pressure, breathing rate, pulse recovery times and cardiovascular efficiency. There are other ways of measuring fitness:

> strength, using amount of weight lifted

> flexibility, using amount of joint movement

> stamina, using time of sustained exercise

> agility, using ability to change direction many times

> speed, in a sprint race.

FIGURE 6: Working out in a gym. What fitness levels will Joe's mum be improving?

Questions

4 What does mmHg mean?

5 What is the systolic reading for Joe?

6 Why does regular exercise help to lower blood pressure?

7 Why is it important to keep fit?

The importance of measuring blood pressure

Joe's mum has high blood pressure. If left untreated, she would be at increased risk of small blood vessels bursting, because of the high pressure exerted on the blood vessel walls.

> If a small blood vessel bursts in the brain, it is called a **stroke**. Brain damage from a stroke can result in some paralysis and loss of speech.

> If a small blood vessel bursts in a kidney, the kidney can fail.

Joe's mum's doctor prescribes a drug to lower her blood pressure. He must be careful not to lower her blood pressure too much. He knows that low blood pressure can cause problems such as poor circulation, dizziness and fainting. The blood will not be at a high enough pressure to carry the required amount of food and oxygen to muscles and other body cells.

Joe's mum enjoys dancing. She devises a fitness programme to match the demands of her dancing. She concentrates on measuring her flexibility, agility and stamina, rather than strength and speed.

Joe realises that different methods of measuring fitness are testing different aspects of fitness. If he wants to improve his fitness for cross-country running, he would concentrate on strength and stamina. However, he would not be very fit for table tennis, which requires flexibility, agility and speed.

Questions

8 How could Joe's mum avoid taking medication for high blood pressure?

9 Explain how low blood pressure can cause poor blood circulation.

Heart disease

You will find out:

> what cholesterol is and how it can cause problems

> what thrombosis is

Joe's family want to keep fit and healthy. They know that heart disease is the main cause of death in the UK.

Heart disease kills more than 110 000 people in England every year.

One adult in the UK dies from heart disease every 3 minutes.

Their doctor explains that their chances of developing heart disease are increased by factors such as:

> having high blood pressure

> smoking

> having a lot of salt in their diet

> eating a lot of saturated fats.

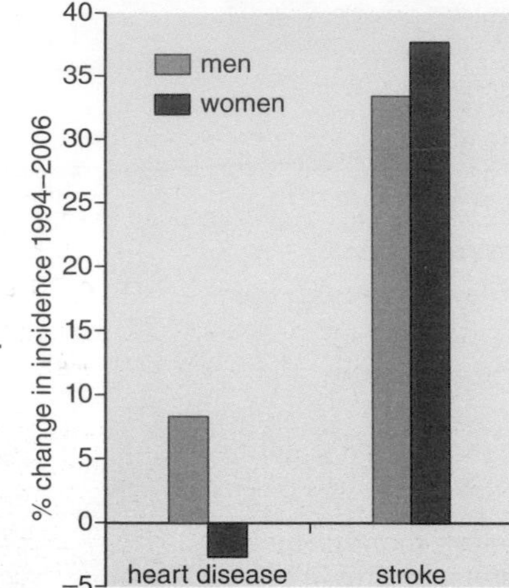

FIGURE 7: Percentage change in incidence of heart disease and stroke in the UK between 1994 and 2006.

Did you know?

The human heart pumps out about 300 litres of blood every hour. That's over 2½ million litres a year.

Cholesterol

Blood flows through arteries at high pressure.

Saturated animal fats such as **cholesterol** can stick to the walls of the arteries. This build up is called a **plaque**. Plaques can slow down or block the flow of blood.

If this happens in the arteries supplying the heart, it can cause a heart attack.

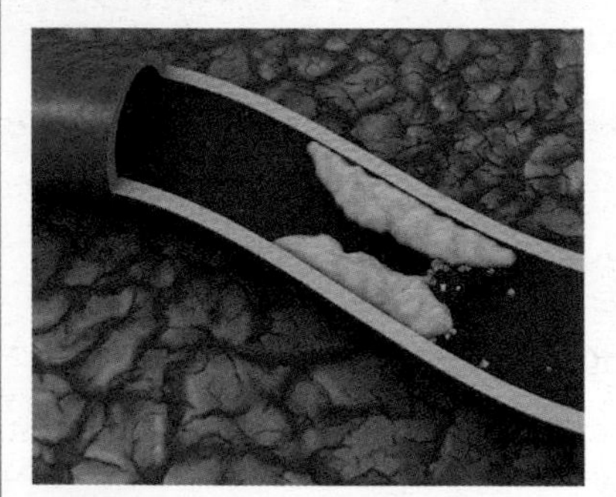

FIGURE 8: The inside of an artery. What is the yellow material attached to the wall?

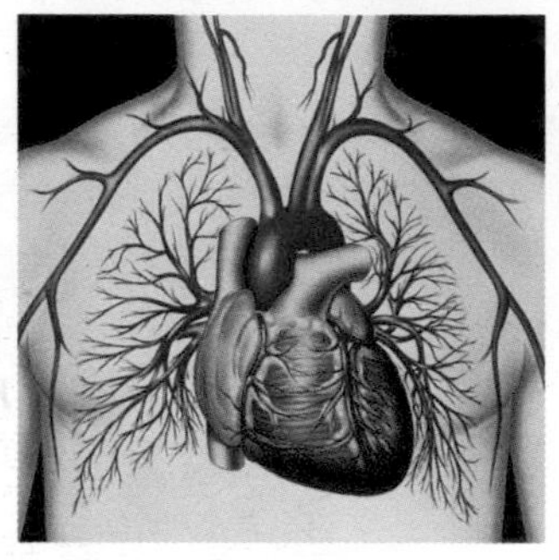

FIGURE 9: The heart and main blood vessels. Why does the heart have its own blood supply?

Questions

10 What are the changes in the incidence of heart disease in the UK between 1994 and 2006?

11 Name two ways of increasing the chance of developing heart disease.

12 What can block an artery by sticking to its walls?

13 What is a plaque?

Diet and blood pressure

A number of studies were carried out in Finland on men. The studies looked at the cholesterol levels in their blood and deaths from heart disease. The results are summarised in Figure 10.

This shows that:

> men with blood cholesterol levels below 5 made up about 10% of the population and less than 5% of coronary heart disease deaths.

> men with blood cholesterol levels above 8 also made up about 10% of the population but about 25% of coronary heart disease deaths.

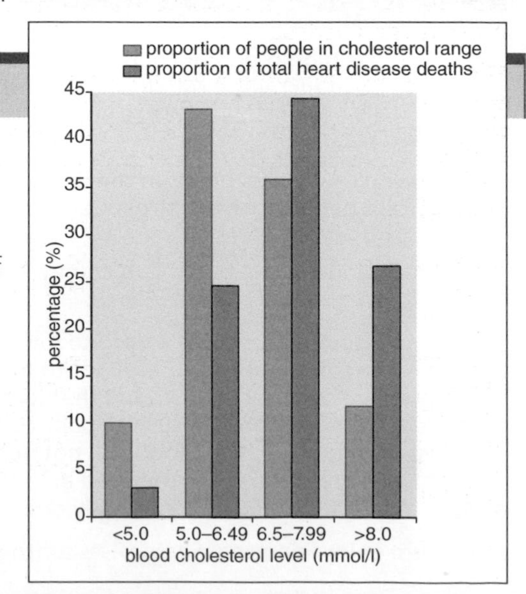

FIGURE 10: Percentage of heart disease deaths by cholesterol group.

animal fats cholesterol salt and your diet

A diet containing a lot of saturated fats and salt increases blood pressure and the risk of heart disease.

> Fats and water do not mix. This causes the cholesterol to build up on the inside of artery walls. Therefore a diet high in saturated fats can result in more plaques in arteries.

> Salt is one of the oldest-known food additives. Its chemical name is sodium chloride. When salt levels are too high, the body retains too much water. This causes a higher volume of blood to be pumped by the heart and so increases the blood pressure.

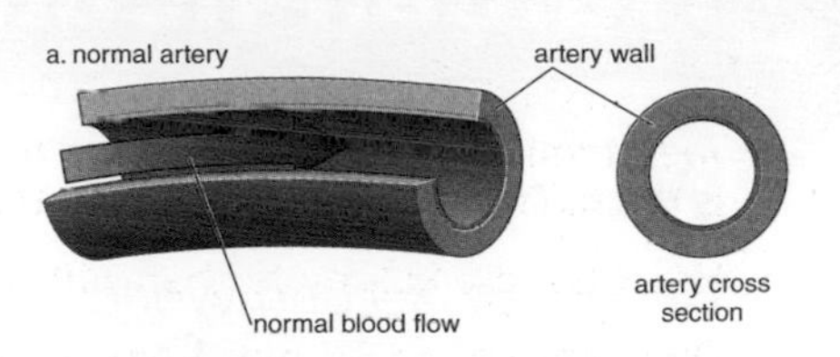

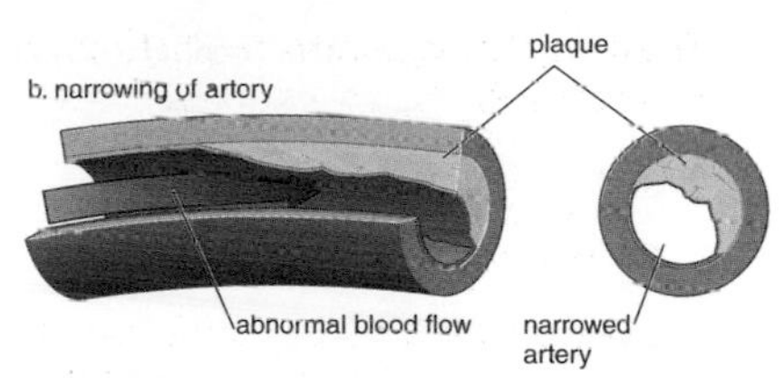

FIGURE 11: A healthy artery and an unhealthy artery. What causes the reduced blood flow in the unhealthy artery?

Smoking and blood pressure

Smoking increases blood pressure. Many chemicals, such as carbon monoxide and nicotine, are in cigarette smoke.

> Carbon monoxide reduces the amount of oxygen carried by the blood. This causes the heart rate to increase to make sure that enough oxygen gets to all parts of the body and puts extra stress on the heart.

> Nicotine is a type of drug called a stimulant. It stimulates the heart to make it beat faster and also makes blood vessels narrower.

Questions

14 Why do we eat salt when it can cause heart problems?

15 Which two chemicals in cigarette smoke affect blood pressure?

Did you know?

The average salt intake is about 9 g a day. The Government recommends a maximum of 6 g a day. This would prevent 70 000 heart attacks and strokes each year.

Coronary artery disease

The coronary artery supplies the heart muscles with oxygen and food. If the artery is narrowed or blocked, the heart muscles are not supplied with energy. This can result in a heart attack, and some of the heart muscle cells die. The artery can be bypassed using a vein transplanted from other parts of the body.

A **thrombosis** is the clotting of blood in a blood vessel causing a blockage. It can happen in both arteries and veins. If it happens in a coronary artery, it can cause a heart attack.

The most common thrombosis is deep vein thrombosis (DVT), with one in every 1000 people in the UK being affected. Travellers on long-distance flights are advised to do leg exercises to prevent DVT.

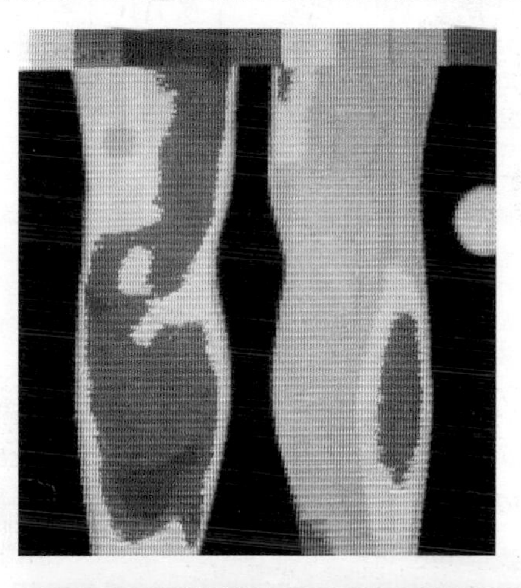

FIGURE 12: A thermogram showing DVT in one leg. The red colour shows heat. What caused one leg to be hotter than the other?

Carbon monoxide

Carbon monoxide is a dangerous gas. It is in cigarette smoke and is also produced by inefficient burning of gas in faulty gas boilers.

Red blood cells contain a pigment called **haemoglobin**. Haemoglobin combines with oxygen so it is carried around the body. However, carbon monoxide combines more readily with haemoglobin, so that the blood carries much less oxygen. This can be fatal.

Questions

16 Why is it so dangerous if there is a blockage in a coronary artery?

17 Explain how a thrombosis is different from a blockage caused by cholesterol.

18 Explain how leg exercises can help to avoid DVT.

Human health and diet

You will find out:

> why we eat food

> the importance of a balanced diet

> what foods are made of

> how foods are stored

War rations

During the Second World War (1939–45), Britain introduced food rationing. This was to make sure that any available food was shared out. Even sweets were rationed.

A 'national loaf' of brown bread was the only available bread. Most people ate less meat, fat and sugar and more vegetables.

Because their diet was better, the health of children improved. They were taller and heavier than before the war.

FIGURE 1: Do you have any relatives who remember rationing during the war?

Shopping for food

Rebecca sees a lot of information on the packets of food in the supermarket. Rebecca knows that the food she eats is called her **diet**. To get a healthy balanced diet she must eat all the foods and nutrients listed below:

> carbohydrates and fats for lots of energy

> protein for growth and repair

> vitamins such as vitamin C to prevent scurvy

> minerals such as iron to make haemoglobin

> fibre to prevent constipation

> water to prevent dehydration.

If she eats too many carbohydrates and fats and does not exercise enough she could become very overweight (**obese**). If she becomes obese she will have a higher risk of suffering from arthritis, heart disease, diabetes and breast cancer.

FIGURE 2: Shopping for a balanced diet. Why is a balanced diet important?

Questions

1 Why did children's health improve during the Second World War?

2 Why do we eat carbohydrates and fats?

3 What does 'obese' mean?

A balanced diet

A person's balanced diet will depend on their age, gender and how active they are. Rebecca knows that:

> she will need more energy than her parents because she is more active

> her brother will need more energy than her because he is much bigger

> they both need a high protein diet because they are growing fast.

Rebecca is a **vegetarian**. She does not eat meat or fish. She thinks it is wrong to kill and eat animals. Her brother is a **vegan** – he does not eat any foods from animals, including milk, cheese and eggs.

Rebecca knows that she must ensure that she eats a sensible diet. She also knows that most people do not have a perfect body shape, so she should not be influenced by the 'perfect images' she sees in magazines and on TV. She has a positive self-image. A poor self-image could lead to a poor diet and health problems.

Some of Rebecca's friends have special diets.

> Shakirah is a Muslim. She eats halal meat.

> Gina is Jewish and does not eat pork.

> Emily has a nut allergy and must not eat nuts.

FIGURE 3: What sort of diet is shown here?

balanced diet food rationing obesity

Food chemistry

Foods are made up of simple chemicals:

> carbohydrates are made up of simple sugars such as glucose

> proteins are made up of amino acids

> fats are made up of fatty acids and glycerol.

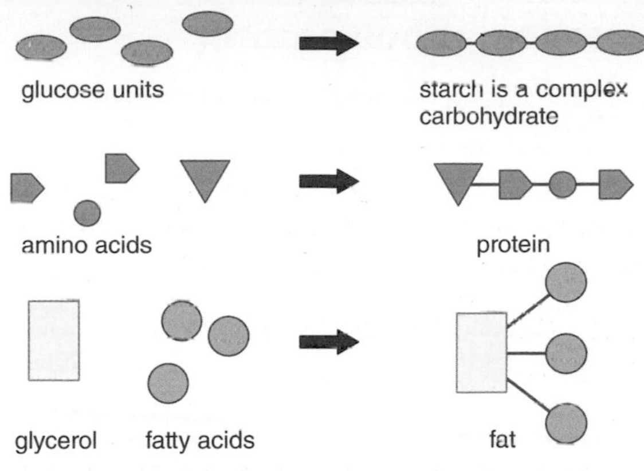

FIGURE 4: Chemistry of foods. Why are carbohydrates, proteins and fats different from each other?

Did you know?

The world's heaviest man is Paul Mason, who lives in Ipswich. He weighs 982 kg (70 stone). In 2009 he needed life-saving weight-loss surgery. A window and wall had to be removed to get him out of his house.

Did you know?

Even astronauts have to eat, despite floating in midair. A typical meal would be shrimp cocktail, chicken, vegetables and butterscotch pudding.

Their meals provide about 14 000 kJ (kilojoules) of energy each day.

Questions

4 How is a vegan diet different from a vegetarian diet?

5 What is a nut allergy?

6 What is starch made of?

Food storage

If you eat too much fat and carbohydrate, they are stored in the body.

> Carbohydrates are stored in the liver as glycogen or converted into fats.

> Fats are stored under the skin and around organs as adipose tissue.

Although proteins are essential for growth and repair, they cannot be stored in the body. However, some amino acids can be converted by the body into other amino acids. Amino acids that can not be made in this way are called essential amino acids

Proteins from meat and fish are called **first class proteins**. They contain all essential amino acids that cannot be made by the body. Plant proteins are called **second class proteins** as they do not contain all the essential amino acids.

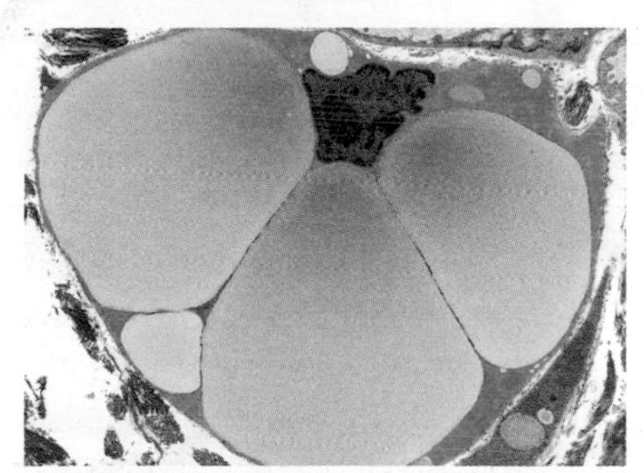

FIGURE 5: An adipose cell containing fat. Where are you most likely to find these cells?

Questions

7 What happens to excess carbohydrates?

8 Suggest what will happen to the glycogen in the liver if the body needs more energy.

9 What type of food contains first class proteins?

adipose tissue first class proteins self-image

Information about protein

Rebecca uses the internet to find out about protein.

She finds out the following information.

> Protein is used to supply energy only in an emergency such as starvation when fats and carbohydrates are not available.

> Protein is part of a balanced diet. It is needed for growth and repair of body tissue such as muscle.

> A high protein diet is necessary for growing teenagers.

> People in many parts of the world do not get enough protein in their diet. Extreme weather conditions can cause crop failure so there are fewer protein-rich foods from plants and animals. Many people may also not have enough money to buy such expensive foods.

You will find out:

- > how to calculate your body mass index
- > about the lack of protein in some countries
- > how to calculate how much protein you need

Did you know?

Each year, 8.3 million tonnes of waste food are thrown away in the UK.

Going food shopping

Rebecca looks for a variety of foods when she shops. This means she can eat a balanced diet.

FIGURE 6: Recommended amounts of different types of foods. Which two food types should we eat the most?

Rebecca looks at some fresh beef and some mycoprotein. Mycoprotein is like beef but made from a fungus.

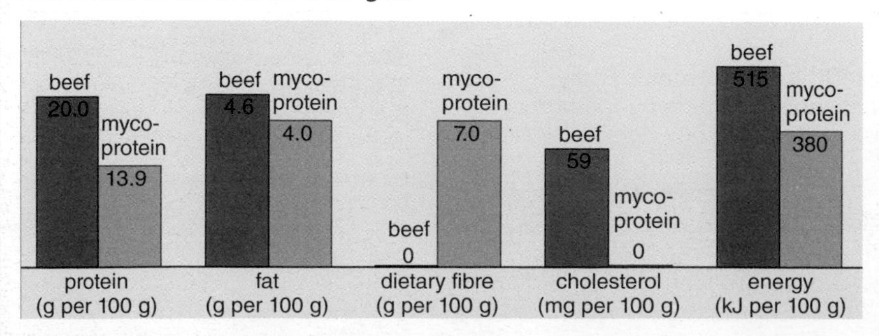

FIGURE 7: A comparison of beef and mycoprotein. Which has the highest amount of protein? Which has the highest amount of cholesterol? Which has the highest amount of fibre?

Questions

10 Why do teenagers need a lot of protein?

11 Rebecca thinks the fresh beef is best. Use only the information in Figure 7 to explain why this might be.

12 Rebecca's friend thinks mycoprotein is best. Use only the information in Figure 7 to explain why.

beef nutritional information starvation muscle

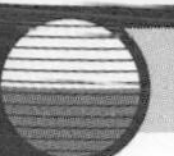

Eating the right amount of food?

Rebecca finds out that 66% of men and 53% of women in the UK are overweight.

She calculates her body mass index (BMI) using the formula:

$$\text{BMI} = \frac{\text{mass (in kilograms, kg)}}{\text{height (in metres, m)}^2}$$

She uses the BMI chart to find out if she is underweight, normal, overweight or obese.

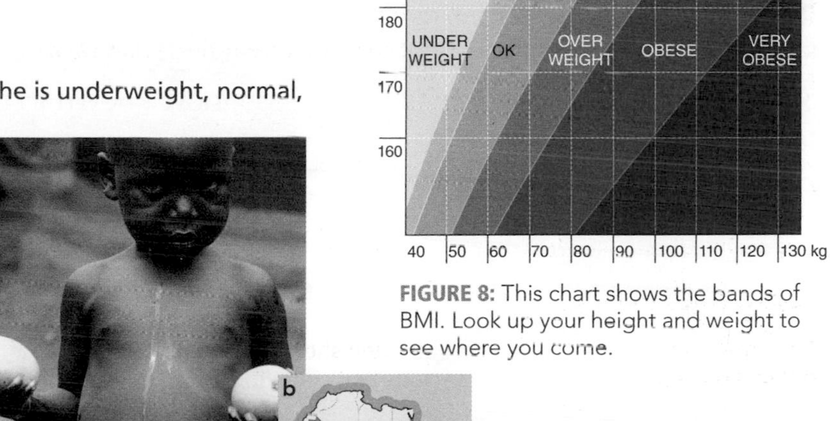

FIGURE 8: This chart shows the bands of BMI. Look up your height and weight to see where you come.

Problems with protein

Famines occur in countries that are overpopulated and where crops have failed. There is little investment in agriculture. Starvation affects thousands of people. One sign of starvation in children is when their bellies are very swollen. This means that they are desperately short of protein. With very little protein in their blood, their body cannot absorb excess water from body tissues and their abdomen swells up with the water. They have **kwashiorkor**.

Rebecca finds out that there is an estimated average requirement (**EAR**) for her amount of required protein.

The EAR is calculated using the formula:
EAR in g = body mass in kg × 0.6.

FIGURE 9:
a What causes this child's abdomen to be swollen?
b This map of Africa shows countries with famines in green.

Questions

13 Which two measurements does Rebecca need to calculate her BMI?

14 What causes kwashiorkor?

15 Rebecca has a mass of 60 kg. What is her EAR of protein?

Why is my EAR different?

Rebecca looks at the label on a tin of beans.

She knows her EAR but realises that the label has other information about protein. She finds out that the EAR is an estimated figure for an average person of a certain body mass.

The multiplication figure of 0.6 in the EAR will vary depending on:

> age, because of different rates of growth and repair
> pregnancy, because the foetus will be growing
> lactation (breast-feeding), because of needing to provide milk.

Amount of protein needed in g

adult man	adult woman	child up to 10 years
55	45	24

Nutritional information

food type	amount in tin in g
carbohydrate	53.4
protein	20.0
fat	0.8

FIGURE 10: A label on a tin of beans. Why is information like this on tins?

Questions

16 If a child eats all the beans from this can, what proportion of their EAR for protein will it be?

17 Explain why a woman breast-feeding her baby will need a high amount of protein.

body mass index   estimated average requirement

Staying healthy

You will find out:

> what causes diseases
> why some diseases are infectious
> how your body defends itself against diseases
> how vectors spread disease

Infection and disease

"Tishoo, tishoo – we all fall down." The words of the nursery rhyme are believed to describe conditions in London in 1665 during the Great Plague. The plague killed 60 000 people in London.

Sneezing was one symptom of the disease, now known to be bubonic plague.

People carried posies (bunches of flowers) to hide the smell of dead bodies.

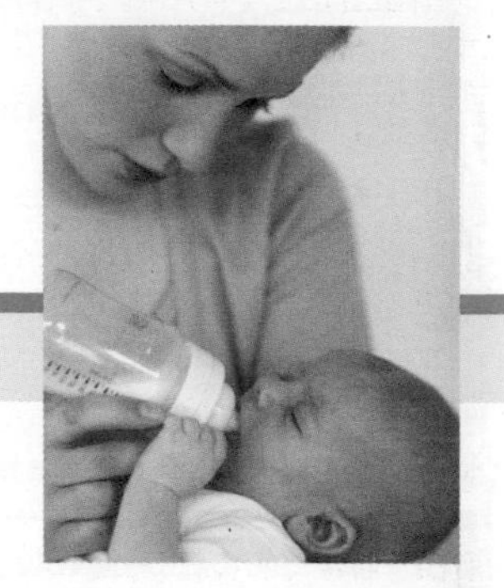

FIGURE 1: What special care must Amanda take to protect Ben from infections?

Infection in humans

Amanda knows she must sterilise baby Ben's feeding bottles.

She wants to protect Ben from **microorganisms**, which cause disease.

These are called pathogens.

There are different types of **pathogen**:

- fungi – athlete's foot is caused by a fungus
- bacteria – cholera is caused by a bacterium
- viruses – flu (influenza) is caused by viruses
- protozoa – malaria is caused by a protozoan.

These pathogens cause **infectious diseases.** These are diseases that spread to other people. If the disease spreads directly from person to person, it is described as **contagious**.

Cancer and inherited disorders, such as red–green colour deficiency, are not caused by pathogens. Diseases that are not caused by pathogens are not infectious.

Questions

1 Why does Amanda sterilise Ben's feeding bottles?

2 What type of pathogen causes flu?

3 How does your skin protect you from pathogens?

4 If your school had 1200 students, how many would have died of flu in 1918?

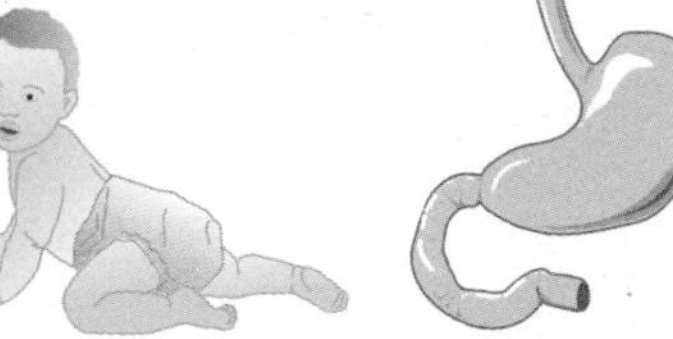

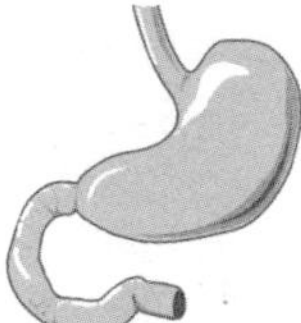

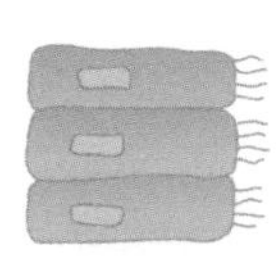

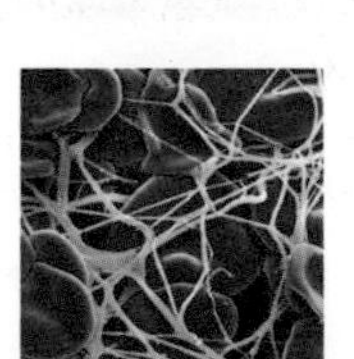

FIGURE 2: In what ways does the body protect itself from pathogens?

Did you know?

This is a virus that causes flu. It has been magnified (made bigger) under a microscope.

In 1918, a flu pandemic swept the world, killing millions of people.

In Australia, it killed half the population. In India, it killed one person in 20. In England it killed one person in 200.

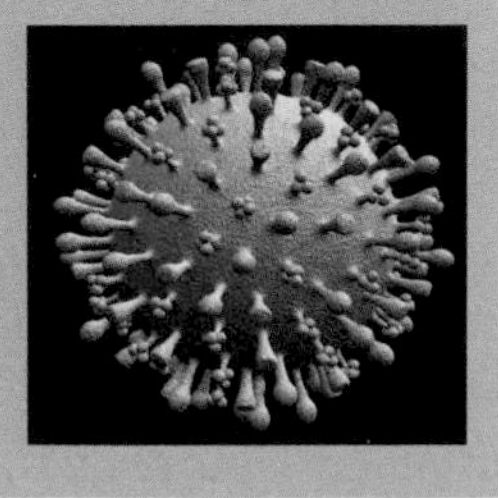

Remember!

Flies and mosquitoes do not carry diseases; they carry the pathogens that cause diseases.

cilia infectious diseases pathogens

How do vectors spread disease?

Some animals, such as the house fly and mosquito, carry microorganisms that cause disease while not suffering from the disease themselves. These animals are called **vectors**.

Amanda knows that mosquitoes are vectors for malaria.

> Malaria is caused by a protozoan called *Plasmodium falciparum*, which is carried by female mosquitoes. *Plasmodium* is a **parasite**, as it lives in a **host**, causing it harm.

> Female mosquitoes feed on human blood using sharp mouthparts to pierce the skin. Some of the *Plasmodium* parasites are left in the host's blood. Here they feed on red blood cells, causing fever in the host which is often fatal.

> Malaria-carrying mosquitoes are not currently a problem in the UK, but every year about 2000 people from the UK get malaria on holiday abroad. In 2008, six people in the UK died from malaria.

Amanda finds out that Ben could suffer from other disorders. She makes sure that:

> Ben's diet includes vitamin C to avoid scurvy

> Ben's diet includes iron to avoid anaemia

> Ben has a healthy lifestyle and a balanced diet so there is less risk of him developing diabetes and some cancers.

Not smoking reduces the risk of lung cancer, eating more vegetables and fruit reduces the risk of bowel cancer and using sunscreen reduces the risk of skin cancer.

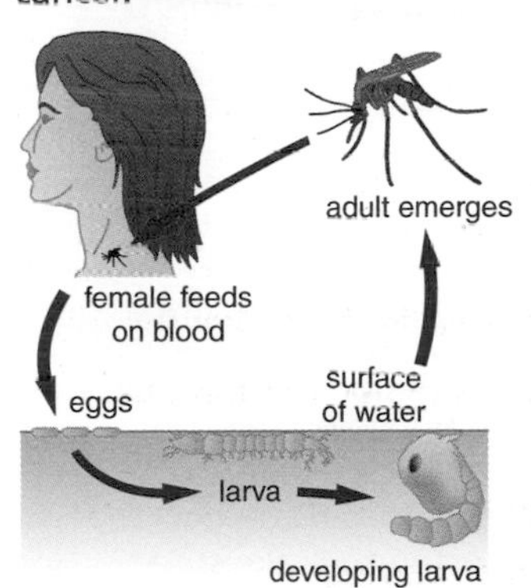

FIGURE 3: The mosquito's life cycle.

Questions

5 Why are house flies and mosquitoes called vectors?

6 Why is *Plasmodium* called a parasite?

How to help the body fight disease

Targeting vectors

Knowledge of the life cycles of the vector and the pathogen is useful in finding ways to control and prevent the spread of disease.

The mosquito has a complicated life cycle. The mosquito larva (young stage) lives in water. Mosquitoes can therefore be controlled by draining stagnant water, or putting oil on the water surface to prevent the larvae from breathing.

The adult can be killed by spraying insecticide. Some drugs, such as Lariam, can be taken by people to kill the protozoan in their blood.

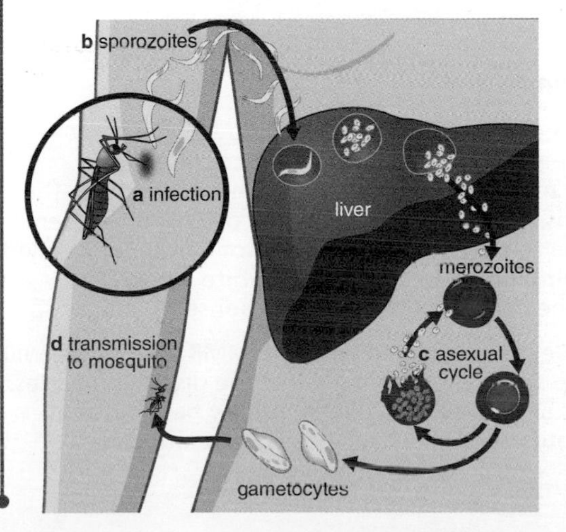

FIGURE 4: The *Plasmodium's* life cycle in diagrammatic form.

Types of cancer

Cancer is a result of cells dividing out of control. The new cells may be odd shapes and form tumours. Benign tumour cells, such as in warts, are slow to divide and are harmless. Malignant tumours are cancerous. The cells divide out of control and spread throughout the body.

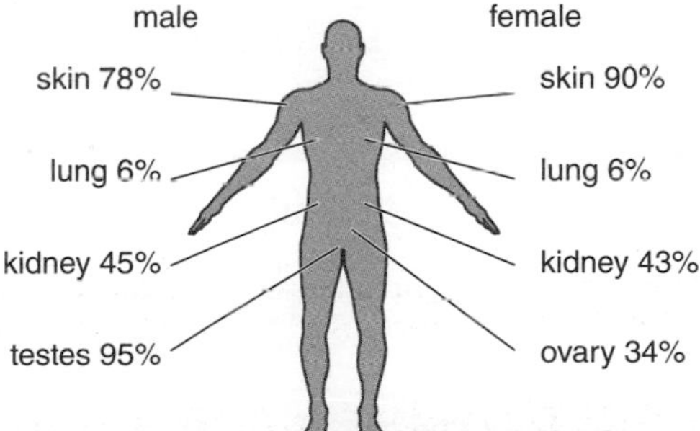

FIGURE 5: Which type of cancer has the best five-year survival rate for: **a** males; **b** females?

Question

7 Suggest why survival rates for skin cancer are different for males and females.

vectors of disease   mosquito   malaria

Fighting sickness

You will find out:

> how your immune system works

> the difference between passive and active immunity

> about the dangers of diseases around the world

> why new drugs are tested

Ben feels hot. Amanda takes his temperature. The temperature should be about 37 °C. It is higher than this. Ben has an infection.

Many pathogens that enter the body are destroyed by white blood cells.

These cells:

> surround and engulf pathogens such as bacteria

> make antibodies that stick the pathogens together.

White blood cells are part of the body's immune system to protect against pathogens.

Amanda's doctor reminds her that her baby should be immunised (vaccinated) against certain diseases such as measles, mumps and rubella.

Amanda asks the doctor if any new drug could help Ben. She knows that a new drug must be thoroughly tested before it can be used. The new drugs are tested to make sure they work and are safe to use.

Amanda knows that the risk of getting some diseases is linked to a country's climate and wealth.

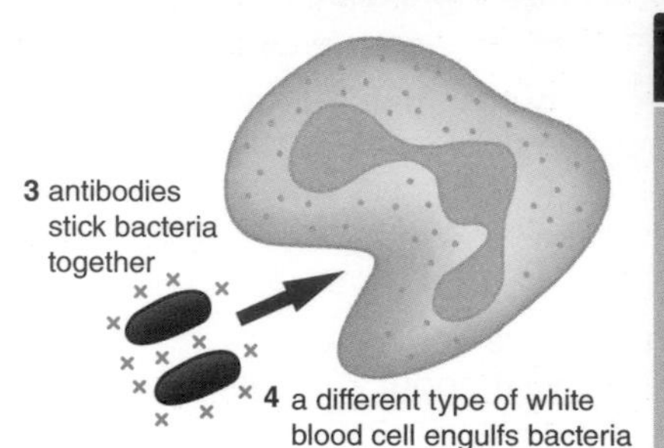

FIGURE 6: How do these white blood cells work?

Did you know?

You sneeze when the lining of your nose is irritated. Your diaphragm contracts violently. It sends out air, dust and bacteria at about 100 mph, spreading almost 10 million bacteria per sneeze.

Questions

8 What does a high body temperature show?

9 What type of blood cell destroys pathogens such as bacteria?

10 What do white blood cells make to stick pathogens together?

11 Why should Amanda's baby be immunised?

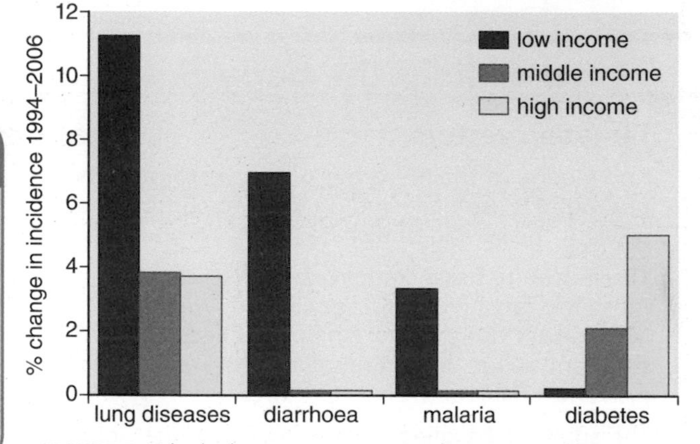

FIGURE 7: Which disease increases with more wealth?

The fight against illness

As Ben has a fever, the doctor gives him an **antibiotic**. Antibiotics are drugs that are used to treat bacterial or fungal infections.

An antiviral drug would be given to treat an infection caused by a virus. Like antibiotics, their action is specific. However, an antibiotic destroys the pathogen, whereas an antiviral drug slows down the pathogen's development.

Ben's fever is a symptom of infection. His body reacts:

> to antigens on the surface of a 'foreign invader'

> to poisonous chemicals, called **toxins**, that are produced by pathogens

> to damage to his cells caused by the pathogen.

Immunity

If the same type of pathogen invades Ben's body again, his white blood cells will recognise it and quickly make lots of **antibodies**. These lock on to the antigens. The pathogen is quickly destroyed, so Ben is immune to that disease. This sort of immunity is called **active immunity**, because Ben's white blood cells make their own antibodies in response to antigens.

When Ben is vaccinated with the MMR vaccine, he will develop active immunity against the diseases measles, mumps and rubella. Active immunity has a lasting effect but can take a few weeks to be effective.

active and passive immunity antibodies white blood cells

Some vaccinations, such as those protecting against a snakebite, give **passive immunity** by providing antibodies. An animal such as the rabbit provides the antibodies. Passive immunity is fast acting but lasts only a short time.

Testing drugs

Amanda knows that drugs, such as antibiotics, are tested in various ways. The drug must not only work, but also must not cause any further damage to the body.

Drugs can be tested on animals, on specially grown human tissue or using computer models.

Some of Amanda's friends are not happy with the idea of using animals in such tests. They say it is cruel and causes unnecessary suffering to the animals.

Questions

12 Which types of pathogens can be controlled by antibiotics?

13 Which type of immunity is needed in an outbreak of disease?

14 Suggest why people may object to testing drugs on animals.

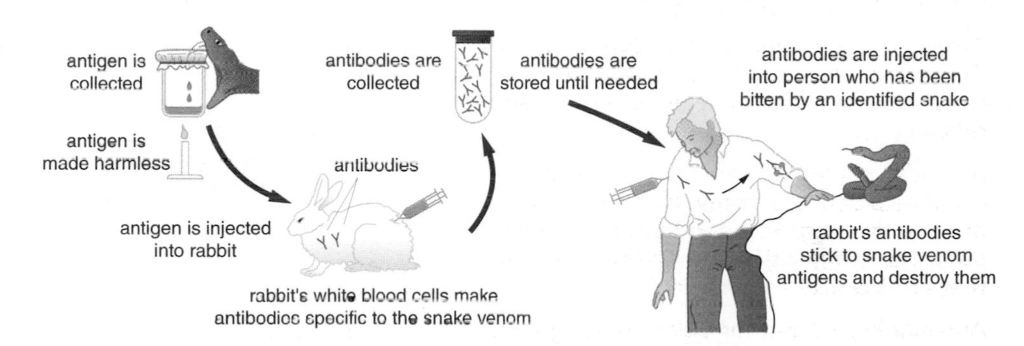

FIGURE 8: Preparation of a vaccination against snake venom.

Is vaccination the complete answer?

Each pathogen has its own **antigens** so specific antibodies are needed to protect against different diseases.

Vaccination (the immunisation process) starts with injecting a harmless pathogen. The antigens it carries trigger an immune response, with white blood cells producing antibodies.

Memory cells, a type of T-lymphocyte cell, remain in the body for a long time after meeting a pathogen. They can then produce a faster response.

Immunisation, like any medical treatment, sometimes carries a small risk to the individual, such as an adverse side effect. However, once the majority of the population has been immunised against an infection, there is little chance of the disease spreading.

Doctors are worried that some antibiotics no longer work. Excessive use of antibiotics has meant that resistant forms of bacteria are more common than the non-resistant forms. With less competition, these new strains thrive. An example is the 'superbug' MRSA (methicillin-resistant *Staphylococcus aureus*).

In testing a new drug, doctors use groups of volunteers. Some of the volunteers take the drug and some a harmless pill called a **placebo** or the best existing drug. In some trials the volunteers do not know which treatment they are receiving **(blind trial)**. In other trials the doctors do not know which treatment is used either (double-blind trial). The information is kept by other doctors.

Research has shown that knowledge of receiving a drug positively affects volunteers' reactions irrespective of whether the drug actually works. This psychological effect is called the 'placebo effect'.

Questions

15 Explain why MRSA has thrived.

16 Suggest why placebos are rarely used in human trials.

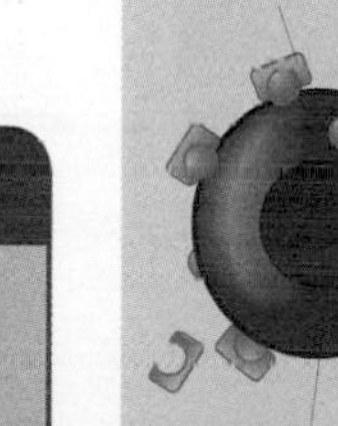

FIGURE 9: What does the body make to destroy 'foreign' antigens?

The nervous system

You will find out:
- about the body's sense organs
- how the eye works
- about problems with your vision
- about your field of vision

Guide dogs

Guide Dogs for the Blind was set up in 1931 to help blind and partially sighted people. Now there are about 5000 working guide dogs in the UK. Hearing Dogs for Deaf People trains dogs to alert people to sounds such as an alarm clock, smoke alarm, telephone or baby crying.

FIGURE 1: A guide dog leads its blind owner by applying pressure to the skin. In what other ways might a dog be a replacement for the senses?

Understanding vision

Rashid has some pieces of grit in his eye. A doctor removes the grit and bandages his damaged eye. Rashid can now only see out of one eye.

Vision in one eye is **monocular vision**. Rashid can't ride his bike because he will not be able to judge distances. When Rashid can use both eyes again he will have **binocular vision**.

Animals with mainly monocular vision have a wider field of view but are poor at judging distance. Those with mainly binocular vision have a narrower field of view but are better at judging distance.

Rashid's sight is later checked at hospital to see if he has any problems such as long sight, short sight or red–green colour blindness.

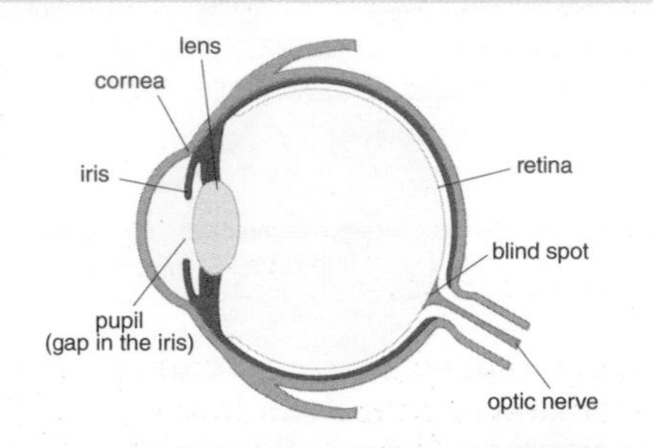

FIGURE 2: What is the coloured part of your eye called?

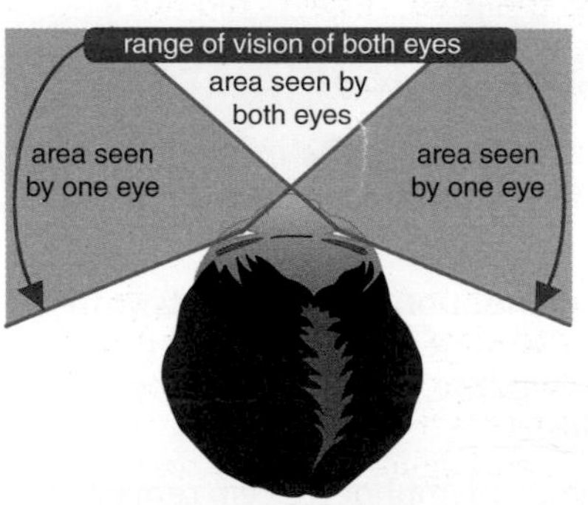

FIGURE 4: Monocular and binocular vision.

Senses

Sense organs (receptors) keep us informed of what is happening. They detect different stimuli and generate nerve impulses to transmit information to the brain.

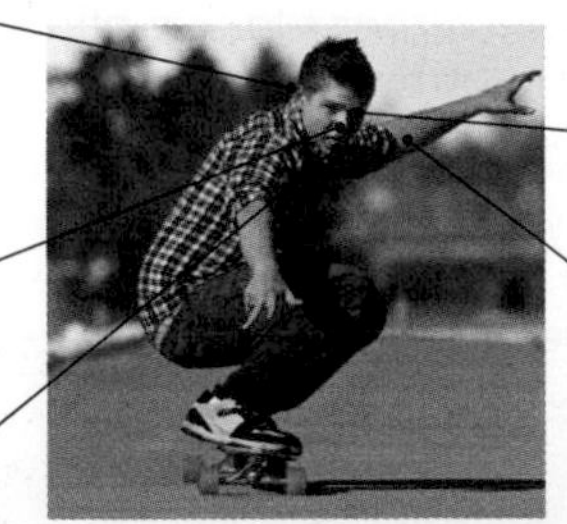

FIGURE 3: How many different sense organs do you have?

Questions

1. What is monocular vision?
2. Why is it dangerous to ride a bike with one eye covered up?
3. Which sense organ reacts to a change in temperature?

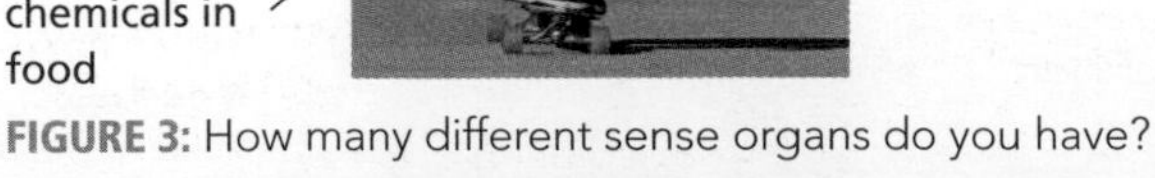

How do eyes work?

Binocular vision

Binocular vision produces two images. The brain compares both images. The more similar they are, the further away is the object they are looking at.

Seeing clearly

Rashid finds out that light rays are refracted (bent) as they go through his cornea and lens. An image is formed on his retina.

eye blind spot field of vision sense organs

The retina has special light receptors, which react to light. Some receptors react to different colours so we have colour vision. Nerve impulses are sent from the retina, through the optic nerve, to the brain. The amount of light entering the eye is controlled by the iris (the coloured part of the eye), which can change the size of the pupil.

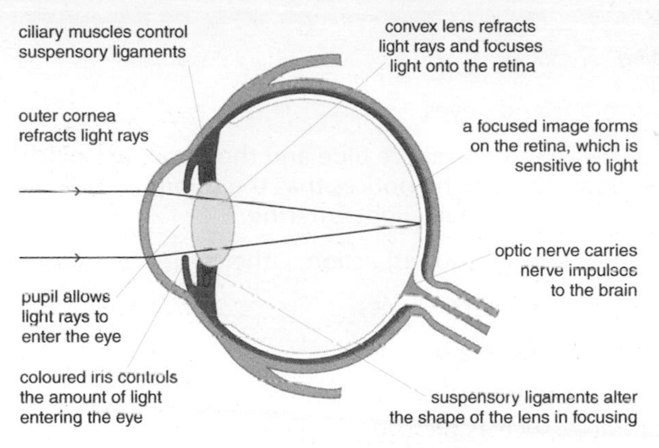

FIGURE 5: What do the cornea and lens do to light rays as they enter the eye?

Vision problems

At the hospital, Rashid finds out he is slightly short-sighted. He cannot see distant things well. His eyeballs or lenses are the wrong shape. Long-sighted people also have eyeballs or lenses that are the wrong shape.

After looking at special charts Rashid knows his colour vision is normal. Some people lack specialised cells in their retinas. This causes red–green colour blindness, which is inherited.

Questions

4 What could cause Rashid to be short-sighted?

5 Suggest why red–green colour blindness cannot be cured.

Elderly eyes in senior citizens

Rashid's grandad nearly stepped out in front of a bus. He had been looking in a shop window, then turned to cross the road. His eyes did not change focus (**accommodate**) fast enough so he did not see how near the bus was.

Since the lens is elastic it can be stretched (becoming thinner) and then return to a rounded shape.

The ciliary muscles and suspensory ligaments in the eye act together to alter the shape of the lens. The fatter the lens, the more the light is refracted. In senior citizens, like Rashid's grandad, the muscles and ligaments become less flexible.

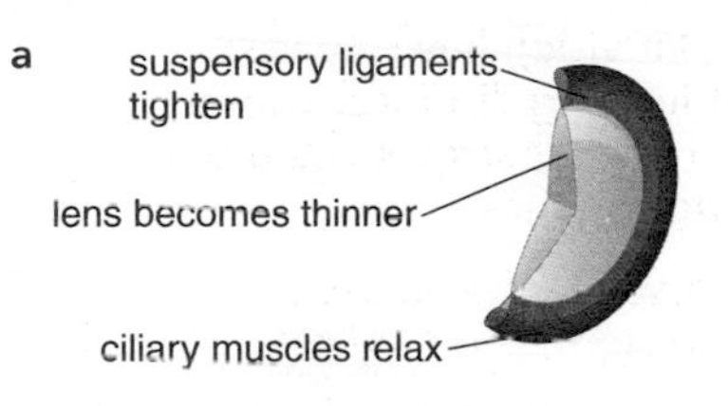

FIGURE 6: How the lens is altered for: a distant objects; b close objects. What vision problem would you have if the lens was not elastic?

Correcting short- and long-sight problems

Rashid is short-sighted. His optician recommends glasses or contact lenses with concave (curving in) lenses. Long-sighted people need convex (bulging out) lenses.

Corneal surgery is now common. Lasers alter the curvature of the cornea to correct focusing problems.

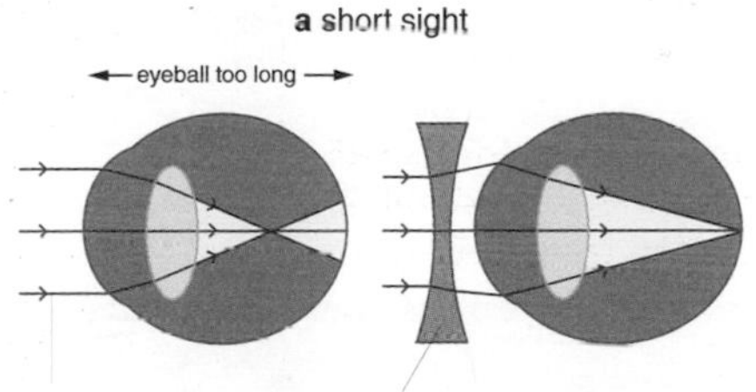

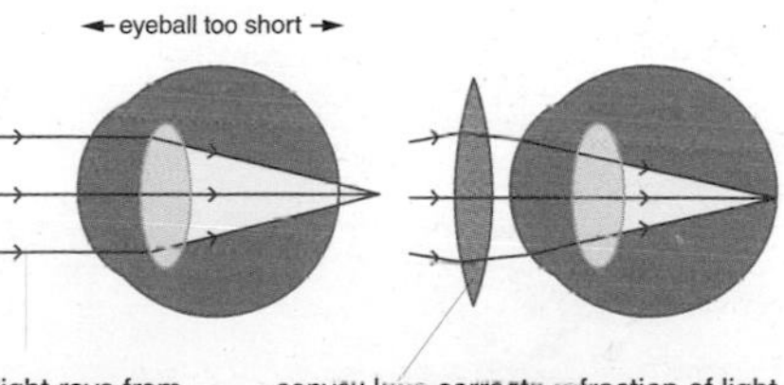

FIGURE 7: How do concave and convex lenses correct short and long sight?

Questions

6 Why might elderly people have problems with their vision?

7 Why do changes take place in the eye when a person goes from a dark room to daylight?

Reflexes

Rashid looks into a friend's eyes.

He notices that his friend's irises are blue and the pupils are black. When a bright light goes on, he notices that the pupils immediately get smaller to prevent too much light entering.

This is a **reflex** (not thought about) action. Other reflex actions are:

> knee jerk (tapping below the knee)

> taking a hand away from a hot plate.

Reflex actions protect us because they are fast and automatic. Voluntary responses, such as picking up a pen, are under conscious control of the brain.

Information is carried round the body as nerve impulses (electrical signals) in nerve cells called neurones. These form our nervous system.

> The **central nervous system (CNS)** is made up of the brain and spinal cord.

> The **peripheral nervous system** is made up of nerves to and from the brain and spinal cord.

You will find out:

> about the parts of your nervous system

> how reflex actions work

> how nerve impulses are transmitted

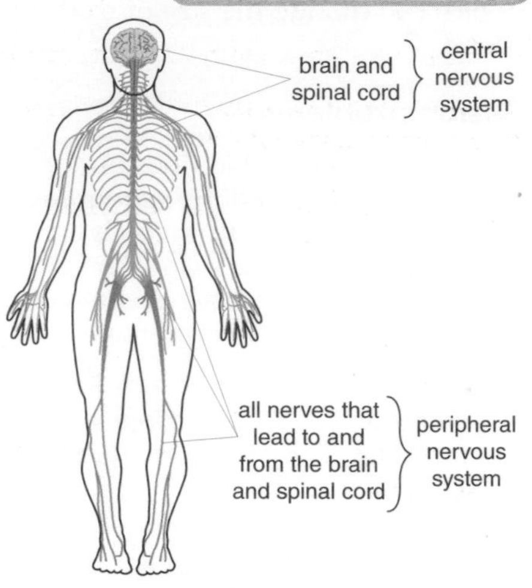

FIGURE 8: How is information carried in the nervous system?

Did you know?

Sprinters are allowed 0.1 seconds to react. A faster time means they have 'jumped the gun' and the race has to restart.The first false-starting sprinter is disqualified and the race restarted.

Questions

8 What happens to your pupils in bright light?

9 Why do your pupils need to be able to change size?

10 Name two other reflex actions.

11 What do neurones carry?

Spinal reflex

A spinal reflex has a number of stages.

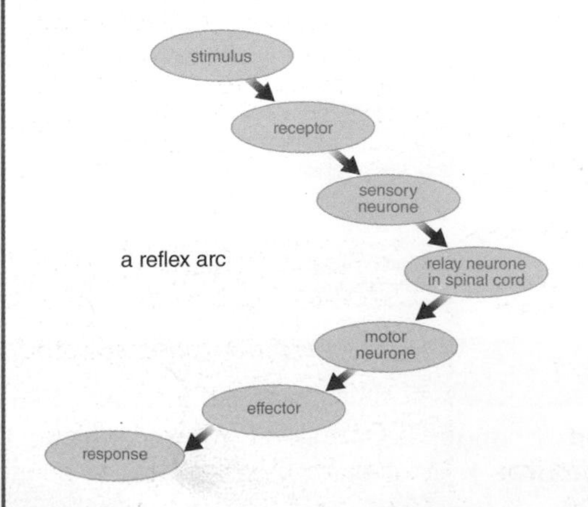

FIGURE 9: Is the brain involved in a spinal reflex?

Rashid helps at lunchtime. Unfortunately he picks up a hot plate. By reflex he immediately drops it and then examines his fingers. The high temperature was the stimulus; dropping the plate was the response.

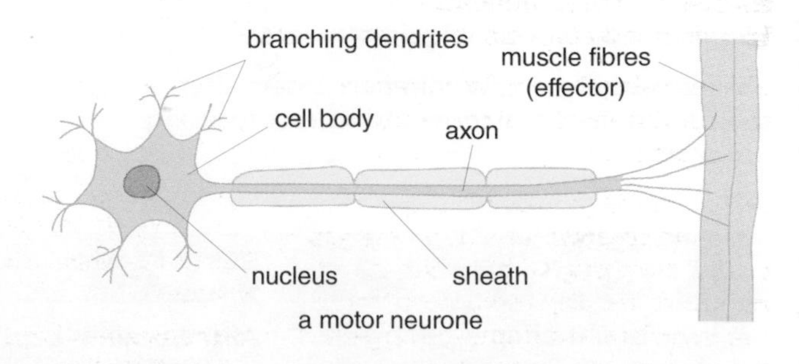

FIGURE 10: In which direction do impulses travel in motor neurones?

neurone cell reaction times reflexes

Dropping a hot plate is an example of a spinal reflex. The reaction takes place without the brain thinking about it. Rashid's brain also receives nerve impulses from pain receptors in his fingers so he examines them to see if they are damaged. The nerve impulse in a reflex arc is carried in neurones.

> **Sensory neurones** carry impulses away from sensory receptors.

> **Motor neurones** carry impulses to an effector (muscle or gland).

So when Rashid dropped the plate, impulses travelled through **axons** of his neurones in a spinal reflex.

Did you know?

Nerve impulses can travel about 100 m/s. A nerve impulse can therefore get to all parts of the body in about 0.02 seconds.

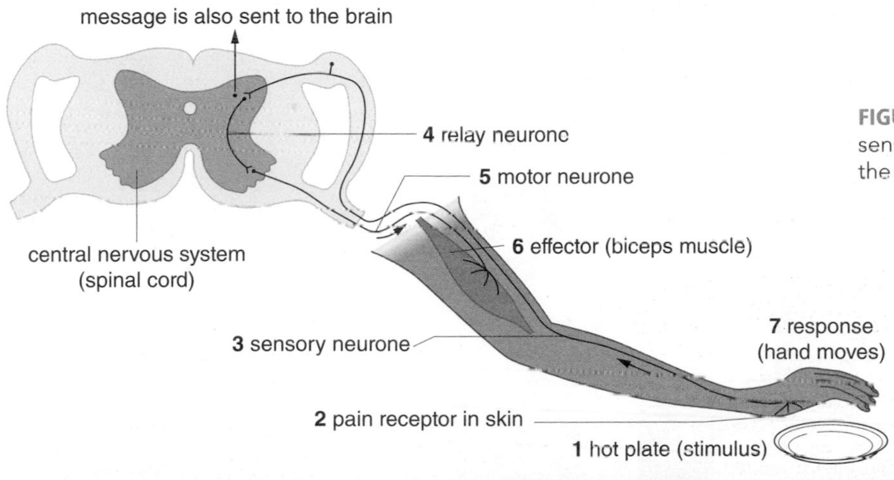

FIGURE 11: A relay neurone connects the sensory neurone to the motor neurone in the spinal cord.

Questions

12 Why are reflex actions very fast?

13 Which part of a neurone carries the nerve impulse?

14 Why is it important for the brain to be kept informed of a reflex action?

Neurones

Neurones are adapted to carry and pass on nerve impulses quickly.

Neurones:

> are long (some nearly 2 m long)

> have branched endings (dendrites) to pick up impulses

> are insulated by a fatty sheath, so that the electrical impulses do not cross over.

Signals have to pass from one neurone to another. There is a gap called the **synapse** between two neurones to allow some control of the impulses. An impulse triggers the release of transmitter substances, such as **acetylcholine**, which diffuse across the synapse.

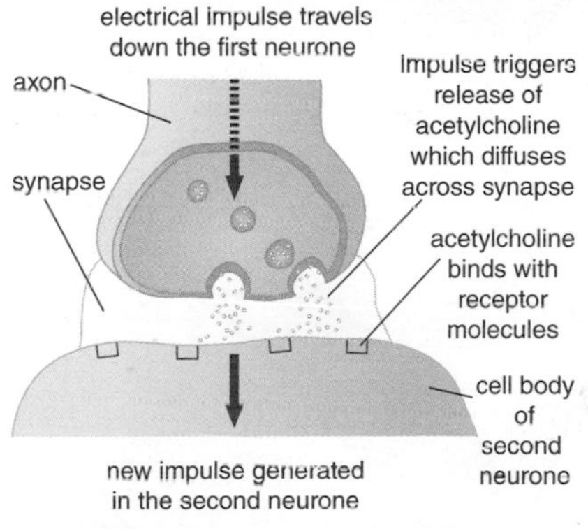

FIGURE 12: What is the purpose of synapses?

As the brain is kept informed during a reflex action, it is possible to modify or 'condition' some reflexes.

Did you know?

Remember sensory neurones carry impulses away from a sensory receptor; motor neurones carry impulses to effectors, such as muscles, so you move.

Questions

15 How are neurones adapted to their function?

16 Describe how an impulse travels from a receptor to an effector.

Preparing for assessment: Applying your knowledge

To achieve a good grade in science, you not only have to know and understand scientific ideas, but you need to be able to apply them to other situations and investigations. These tasks will support you in developing these skills.

The polio vaccination

Polio is a disease that can be prevented but not cured. Teams of scientists came up with vaccines in the 1950s and since then it has been virtually eradicated in North America and Europe.

The vaccination meant that it was then almost impossible for Alice to catch polio, even if the virus entered her body.

Alice is vaccinated against polio.

The vaccine goes into the intestines. It contains a virus, which is multiplying and entering the bloodstream.

The virus enters the bloodstream. It is the job of the white blood cells to kill the virus, but they are caught unprepared. They have never come across anything like this before. Alice's current antibodies are useless.

This type of polio virus doesn't cause the disease though, and soon Alice's white blood cells are busy making new specific antibodies to deal with it.

Now Alice's body is ready to fight any dangerous polio virus that attacks her – she has the right type of antibodies ready and waiting.

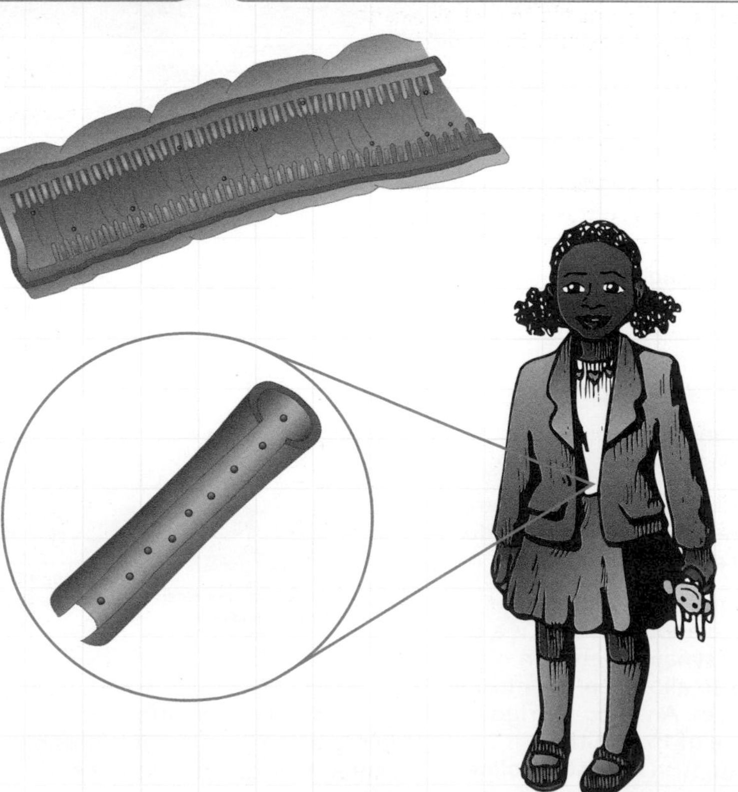

Task 1

What would happen if the polio virus got into Alice's body after she had been vaccinated?

Task 2

Vaccination involves putting a 'live' virus into the body. How does the body recognise a virus?

Task 3

How does the body respond to a virus?

Task 4

Why is the body unable to use its existing antibodies to deal with this virus?

Task 5

How is the body now protected?

Maximise your grade

Answer includes showing that you can...
Use the term immunisation.
Recall that immunisation gives protection against certain pathogens.
State that receiving the polio vaccine means that the person is protected from the disease in the future.
Describe how the body reacts to the polio virus by producing antibodies.
Describe how immunity to polio comes from prior infection.
Explain how being infected by polio causes the production of specific antibodies.
Explain that polio has its own antigens and that specific antibodies are needed.
Explain how immunisation works.
As above, but with particular clarity and detail.

Drugs and you

You will find out:
- > about how drugs can be useful or harmful
- > how and why drugs are classified
- > what harmful drugs do to the body

The discovery of aspirin

In 1897, Felix Hoffman was trying to develop a drug to help his father, who was suffering from arthritis. He succeeded and developed aspirin. Aspirin is a painkiller.

Today over 100 billion tablets of aspirin are made each year.

In recent years scientists have discovered other uses for aspirin. It reduces the risk of some cancers and helps to prevent heart attacks.

It has earned the name 'wonder drug'.

FIGURE 1: Why are soluble pills a good way of taking drugs?

Harmful and useful drugs

Useful drugs help the body. Insulin helps diabetics. Penicillin kills bacteria. You need a doctor's prescription to get many useful drugs. However, misusing these drugs could harm you.

Different types of drugs and their effects

Harmful drugs change how you behave and feel. They damage your health.

Alcohol and cannabis are harmful drugs.

> **Depressants** slow down the workings of the brain.

> **Hallucinogens** distort what a person sees and hears.

> **Painkillers** stop nerve impulses so no pain is felt.

> **Performance enhancers** develop muscles.

> **Stimulants** speed up the workings of the brain and help people with depression.

A person can become **addicted** to a drug, such as nicotine in tobacco. This means it is very hard to give it up.

When a person has been taking a drug for a long time they find they need to take larger doses of the drug to get the same effects. This is called tolerance.

When a person decides to try to give up a drug addiction they can join a rehabilitation group that helps them to stop taking the drug.

The person suffers **withdrawal symptoms**, such as being bad tempered, if they do try to give the drug up.

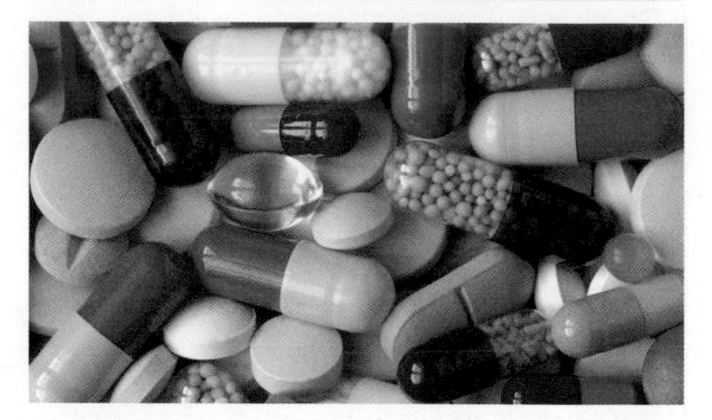

FIGURE 2: Suggest why prescription drugs are different shapes, sizes and colours.

Questions

1 Name three things that harmful drugs do to you.

2 Penicillin is used to treat some diseases. What do you need before you can get this drug?

3 What type of drug would you take for a toothache?

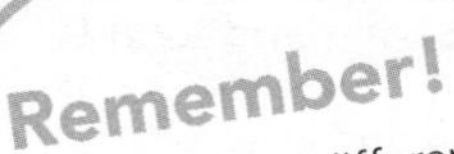

Remember!

There are many different drugs. Some are useful and some are harmful.

Did you know?

The NHS spends about £22 million a day on prescription drugs. We each pick up on average 16 prescriptions a year – twice as many as 20 years ago.

Drug classification

Drugs are classified by law based on how dangerous they are.

CLASS A	CLASS B	CLASS C
7 years in prison* and a fine for possession	5 years in prison* and a fine for possession	2 years in prison* and a fine for possession
life in prison for supplying	14 years in prison for supplying	14 years in prison for supplying
heroin, methadone, cocaine, ecstasy, LSD, magic mushrooms	amphetamines, barbiturates, cannabis	anabolic steroids, Valium, temazepam

*These are the maximum prison sentences.

Examples of drugs

FIGURE 3: Depressants such as alcohol, solvents and temazepam

FIGURE 4: Hallucinogens such as LSD.

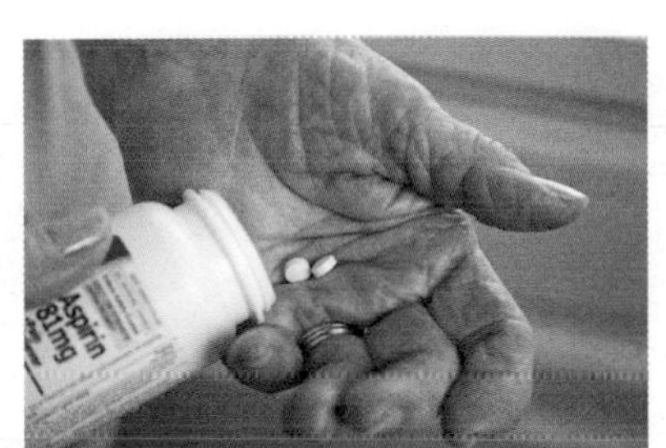

FIGURE 5: Painkillers such as aspirin and paracetamol.

FIGURE 6: Performance enhancers such as anabolic steroids.

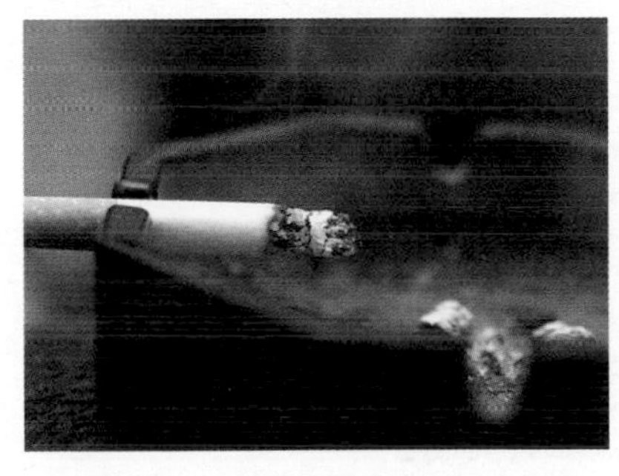

FIGURE 7: Stimulants such as nicotine, ecstasy and caffeine.

Questions

4 Which class of drug (A, B or C) is the most dangerous?

5 Aspirin and heroin are both painkillers. Which one is illegal?

6 Suggest why the penalties for supplying drugs are greater than for possession.

Effect of drugs on the nervous system

Nicotine is a stimulant. It affects synapses (junction of two neurones) by causing more neurotransmitter substances to cross to the next neurone and bind to the receptor molecules.

Alcohol is a depressant. Depressants bind with receptor molecules so blocking the transmission of impulses. See Figure 12 on page 25 for more detail on nerve impulse transmission across a synapse.

Questions

7 How does nicotine affect the nervous system?

8 Explain why the effects of alcohol are different from the effects of nicotine.

What is in tobacco smoke?

The main chemicals in tobacco smoke are:

> carbon monoxide – stops red blood cells from getting oxygen and increases risk of heart disease

> nicotine – is an addictive drug

> tar – irritates and causes cancer

> particulates – collect in lungs, blocking them.

Many diseases are linked to smoking.

Some chemicals get into the blood and can cause heart disease, emphysema, bronchitis and cancer of the mouth, throat, oesophagus or lungs.

In February 2003, a ban was placed on advertising tobacco products in newspapers, magazines and posters. In July 2007, another ban was placed on smoking in work and public places in England.

You will find out:

> about the dangers of smoking and drinking alcohol

> about damage caused by smoking

> how alcohol affects behaviour

FIGURE 8: This is a breathalyser. What is it measuring?

What are the effects of drinking alcohol?

Drinking alcohol has short-term effects.
A person who has drunk alcohol may:

> do silly things and find it hard to talk clearly

> find it difficult to coordinate their movements and easily lose their balance

> have blurred vision and feel sleepy

> have increased blood flow to skin.

No wonder there is a legal limit to the amount of alcohol car drivers and pilots can have in their bloodstream.

There are long-term effects too. A person may suffer from damage to the liver and brain.

Did you know?

The legal UK limit is 80 mg alcohol/100 ml of blood. Finland, Hungary and Romania have a zero limit…no alcohol at all.

Questions

9 What is the legal limit for blood alcohol levels for drivers?

10 Suggest why it is illegal to advertise tobacco products in magazines, newspapers and on posters.

11 Name the four main chemicals in tobacco smoke.

12 What chemical makes a person addicted to smoking?

The effects of tobacco smoke and alcohol

Smoker's cough

Cells that line the trachea, bronchi and bronchioles are called epithelial cells. Some cells have tiny hairs called cilia and others make sticky mucus.

Cigarette smoke stops the cilia from moving and makes the goblet cells produce extra mucus. Dust and particulates collect and irritate the cells, causing a 'smoker's cough'. Smokers cough to move this mess upwards so it can be swallowed.

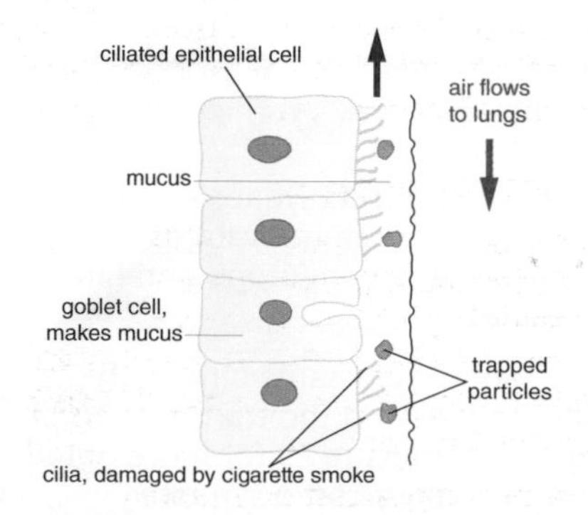

FIGURE 9: What causes smoker's cough?

Alcoholic drinks

The alcohol content of a drink is measured in **units of alcohol**.

Different drinks contain different amounts of alcohol. One unit is equivalent to 10 ml of pure alcohol.

The legal limit of alcohol when driving is 80 mg of alcohol per 100 ml of blood. A breathalyser is used as a roadside test to see if a more accurate blood test is required.

Data on alcohol-related vehicle accident deaths in the UK show a welcome reduction from 1550 to 480 in the last 20 years. This could be due to the use of breathalysers or better car design.

The information on ambulance call-outs linked to alcohol-related injuries is also interesting (see Figure 11).

Drinking alcohol also increases reaction times. The car travels further before the driver reacts. Researchers have found that the average reaction time increases by 0.15 s.

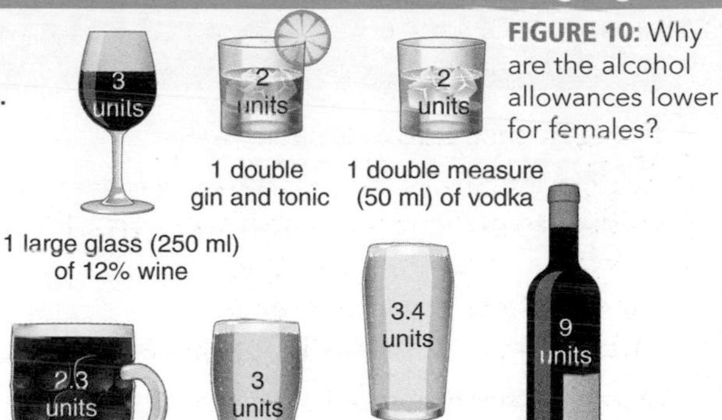

FIGURE 10: Why are the alcohol allowances lower for females?

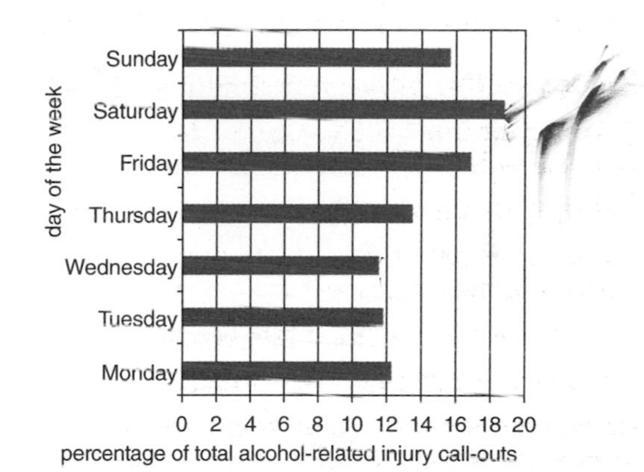

FIGURE 11: On which day do ambulance crews receive the fewest alcohol related injury call-outs?

Questions

13 Why are cilia in the trachea important?

14 George drinks beer (ale). What is his maximum recommended amount per week?

15 Why do alcohol-related injury call-outs vary during the week?

16 A car is travelling at 31.3 m/s. When stopping, how much further will the car travel when the driver's reaction time is increased by 0.15 s?

The effects of alcohol on the liver

The liver performs about 1500 different functions in the body. One important function is to break down (oxidise) harmful substances such as alcohol.

Enzymes in the liver break down alcohol into other toxic chemicals, acetaldehyde and acetic acid. Eventually these are further broken down into carbon dioxide and water.

Alcohol and its toxic breakdown products damage the liver, forming scar tissue. This is called alcoholic cirrhosis of the liver and it affects other useful liver functions. This can lead to liver failure and death.

Data on the effects of smoking

The idea of smoking being a harmless habit changed about 50 years ago. A report was published linking smoking with lung cancer and other diseases. It showed that in the UK nearly 140 men (who smoked only cigarettes) in every 100 000 died from lung cancer each year.

In 1970 about 50% of the population smoked cigarettes. It is now about 22%.

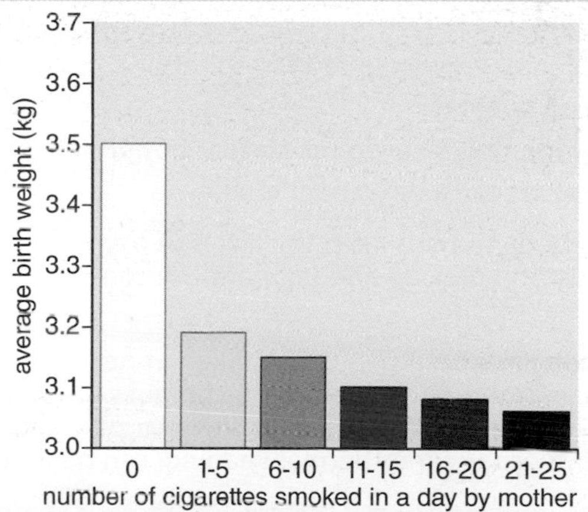

FIGURE 12: What link is there between smoking and birth weight?

Questions

17 Why can alcohol cause a lot of damage to the liver?

18 Suggest why the percentage of smokers in the UK has dropped from 50 to 22 since 1970.

Staying in balance

You will find out:
> how the insides of our bodies are kept in balance
> about homeostasis
> about the dangers of being too hot or cold

Weather warning

In 2005, the Government published guidelines to keep people safe in very hot weather.

People were advised to:
> avoid going out at midday
> take cool showers
> stay in the shade
> eat cold salads.

The heatwave in 2003 caused 27 000 deaths in Europe. In England, 2000 people died.

Did you know?

The highest ground temperature of 93 °C was recorded at Furness Creek in Death Valley, USA in 1972.

FIGURE 1: Why are travellers advised to carry at least 2 litres of water in their car when driving in Death Valley?

Keeping steady

Evie is taking part in the Duke of Edinburgh's Gold Medal Award. She is training for a 50-mile walk.

She could be in danger if she gets too hot or too cold as enzymes would stop working in her body.

Her core body temperature is normally 37 °C. Body cells work best at this temperature.

Evie loses more heat by:
> sweating
> more blood flow near to the surface of her skin.

Evie gains or keeps heat by:
> exercising
> shivering
> releasing energy from food (respiration)
> reduced sweating
> less blood flow near to the surface of her skin
> putting on extra or warmer clothes.

Various body systems keep the levels of oxygen, water, carbon dioxide and temperature constant. This is essential for life.

Evie's body temperature can be taken at her ear, fingers, mouth or anus. Various techniques, such as using a clinical thermometer, sensitive strips, digital recording probes and thermal imaging, can be used.

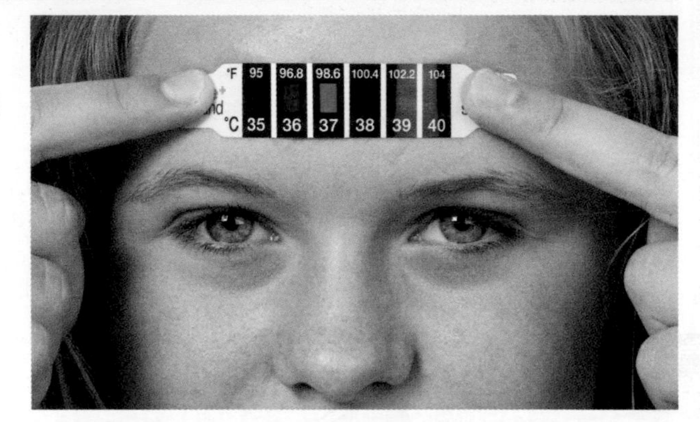

FIGURE 2: A heat-sensitive strip uses colour to show temperature.

Questions

1 How many people died in England as a result of the heatwave in 2003?

2 What is Evie's normal body temperature?

3 Write down one way in which Evie can lose heat.

4 Describe one way of measuring body temperature.

What is homeostasis?

Keeping a constant internal environment by balancing body inputs and body outputs is called **homeostasis**. Evie's body has automatic control systems to keep her temperature, water content and carbon dioxide at steady levels. This makes sure that all her cells can work at their optimum (best) levels.

body temperature   why do we shiver?   why do we sweat?

Sweating

Evie knows she sweats all the time. Sweat comes from sweat glands in her skin.

She sweats even more when she exercises. The water in the sweat evaporates – it uses heat from her skin to change from a liquid to a gas. Therefore the sweat cools the skin, and the blood flowing through it, when it evaporates.

Evie wears anti-perspirant and showers after exercise. This stops bacteria from living on the sweat and causing smells. When she steps out of the shower she feels very cold as the shower water evaporates from her skin.

Dangers of high and low temperatures

It is important to keep the body temperature at 37 °C. This is the optimum temperature for many enzymes.

Evie knows that she must wear the correct clothing and be well equipped for her expedition to help her body keep a steady internal environment.

If she gets too hot on her walk she could suffer from:

> **heat stroke**, when sweating is ineffective and her skin becomes cold and clammy and her pulse is rapid and weak

> **dehydration** because she has lost too much water.

If these conditions are left untreated she could die.

Evie must not get too cold. If the amount of heat generated is too low and her temperature falls she could suffer from **hypothermia**. Her pulse rate slows down and she shivers violently.

She could also die from this condition if it is untreated.

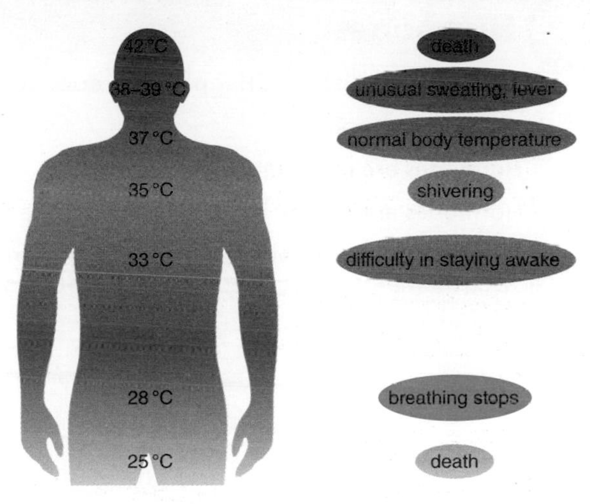

FIGURE 3: Suggest why shivering helps a person's body temperature to rise.

Questions

5 Why does Evie feel cold when she steps out of the shower?

6 What are the symptoms of hypothermia?

7 At which two temperatures are people likely to die?

Feedback controls

Premature babies are put into an incubator.

The temperature of an incubator is controlled by a feedback mechanism.

If the temperature is too low, the heater is switched on. This is called a **negative** feedback, as turning on the heater negates (cancels out) the decreasing temperature. Negative feedback systems are used in the human body.

The hypothalamus

A small gland in the brain, called the hypothalamus, helps to keep the body in balance by detecting when the blood is too hot or too cold.

The hypothalamus then triggers protective measures such as shivering or sweating, or changing the size of small blood vessels in the skin (**vasoconstriction** and **vasodilation**). These changes are under the control of the nervous and hormonal systems.

Question

8 Explain how the hypothalamus helps to maintain a constant body temperature.

Vasoconstriction. When the body is too cold small blood vessels in the skin constrict and so less blood flows through them, reducing heat loss.

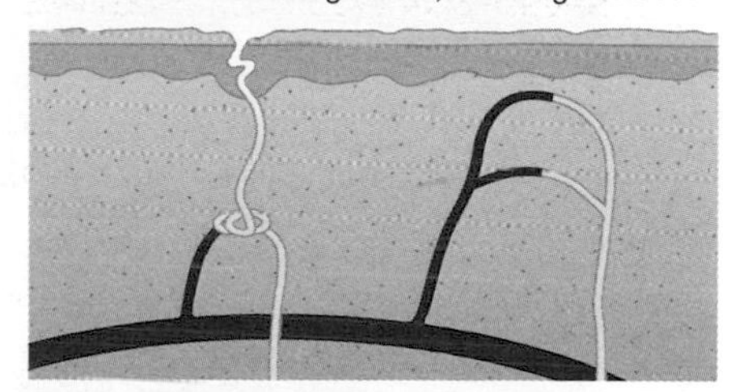

Vasodilation. When the body is too hot small blood vessels in the skin dilate and so blood flow increases, bringing more blood to the surface, where it loses heat.

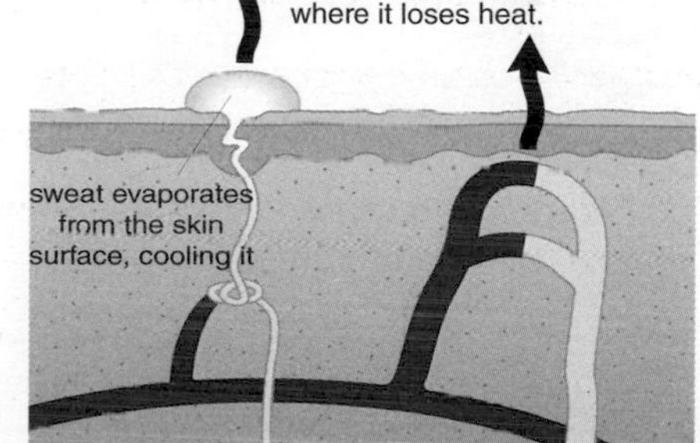

FIGURE 4: Vasoconstriction and vasodilation.

dehydration hypothermia negative feedback

Hormones

You will find out:

> where hormones are produced

> how blood sugar levels are controlled

> why a lack of insulin causes a problem

Evie has Type 1 diabetes. Her pancreas does not produce enough insulin.

Insulin is a **hormone**.

> Hormones are made in special glands.

> Hormones are carried in the blood to where they have an effect.

> As the hormones travel in the blood, hormonal responses are usually slower than those controlled by the nervous system.

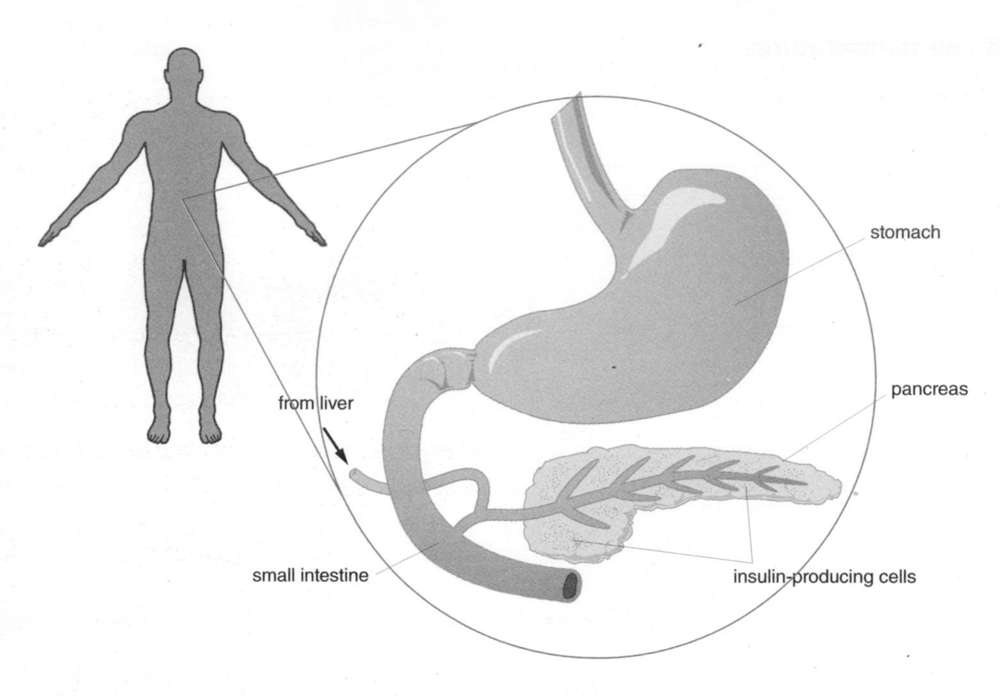

FIGURE 5: What hormone is produced in the pancreas?

Questions

9 How do hormones get around the body?

10 What sort of chemical is insulin?

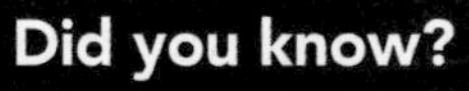

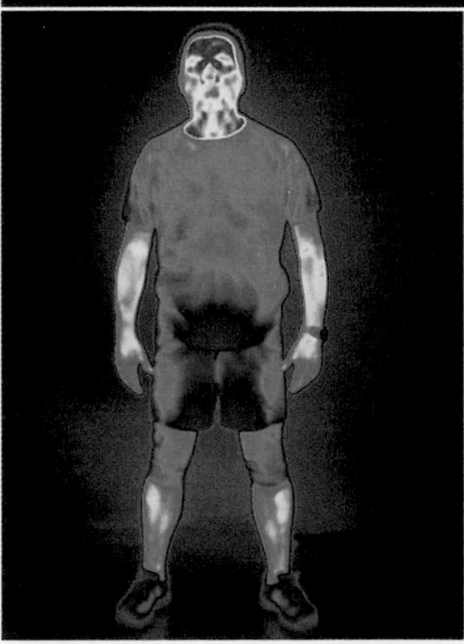

Hospitals use thermal imaging to show the temperature of parts of the body.

Hotter parts are shown in red, colder parts in blue.

Insulin and diabetes

> Insulin controls the level of glucose in the blood. If the level of glucose in Evie's blood falls too low she could go into a coma.

> The level of glucose in the blood is called the **blood sugar level**.

> Insulin is produced in the pancreas and carried by the blood to the liver, where it affects the glucose level. The liver is therefore the target organ for insulin. The effects are slower than if the process was controlled by nerves.

Types of diabetes

Evie has Type 1 diabetes. This is when the body is unable to make any insulin. It usually occurs in teenagers and affects about 15% of the population. Type 1 diabetes needs to be treated by injections of insulin.

Type 2 diabetes is different. This occurs when the body either produces too little insulin, or the cells do not react to it. It is linked to obesity and is more common in older people. Type 2 diabetes can often be controlled by altering the sugar content in the diet rather than by injections of insulin.

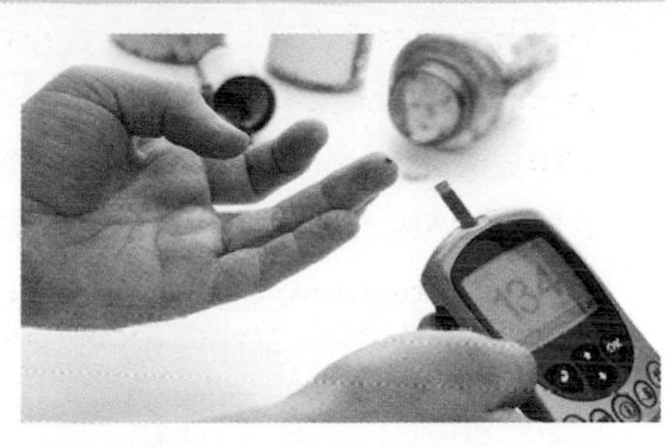

FIGURE 7: Why must a diabetic person frequently test their blood?

Questions

11 What does insulin control?

12 Why are the controlling effects of insulin slower than a reflex action?

13 What is the target organ for insulin?

Control of blood sugar levels

After a meal containing carbohydrates, the digestive system breaks the large carbohydrate molecules down into simple sugars such as glucose. Glucose is carried in the blood plasma.

When the levels of glucose in the blood are too high, insulin converts the excess glucose into glycogen, which is stored in the liver.

As Evie does not have enough insulin, a lot of her glucose is removed from her blood by her kidneys. Glucose is excreted in her urine.

Her body therefore does not have a store of glycogen to use later on.

She needs to inject herself with insulin to keep her insulin levels as close to normal as possible. The size of the insulin dose that Evie needs depends on her diet and how much exercise she takes.

Using an insulin pump

Instead of checking blood sugar levels and then injecting the required amount of insulin, a new technique has been developed. It uses an insulin pump. The blood sugar levels are continually checked and the appropriate amounts of insulin are pumped into the blood.

FIGURE 8: How is the insulin pump better than having insulin injections?

Questions

14 Explain why the process of blood sugar regulation is described as negative feedback.

15 Suggest which foods to avoid to prevent too high a level of blood sugar.

16 Explain why the insulin dosage for a person with Type 1 diabetes depends on the amount of physical activity they undertake.

Controlling plant growth

You will find out:

> about plant hormones

> how plants respond to light and gravity

Plant growth and ripening

At the end of the summer many gardeners are faced with handfuls of unripe green tomatoes.

Some gardeners either throw them away or make tomato chutney.

The alternative is to put the tomatoes in a bag with a ripe banana. Over-ripe bananas release a hormone that ripens fruit.

FIGURE 1: Many fruits take time to ripen. Are there commercial advantages to fruits that ripen slowly?

Plant hormones

Plants as well as animals respond to changes in their environment.

Plants make special chemicals called **plant hormones**. Plant hormones control different processes in a plant:

> growth of shoots towards light

> growth of roots downwards so they will find more water (in response to gravity)

> growth of flowers

> ripening of fruit.

When shoots grow towards the light the plant has a better chance of survival. The leaves will get more light and produce more food from photosynthesis.

Growing cress

Rosa grows cress seeds. She puts some seedlings in a box with a small opening. These seedlings get light from only one direction.

She puts some seedlings in a box on top of a clinostat. The clinostat acts as a turntable, slowly turning the dish so the seedlings get light from all directions.

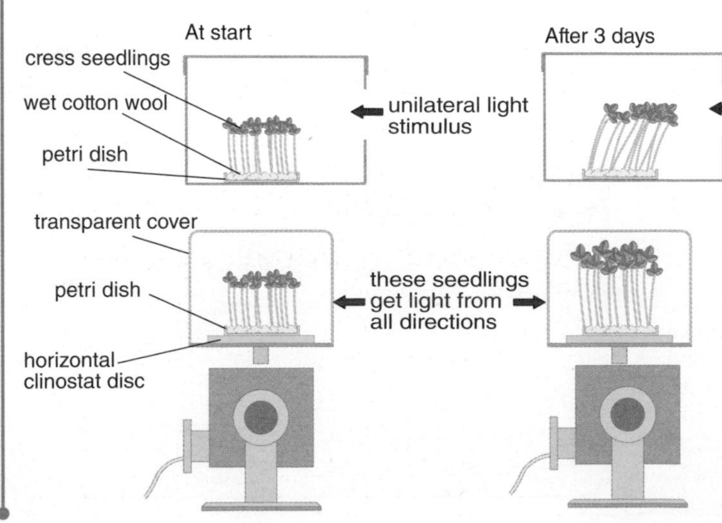

FIGURE 2: What happens to the cress seedlings that get light from only one direction? What happens to the cress seedlings that get light from all directions?

Questions

1 What name is given to chemicals inside plants that control growth?

2 Do plant shoots grow towards or away from light?

3 Which part of the plant responds to the force of gravity by growing downwards?

4 How does a hormone from a banana affect a green tomato?

5 Explain one advantage of a plant shoot growing towards light.

6 Explain one advantage of a plant root growing down into soil.

Positive and negative

A plant is sensitive. It responds to different stimuli. A plant's response can be:

> negative – it grows away from a stimulus

> positive – it grows towards a stimulus.

One group of plant hormones is called **auxins** and they control the response. Auxin is made in the tips of roots and shoots of a plant. It travels through a plant in solution. The roots and shoots respond to auxin in different ways.

Phototropism

When a plant responds to light it is called **phototropism**.

> Plant shoots grow towards light. This response is called positive phototropism.

> Plant roots grow away from light. This is negative phototropism.

Geotropism

When a plant responds to gravity it is called **geotropism**.

> Plant shoots grow away from the pull of gravity. This response is called negative geotropism.

> Plant roots grow with the pull of gravity. This is positive geotropism.

Question

7 Name two responses in plants that involve auxin.

Remember!
The 'positive' and 'negative' terms are difficult to remember.
If it's positive it grows towards the stimulus. Imagine being positively attracted to someone.

How auxin works

When the tip of a shoot is cut off it stops growing. If the tip is replaced on the stem it starts to grow again. Removing the tip removes the source of auxin and stops growth.

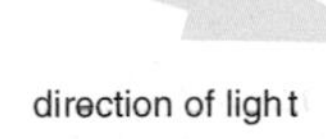
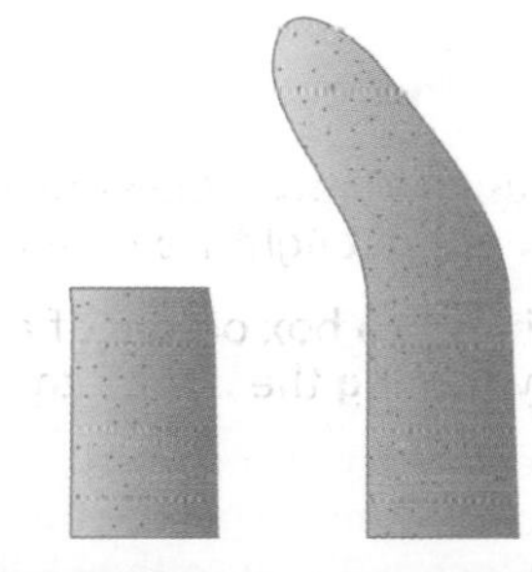

FIGURE 3: Growth of plant shoot tips in response to a light source. Why is only one shoot growing towards the light?

Question

8 Look at the experiment shown in the diagram.

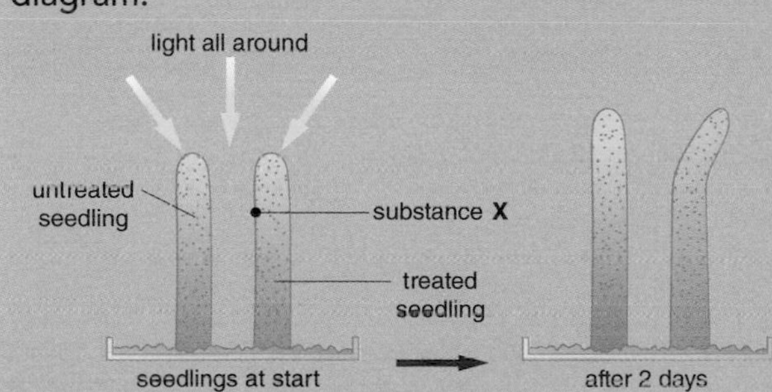

a Describe the results of the experiment after 2 days.

b Name substance X.

c Explain the effect of substance X on the shoot.

Speeding up and slowing down growth

Farmers can spray crops such as fruit trees with hormones to speed up the growth and development of the fruit.

Slowing down growth

Farmers can also use hormones to slow down growth. This stops the fruit from falling off the tree before the harvest.

FIGURE 4: Apples on a tree. What do farmers spray on trees to make fruit grow?

FIGURE 5: How do farmers stop fruit from falling off trees?

Questions

9 Copy and complete the following sentence.

Farmers can use h_________ to speed up or slow down plant g_________.

10 Suggest why farmers need to slow down the growth of fruit on their trees.

You will find out:

> how plant hormones can be used in agriculture

> about commercial use of plant hormones

Did you know?

Apple seeds contain a small quantity of cyanide.

Using plant hormones

Farmers, gardeners and fruit growers can all use plant hormones. They mostly use a human-made auxin called synthetic auxin.

Selective weedkillers

Synthetic auxin is sprayed on crops to kill weeds. The hormone makes weeds grow too fast and they die. The concentration used only affects broadleaved weeds. The narrow-leaved crops are unaffected. Because they kill only certain weeds they are called selective weedkillers.

Growing roots

New plants can be grown by taking cuttings from existing plants. A cutting is dipped into a hormone-based rooting powder. The synthetic auxin in the powder stimulates roots to grow.

Germinating seeds

All seeds are dormant, needing water and oxygen to germinate. Some seeds, such as ginseng, need very cold conditions for a year before they **germinate.** Growers now use a hormone called **gibberellic acid** to force these seeds to germinate, saving time and money.

FIGURE 6: A farmer spraying crops with a selective weedkiller. Why is the weedkiller described as 'selective'?

dormant seeds   using rooting powder   how do selective weedkillers work?

Transporting bananas

Bananas are easily damaged during transport. To avoid this they are harvested before they are ripe. During transport a hormone called **ethene** is sprayed on them. Ethene causes bananas to ripen ready for sale.

Other types of fruit trees can be sprayed with plant hormones to delay fruit ripening. This ensures the fruit reaches its full size and does not drop from the trees.

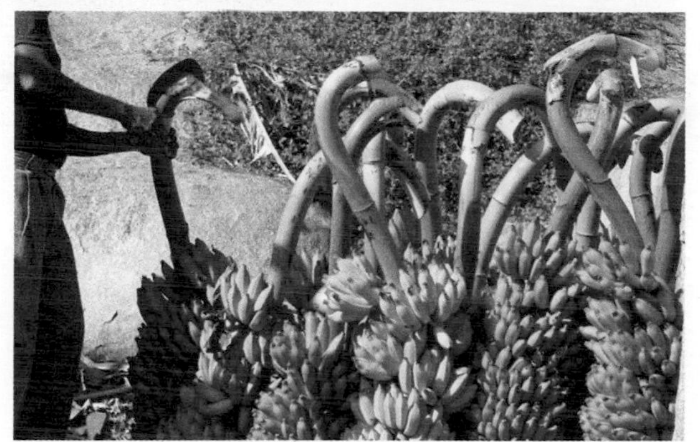

FIGURE 7: How are bananas supplied undamaged and ripe to supermarkets?

Questions

11 How do hormones kill weeds?

12 How does dipping a cutting in rooting powder help it to grow?

13 Bananas are picked when they are green, yet when we buy them they are yellow and ripe. Why is this?

14 One reason bananas are picked before they are ripe is to prevent damage. Suggest another reason.

15 Suggest one reason why some people prefer fruit that has not been sprayed with hormones.

How does phototropism work?

In 1880, Charles Darwin did many experiments on phototropism.

He concluded:

> the tip controls growth and the plant's response to light

> the tip produces a soluble chemical that travels down the shoot

> the shoot grows towards light.

We now know more about phototropism.

> Auxin is made in the tip.

> Auxin moves away from light and collects on the shady side of a shoot.

> Auxin causes cells on the shady side to elongate (grow longer) more than cells on the light side.

> The shady side becomes longer and causes the shoot to bend.

Questions

16 Which of Darwin's experiments showed that the substance produced by the tip was soluble?

17 Which of Darwin's experiments showed that the tip was sensitive to light?

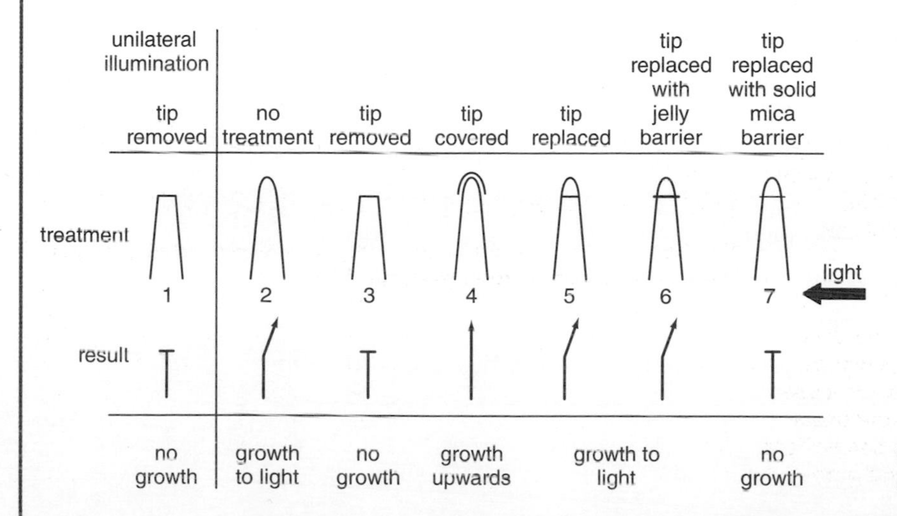

FIGURE 8: Summary of Darwin's experiments.

Variation and inheritance

You will find out:

- about inherited characteristics
- about characteristics that are caused by the environment
- what makes a boy or a girl
- about mutations

The human blueprint

The Human Genome Project to identify all of the 30 000 human genes was finished in June 2000. The sites of many faulty genes are now known.

genes on chromosome 7 cause excess mucus production (cystic fibrosis)

genes on chromosome 4 cause on-going nerve damage (Huntington's disease)

genes on chromosome 11 cause abnormal haemoglobin (sickle-cell anaemia)

genes on chromosome X cause lack of blood clotting factor (haemophilia)

1 2 3 4 5 6 7 8 9 10 11 12 13 14 15 16 17 18 19 20 21 22 XX

FIGURE 1: Paired chromosomes. Chromosomes carry genes. Each chromosome pair is labelled with a number, except for the sex chromosomes, which make a female or a male. These are labelled 'XX' in a female and 'XY' in a male.

Human characteristics

Calvin is a teacher. He draws his new students so he can remember their names. This is Darren. He has red hair, blue eyes, no ear lobes, a large straight nose and freckles. These are examples of characteristics that he has inherited from his parents.

Each inherited characteristic is controlled by his **genes**. His scar and Yorkshire accent are examples of characteristics that are not inherited. They have been caused by his environment.

Other characteristics caused by both inheritance and environment are:

- height
- intelligence
- body mass.

Some human disorders such as red–green colour blindness, sickle-cell anaemia and cystic fibrosis are also inherited.

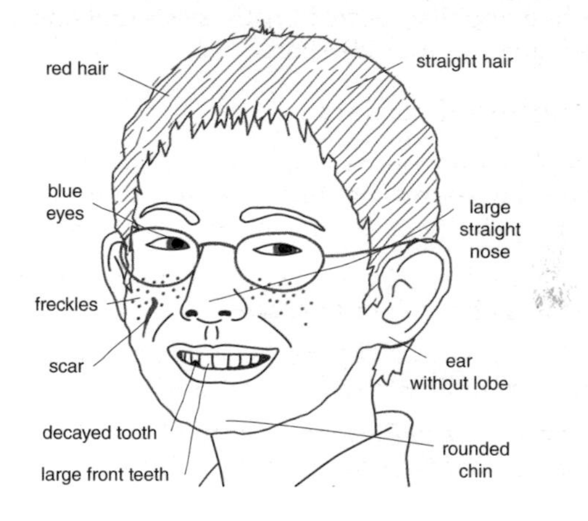

FIGURE 2: Can you suggest another characteristic that Darren has on his face that he has inherited?

Questions

1 Look at the picture of Darren. Write down four of his characteristics that are inherited.

2 What controls these characteristics?

3 Write down three of Darren's characteristics that are caused by his environment.

genes and inheritance

Sex chromosomes

In the human population there are nearly the same number of males as there are females. Why is this? All humans have 46 chromosomes, arranged in pairs, in their body cells. Gametes have only 23 chromosomes.

One pair of chromosomes is called the **sex chromosomes**.

The number of chromosomes varies between species. The worm, *Ascaris* has only two chromosomes in each cell.

In all mammals, including humans, there are two types of sex chromosomes called X and Y. The Y chromosome is only found in males and is shorter and a different shape from the X chromosome. In a female body cell, the sex chromosomes are XX. In a male body cell, the sex chromosomes are XY.

All the students in Calvin's class are unique. Even siblings in the school are not identical. This is called **variation**. Variation is caused by:

> genes being mixed up when **gametes** are formed.

> genes coming from two parents (fertilisation)

> changes in genes or chromosomes called **mutations**.

Faulty genes

Faulty genes are responsible for inherited disorders such as haemophilia, cystic fibrosis, sickle-cell anaemia and red–green colour blindness.

Inherited disorders

People with inherited disorders not only face problems caused to themselves but also carry the responsibility of making decisions about future generations.

> Should they inform their partners, employers and insurers about their disorder?

> Should they accept the risk of passing a genetic disorder to their next generations?

> Should they rely on new technology to select an embryo that does not carry the disorder?

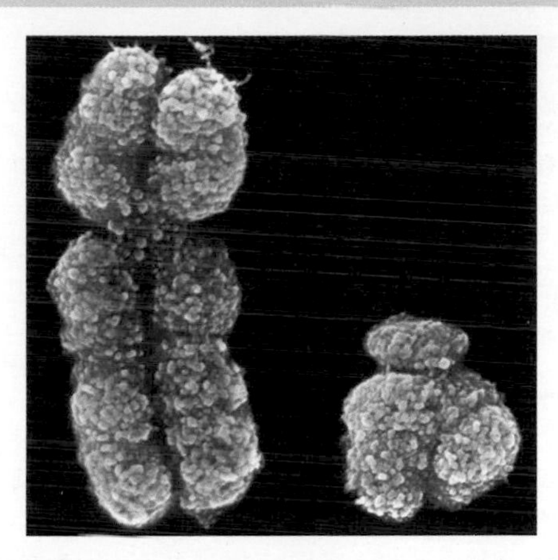

FIGURE 3: X and Y sex chromosomes magnified highly. How is the shape of the X chromosome different from that of the Y chromosome?

Did you know?

Queen Victoria carried a mutant gene for haemophilia. One of her sons, Leopold, who had haemophilia and two of her daughters carried the mutant gene.

One daughter, Alice, married into the Russian royal family. This resulted in the famous person with haemophilia, Alexis, the last Tsarevitch (King).

Questions

4 How many pairs of chromosomes are in a human body cell?

5 How many sex chromosomes do you have in each body cell?

6 What is a mutation?

Genetic issues

Nature versus nurture?

There is an ongoing debate over the balance between genetic (nature) and environmental (nurture) factors in determining human attributes such as intelligence, sporting ability and health.

Could you produce an Olympic champion by long-term extensive training? Or is the ability controlled only by genes?

Scientists are discovering more about genes and how they work by studying twins. The Twin Research Unit now believes that inheritance plays a part in short-sightedness, osteoarthritis, acne and migraines.

Question

7 Write down one piece of evidence for genetic factors controlling sporting characteristics and one piece against.

Characteristics

You will find out:

- > about dominant and recessive characteristics
- > how to work out a genetic cross
- > the chances of inheriting a genetic disorder

Chromosome numbers

The number of chromosomes in a cell is usually an even number. This is because the paired chromosomes separate when eggs and sperm are formed. Humans have 46 chromosomes, arranged in 23 pairs, in each cell. Eggs and sperm (gametes) have only one chromosome from each pair. They combine to make a fertilised egg that has 46 chromosomes.

Calvin teaches a new class every year. One of his new students is called Charlotte.

There are about 7 billion people in the world, but there is only one Charlotte.

Her genes are in chromosomes in the nucleus of every cell of her body except red blood cells.

The chromosomes have all the information to make Charlotte. The chromosomes are arranged in matching pairs.

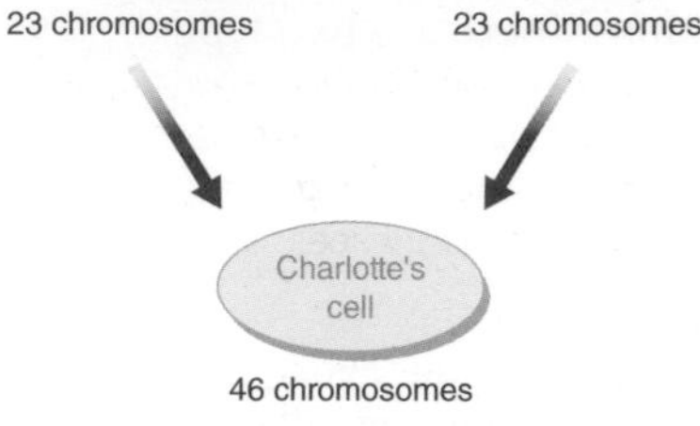

FIGURE 4: What is a female gamete called?

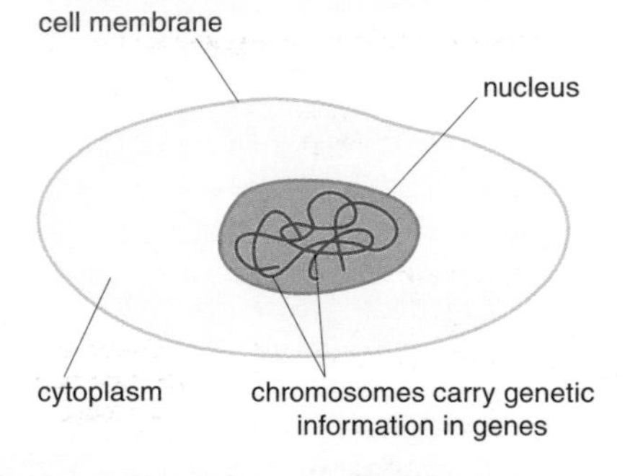

FIGURE 5: What does a nucleus contain?

Questions

8 What do Charlotte's chromosomes carry?

9 Where in a cell are chromosomes found?

10 Explain why body cells of organisms usually have an even number of chromosomes.

11 Does a sperm contain a quarter, half, three-quarters or all of the information to make a baby?

Breeding tomato plants

Calvin's class does breeding experiments with two types of tomato plants.

One type has genes only for green stems, another type has genes only for purple stems. The gene involved has two different versions called alleles.

The two plant types are grown to adulthood and cross-pollinated so that the male pollen and the female ovules (eggs) from both types are mixed. The seeds are collected and grown to form the next generation.

This generation (called the F1) all have purple stems, showing that the characteristic (and allele) for purple colour is dominant over the green colour characteristic (and allele).

The characteristic for green colour is called recessive and will be expressed only in the absence of a dominant allele.

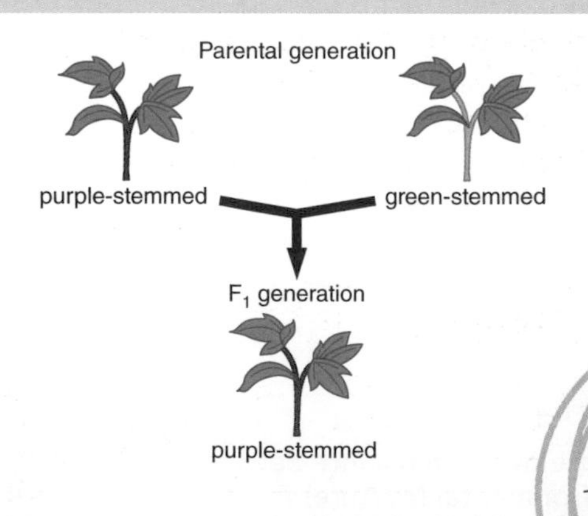

FIGURE 6: Breeding green- and purple-stemmed tomato plants.

Remember!

The study of genetics uses many new words. Make a list of them and their meanings.

Question

12 What does cross-pollinated mean?

dominant characteristics recessive characteristics genotype

A monohybrid cross

Male or female?

Each egg carries one type of sex chromosome (X). Each sperm carries an X or a Y sex chromosome.

There is always a random chance of which sperm fertilises an egg. During fertilisation the sex chromosomes mix.

A **monohybrid cross** involves different **alleles** of a gene. These are carried on a pair of chromosomes: one chromosome could be carrying the dominant allele, the other chromosome could be carrying the recessive allele.

Darren has no ear lobes. Symbols are used to work out the genetic cross. Darren's parents each have two different alleles (E and e); they are **heterozygous**. Their **genotype** (genetic make-up) has both dominant and recessive alleles.

Their **phenotype** (expressed characteristic) is to show lobed ears. Darren has two identical recessive alleles (one type). He is **homozygous**. His genotype has only recessive alleles. His phenotype is to show ears without lobes because the dominant allele is absent.

The combination EE also has two identical alleles so it is homozygous.

Cystic fibrosis

Darren has cystic fibrosis, an inherited condition. He has difficulty in breathing because his body produces too much mucus, which collects in his lungs. It is a recessive characteristic.

Inherited disorders, such as cystic fibrosis, are caused by faulty alleles. Most of these alleles are recessive. When both healthy parents are heterozygous for the condition (they are Cc) there is a one in four chance of their baby being born with cystic fibrosis (cc). Darren will have important decisions to make in later life.

In this case, none of their children will have cystic fibrosis, although they will be carrying the recessive allele.

In the future, Darren would like to start a family. He hopes his partner will be homozygous (CC), without a recessive allele for cystic fibrosis.

		Darren's partner	
		C	C
Darren	c	Cc	Cc
	c	Cc	Cc

FIGURE 9: What would be the phenotype and genotype of these offspring?

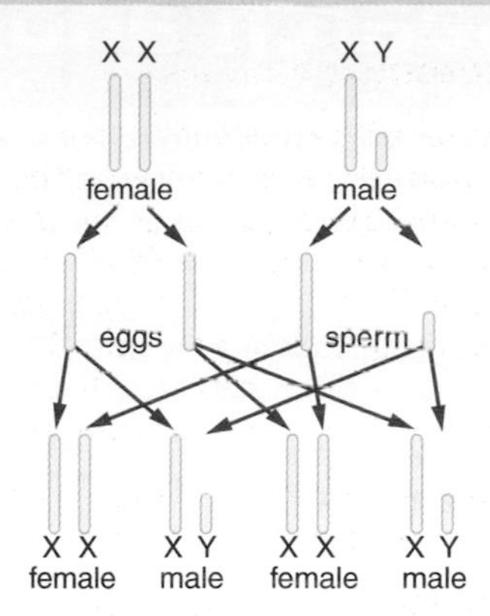

FIGURE 7: There should be equal numbers of males and females in the human population.

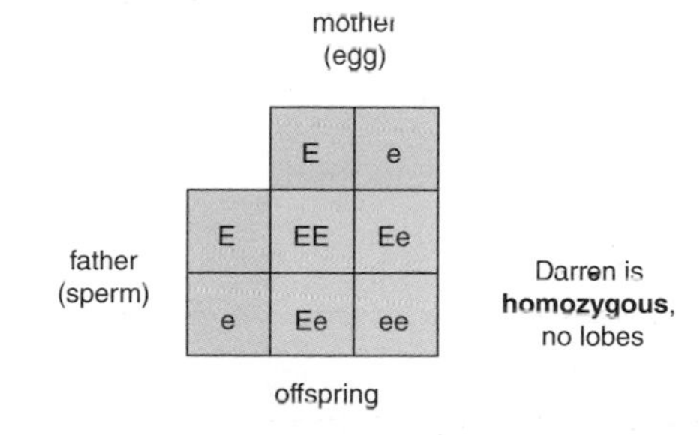

FIGURE 8: A genetic diagram.

Questions

13 Suggest why there are slightly fewer men than women in the world.

14 Using a genetic diagram work out the possible combinations if Darren's parents were Ee and EE in the ear lobes cross.

15 Could parents with gene combinations of CC and Cc produce a baby with cystic fibrosis?

heterozygous homozygous monohybrid cross phenotype

Preparing for assessment: Research and collecting secondary data

To achieve a good grade in science, you not only have to know and understand scientific ideas, but you need to be able to apply them to other situations and investigations. These tasks will support you in developing these skills.

Task

> Which part of the human body is most accurate in testing the temperature of a baby's bath water?

Context

Charlotte has a new baby. She knows she has to bath her baby every day.

She looks up information about the recommended water temperature of the baby's bath.

She knows that a high water temperature is extremely dangerous.

One leaflet tells her to use the back of her hand to test for the right temperature if she hasn't got a thermometer.

A book on baby care says that the elbow should be used to test bath water temperature.

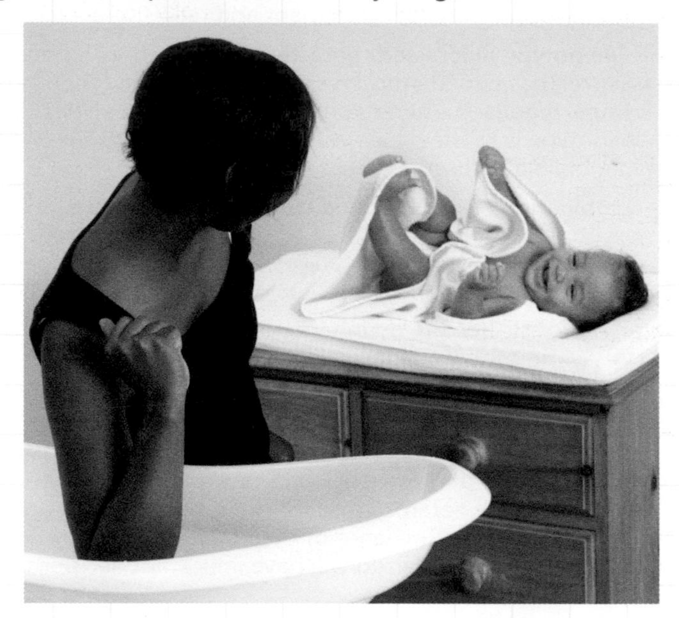

Planning

Plan how you are going to collect this information. You need to:

1. Write down how you found the information.

2. Write down a list of all the sources of your information.

3. Clearly present the information so it could be used to plan an actual investigation.

General rules

1. You may work with other students but your written work should be done on your own.

2. You cannot get detailed help from your teacher.

3. You are not allowed to redraft your work.

4. Your work can be handwritten or word processed.

5. It is expected that you complete this task in two hours.

6. You are allowed to do this research outside the laboratory.

7. You must be aware of and mention any health and safety issues.

What do you need to research?

Where would you find this information?

Find out about skin structure and function.

Find out if some areas of the skin are more sensitive than others.

Use the search terms **human skin structure**, **baby bath water temperature** and **hot water damage to skin**.

What do you find out about?

How do you record all this information?

Put together or print off a final account of your research and hand it to your teacher.

Research and collecting secondary data

To research:

- water being too hot and damaging the skin
- the structure of the skin
- how and why the skin reacts to temperature changes.

Information sources:

- Science textbooks in the school library
- General books on baby care
- The internet

Research found:

- Recommended water temperature
- Skin damage
- Different receptors

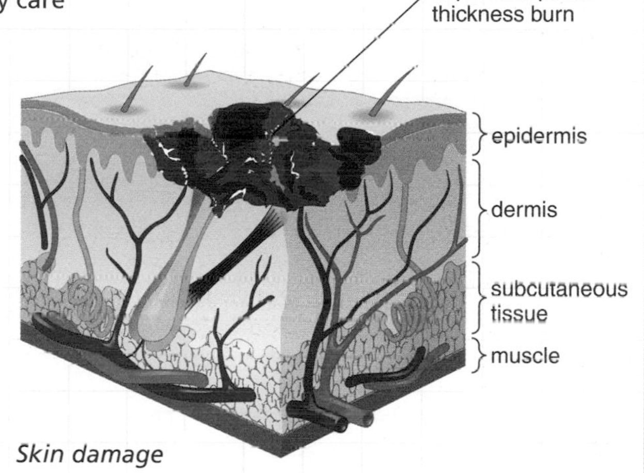

Skin damage

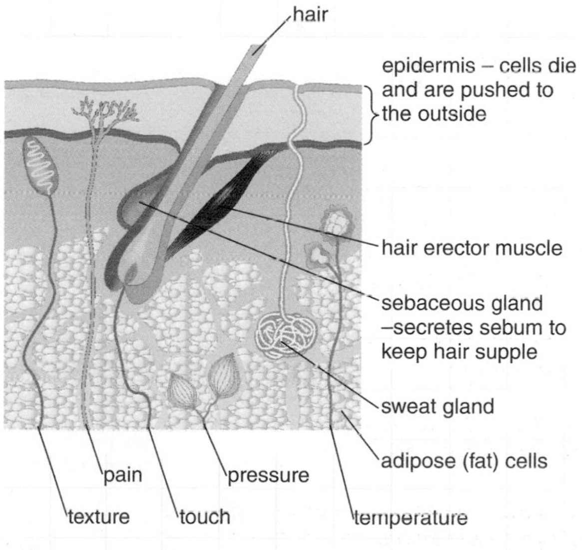

Different receptors

Record:

- Use a notebook to write down the information and where it came from.
- Use a computer to record images, conversation and diagrams.

B1 Checklist

To achieve your forecast grade in the exam you'll need to revise

Use this checklist to see what you can do now. It gives you many of the important points you will need to know. Refer back to the relevant pages in this book if you're not sure and to see if there is anything else you need to know. Look across the three columns to see how you can progress.

Remember you'll need to be able to use these ideas in various ways, such as:

> interpreting pictures, diagrams and graphs
> applying ideas to new situations
> explaining ethical implications
> suggesting some benefits and risks to society
> drawing conclusions from evidence you've been given.

Look at pages 278–299 for more information about exams and how you'll be assessed.

To aim for a grade E	To aim for a grade C	To aim for a grade A
explain why the blood in arteries is under pressure **describe** how cholesterol can restrict or block blood vessels	**describe** the factors that increase blood pressure **describe** the factors that decrease blood pressure **explain** how diet can increase the risk of heart disease	**explain** the consequences of high blood pressure **explain** the consequences of low blood pressure **explain** how narrowed arteries increase the risk of a heart attack **explain** how a thrombosis can increase the risk of a heart attack
explain what a balanced diet should contain **explain** why a high protein diet is necessary for teenagers **recall** that obesity is linked to increased health risks	**recall** what proteins, fats and carbohydrates are made from **explain** factors that affect how a balanced diet will vary **explain** why protein deficiency is linked to developing countries **calculate** EAR of protein **calculate** BMI and understand the results	**describe** what happens to various food types in the body **describe** the differences between first and second class proteins **explain** why the EAR results may vary
recall the causes of infectious diseases **describe** how the body defends itself from pathogens **explain** why new medical drugs are tested before use	**recall** how pathogens cause the symptoms of a disease **explain** the difference between active and passive immunity **describe** how new treatments are tested **understand** objections to some forms of testing	**explain** why each pathogen needs specific antibodies **explain** the process, benefits and risks of immunisation **explain** why blind trials are used in testing new drugs

To aim for a grade E	To aim for a grade C	To aim for a grade A
name and locate the main parts of the eye **describe** the main problems in vision **describe** reflex actions as fast, automatic, protective	**describe** the functions of the main parts of the eye **explain** how long sight and short sight is caused **describe** a reflex arc and the path taken by a spinal reflex	**explain** how the eye focuses light **explain** how long sight and short sight can be corrected **describe** the nerve transmission across a synapse
describe the effects of each drug category: depressants, painkillers, stimulants, performance enhancers, hallucinogens **know** the effects of alcohol and cigarette smoke on the body	**recall** examples of each drug category **explain** the basis of the legal classification of drugs **interpret** information on reaction times, accident statistics and alcohol levels **describe** the effects of cigarette smoke on the lungs	**explain** the action of depressants and stimulants on nerve synapses **describe** the effects of alcohol on the liver **interpret** information on cancer statistics
describe how heat can be gained or retained by the body **recall** that type 1 diabetes is caused by a failure of the pancreas to produce insulin	**understand** homeostasis involves balancing body inputs and outputs using automatic systems **describe** the effects of high and low temperatures on the body **explain** the difference between Type 1 and Type 2 diabetes	**explain** how negative feedback mechanisms maintain a constant internal environment **interpret** how insulin helps to regulate blood sugar levels by converting excess blood glucose to glycogen in the liver
understand that plant growth is controlled by plant hormones **understand** that stems grow towards the light and roots grow downwards in response to gravity **recognise** that plant hormones can be used to speed up or slow down plant growth	**recall** that auxins (plant hormones) move through the plant in solution and are involved in phototropism and geotropism **relate** the action of plant hormones to their commercial use	**explain** how auxin brings about shoot curvature in terms of cell elongation
recall that the nucleus contains chromosomes **recall** that chromosomes carry information in the form of genes **recognise** that some disorders are inherited	**describe** how sex in mammals depends on sex chromosomes **understand** that inherited disorders are caused by faulty genes	**explain** sex inheritance using genetic diagrams **use** genetic diagrams to predict the probabilities of inherited disorders passing to the next generation

B1 Exam-style questions

Foundation Tier

1 Five students were assessed using different ways of measuring fitness. They were graded 1, 2, 3, 4 or 5, with grade 1 being the highest level of fitness.

	strength	flexibility	stamina	agility	speed
Emily	5	4	4	3	3
Lauren	2	1	2	1	1
Kayla	3	2	3	2	2
Nadia	4	5	5	5	4
Alyssa	1	3	1	4	5

AO2 **(a)** Suggest and explain why each student did not show the same level of fitness for each way of assessing fitness. [2]

AO3 **(b)** Which student showed the best overall fitness level? [1]

AO3 **(c)** Comment on the fitness level of Alyssa and suggest a sport or activity to suit her fitness. [3]

[Total: 6]

AO1 **2 (a)** Match up each part of a balanced diet with its correct function.

1. iron	A. prevents scurvy
2. vitamin C	B. makes haemoglobin
3. fibre	C. prevents constipation
4. high protein	D. high energy source
5. high carbohydrates	E. growth

[4]

(b) Chris is 1.8 m tall. He weighs 85 kg.

Look at the BMI chart.

Weight in kilograms

Height in centimeters	45	48	50	53	55	58	60	63	65	68	70	73	75	78	80	82.5	85	87.5	90
145.0	21.4	22.6	23.8	25.0	26.2	27.3	28.5	29.7	30.9	32.1	33.3	34.5	35.7	36.9	38.0	39.2	40.4	41.6	42.8
147.5	20.7	21.8	23.0	24.1	25.3	26.4	27.6	28.7	29.9	31.0	32.2	33.3	34.5	35.6	36.8	37.9	39.1	40.2	41.4
150.0	20.0	21.1	22.2	23.3	24.4	25.6	26.7	27.8	28.9	30.0	31.1	32.2	33.3	34.4	35.6	36.7	37.8	38.9	40.0
152.5	19.3	20.4	21.5	22.6	23.6	24.7	25.8	26.9	27.9	29.0	30.1	31.2	32.2	33.3	34.4	35.5	36.5	37.6	38.7
156.0	18.7	19.8	20.8	21.9	22.9	23.9	25.0	26.0	27.1	28.1	29.1	30.2	31.2	32.3	33.3	34.3	35.4	36.4	37.5
157.5	18.1	19.1	20.2	21.2	22.2	23.2	24.2	25.2	26.2	27.2	28.2	29.2	30.2	31.2	32.2	33.2	34.3	35.3	36.3
160.0	17.6	18.6	19.5	20.5	21.5	22.5	23.4	24.4	25.4	26.4	27.3	28.3	29.3	30.3	31.3	32.2	33.2	34.2	35.2
162.5	17.0	18.0	18.9	19.9	20.8	21.8	22.7	23.7	24.6	25.6	26.5	27.5	28.4	29.3	30.3	31.2	32.2	33.1	34.1
165.0	16.5	17.4	18.4	19.3	20.2	21.1	22.0	23.0	23.9	24.8	25.7	26.6	27.5	28.5	29.4	30.3	31.2	32.1	33.1
167.5	16.0	16.9	17.8	18.7	19.6	20.5	21.4	22.3	23.2	24.1	24.9	25.8	26.7	27.6	28.5	29.4	30.3	31.2	32.1
170.0	15.6	16.4	17.3	18.2	19.0	19.9	20.8	21.6	22.5	23.4	24.2	25.1	26.0	26.8	27.7	28.5	29.4	30.3	31.1
172.5	15.1	16.0	16.8	17.6	18.5	19.3	20.2	21.0	21.8	22.7	23.5	24.4	25.2	26.0	26.9	27.7	28.6	29.4	30.2
175.0	14.7	15.5	16.3	17.1	18.0	18.8	19.6	20.4	21.2	22.0	22.9	23.7	24.5	25.3	26.1	26.9	27.8	28.6	29.4
177.5	14.3	15.1	15.9	16.7	17.5	18.3	19.0	19.8	20.6	21.4	22.2	23.0	23.8	24.6	25.4	26.2	27.0	27.8	28.6
180.0	13.9	14.7	15.4	16.2	17.0	17.7	18.5	19.3	20.1	20.8	21.6	22.4	23.1	23.9	24.7	25.5	26.2	27.0	27.8
182.5	13.5	14.3	15.0	15.8	16.5	17.3	18.0	18.8	19.5	20.3	21.0	21.8	22.5	23.3	24.0	24.8	25.2	26.3	27.0
185.0	13.1	13.9	14.6	15.3	16.1	16.8	17.5	18.3	19.0	19.7	20.5	21.2	21.9	22.6	23.4	24.1	24.8	25.3	26.3
187.5	12.8	13.5	14.2	14.9	15.6	16.4	17.1	17.8	18.5	19.2	19.9	20.6	21.3	22.0	22.8	23.5	24.2	24.9	25.6
190.0	12.5	13.2	13.9	14.5	15.2	15.9	16.6	17.3	18.0	18.7	19.4	21.0	20.8	21.5	22.2	22.9	23.5	24.2	24.9

underweight normal overweight obesity

AO2 **(i)** What is his BMI? [1]

AO3 **(ii)** In which group is Chris? [1]

AO3 **(iii)** Suggest one thing Chris can do so he will be in the normal group. [1]

[Total: 7]

3 The police have issued this poster.

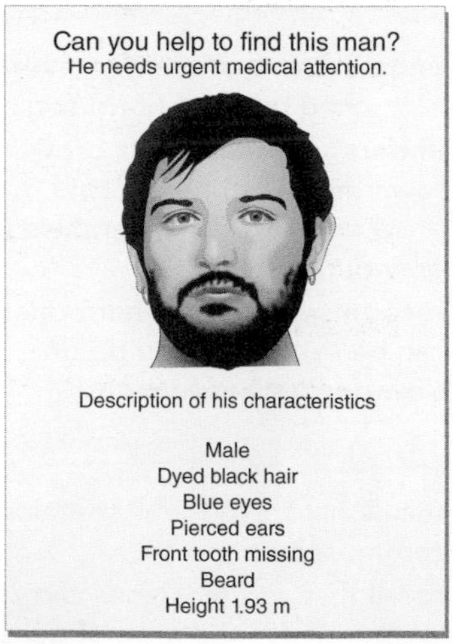

Can you help to find this man?
He needs urgent medical attention.

Description of his characteristics

Male
Dyed black hair
Blue eyes
Pierced ears
Front tooth missing
Beard
Height 1.93 m

AO1 **(a) (i)** Name two of the characteristics due to genetics.

AO1 **(a) (ii)** Name one characteristic caused by both genetics and the environment. [2]

AO1 **(b)** Explain how the man could have caught malaria and why it is important, to him and other people, that he should receive urgent medical attention. [2]

[Total: 4]

4 Dan finds out that cigarette smoke contains carbon monoxide, nicotine and tars.

AO2 Explain why this information should stop him from smoking. [3]

[Total: 3]

AO1 recall the science AO2 apply your knowledge AO3 evaluate and analyse the evidence

Worked Example – Foundation Tier

The body has many sense organs.

(a) Harry uses a skateboard.
How does his body detect changes in the environment?
In your answer write about.
- what these changes are called
- which body structures detect these changes
- how these structures react. [3]

These changes are called stimuli.

Harry will detect things such as lamp posts in his way using his eyes. Eyes are an example of a receptor.

His eyes will send messages to the rest of his body.

(b) Complete the labels to show the structure of the eye. [3]

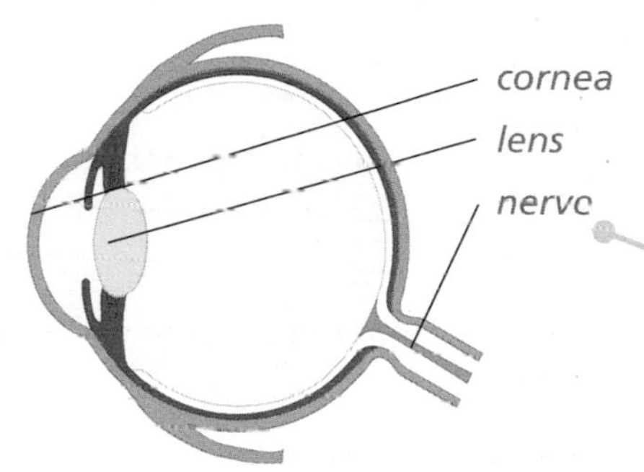

(c) Describe how light from an object reaches the retina. [3]

Light is reflected by the cornea.

The light is focused by the lens.

So it reaches the retina at the back of the eye.

(d) If the nerve from the eye to the brain is cut, blindness would result.

Explain why. [2]

Nerve impulses are stopped.

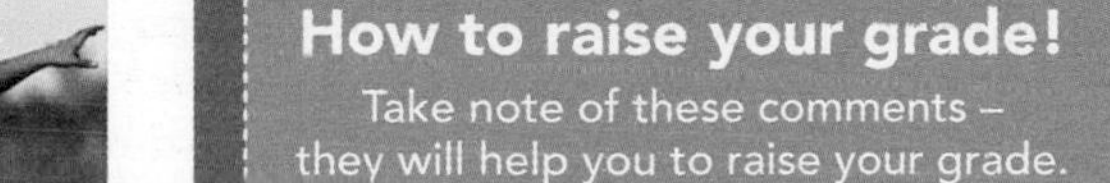

How to raise your grade!
Take note of these comments – they will help you to raise your grade.

The student has correctly identified stimuli. 1/1

The student has correctly identified receptors and given eyes as an example. 1/1

However, the answer "sending messages" is not considered a good enough answer to score a mark. The student should have written *The receptors generate nerve impulses, which are carried by the axons in neurones.* 0/1

This answer is too vague, and would not be credited. Optic nerve is the correct answer. Always try to give a complete answer. 2/3

The light is refracted, NOT reflected. The answer contains only two facts, yet 3 marks are available. The last sentence simply repeats the question. Only one fact is correct. 1/3

The answer is correct but not complete. There are 2 marks available. Reference to the impulse not reaching the brain is also required. 1/2

This student has scored 6 marks out of a possible 11. This is just below the standard of a Grade C. With more care the student could have achieved a Grade C.

B1 Exam-style questions

Higher Tier

1 Jenny is going on holiday to a tropical island. She needs to have certain vaccinations.

AO1 **(a)** Suggest why she needs vaccinations and describe the main stages of how immunisation works. [6]

AO2 **(b)** There are no vaccinations against malaria. Jenny is given this advice:

- carry an insect repellent spray
- wear long-sleeved tops and trousers at night
- sleep under a fine mesh net
- avoid areas of stagnant water.

What does this information tell you about how malaria is spread? [3]

[Total: 9]

2 The shoot and root of a plant react to light.

AO1 **(a)** Complete the sentences using some of these words.

geotropic phototropic photosynthetic negatively positively

Shoots are positively …………… and negatively………

Roots are ………… phototropic and ……………

……………… [2]

AO3 **(b)** The graph shows the effect of auxin concentration on the growth of a plant shoot and root.

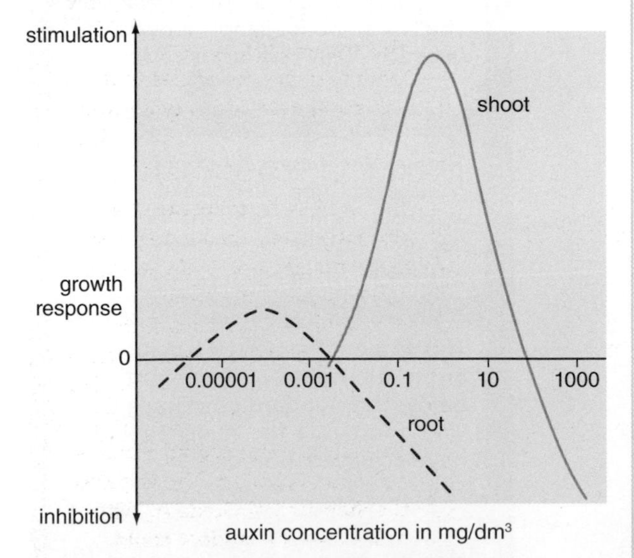

Describe and explain any use of these results. [6]

[Total: 8]

3 Look at the information about smoking.

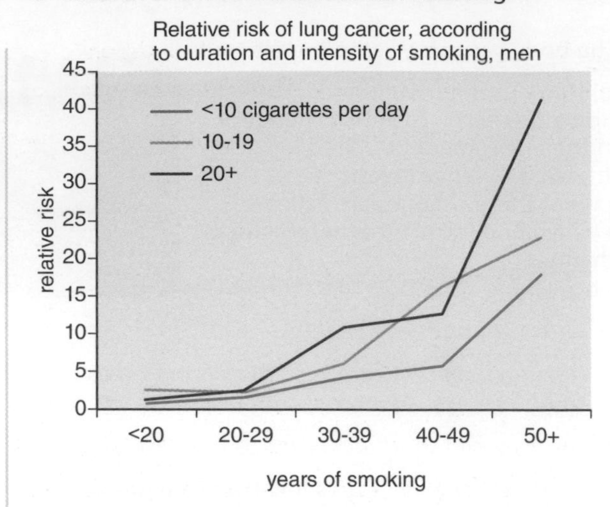

AO2 **(a)** Identify two factors that increase the relative risk of lung cancer. [2]

AO3 **(b)** Which duration of smoking groups and intensity of smoking group show the greatest increase of risk? [2]

[Total: 4]

4 Look at the data about deaths from respiratory diseases and circulatory diseases.

	Death in adults aged 35 and over				
	number of deaths			% of deaths	
	(m)	(f)	total	(m	f)
respiratory diseases such as emphysema	11 800	10 500	22 300	54	40
circulatory diseases including heart disease	13 500	8500	22 000	19	11

AO2 What patterns are shown by these data? [2]

[Total: 2]

5 The diagram shows how a nerve impulse is transmitted.

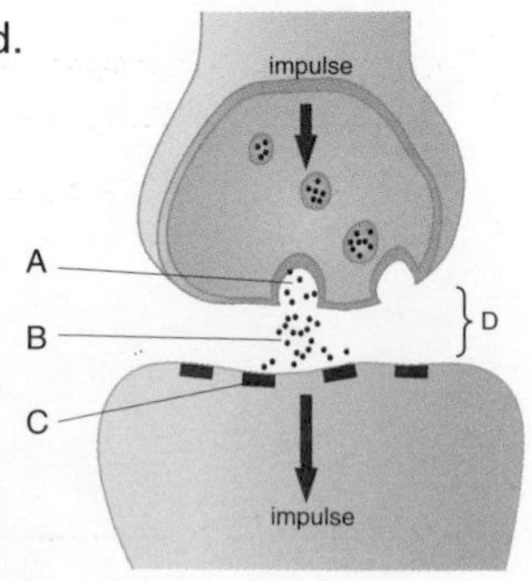

AO1 **(a)** Describe the role of structures A, B, C and D in the transmission of a nerve impulse. [4]

AO2 **(b)** What effect does a depressant drug have on this process? [2]

[Total: 6]

AO1 recall the science AO2 apply your knowledge AO3 evaluate and analyse the evidence

Worked Example – Higher Tier

Marfan's syndrome is a rare inherited condition.

The American President Abraham Lincoln had it.

People with Marfan's syndrome have very long arms, legs, fingers and toes.

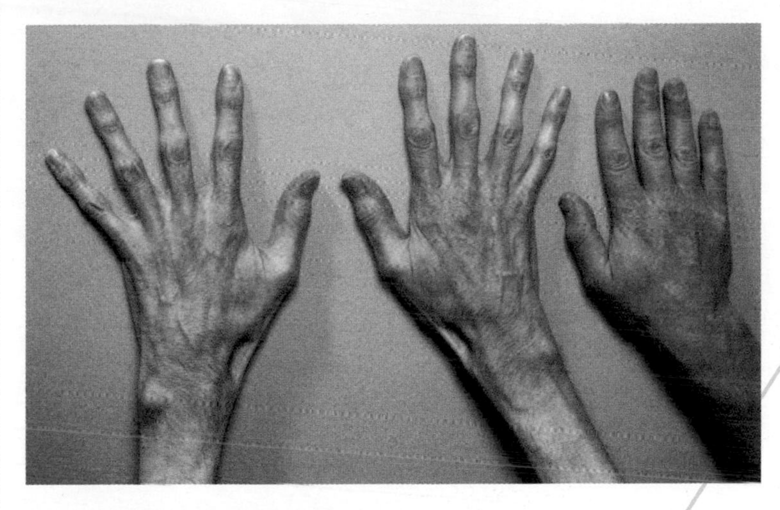

(a) The condition is caused by a faulty recessive allele (r).

What is a **recessive allele**? [1]

An allele that is recessive to a dominant allele.

(b) Two parents have the genotype Rr.

(i) What will be their phenotype? [1]

They will carry a recessive and a dominant allele for Marfan's syndrome.

(ii) Complete the genetic cross for these two parents. [2]

		parent	
		R	*r*
parent	*R*	*RR*	*RR*
	r	*Rr*	*rr*

(iii) Which combination of alleles will result in Marfan's syndrome? [1]

rr

(iv) Discuss the issues raised when two parents discover they both carry a faulty allele. [3]

They will have to decide whether to have children.

How to raise your grade!

Take note of these comments – they will help you to raise your grade.

This is a low level answer and would not score a mark. The correct answer is that it is only expressed in the absence of a dominant allele. 0/1

The student does not understand the meaning of the word phenotype. Learning the new words in genetics is important. The correct answer is *They will both be normal/will not have Marfan's syndrome.* 0/1

The student has been careless. This box should be Rr. Always check your answers, especially in questions on genetics. 1/2

The student has correctly identified the two alleles that will result in Marfan's syndrome. 1/1

This is a low level answer. It is also very brief. The student has made no attempt to discuss the issues, such as the chances (1 in 4) of having a child with Marfan's, or ethical arguments. 1/3

This student has scored 3 marks out of a possible 8. This is below the standard of Grade A. With more care the student could have achieved a Grade A.

B2 Understanding our environment

Ideas you've met before

Classification of organisms

Living things can be classified into different groups.

The toucan is a bird because it has feathers. It is also a vertebrate because it has a backbone.

Other vertebrate groups include mammals, fish, amphibians and reptiles.

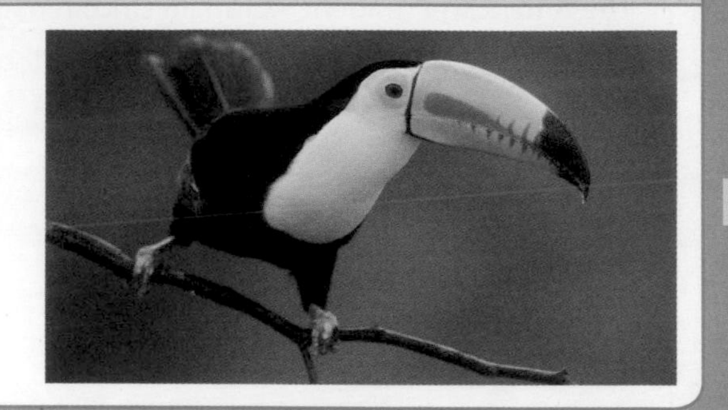

Name the group of organisms that do not have a backbone.

Energy flow and recycling elements

Food chains show what eats what. Energy from the Sun passes from one organism to another.

Some organisms feed on dead matter to release the elements. The recycled elements can then be used by plants.

Name two elements that can be recycled.

Adaptations

Animals and plants are adapted to survive in their habitats.

Dolphins are streamlined and have flippers to help them swim.

How is a bird adapted to fly?

Populations and survival

The population of a species can rise or fall.

When the environment changes, a species must adapt or become extinct.

The number of predators will affect a prey population.

If the number of predators goes up, what will happen to the number of prey?

In B2 you will find out about...

> what makes a species

> why some organisms are difficult to classify

> the characteristics of the arthropod groups

> how pyramids of biomass show the mass of organisms in a food chain

> how energy can be lost from a food chain

> decomposers such as fungi and bacteria

> how animals are adapted to hot or cold environments

> why plants can survive in the desert

> how predators and prey are adapted to survive

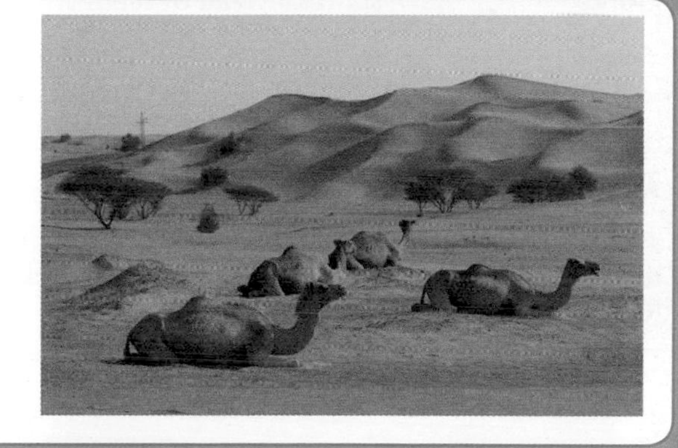

> how man has affected the environment and animal populations

> Darwin's theory of natural selection

> cyclic fluctuations of predator–prey relationships

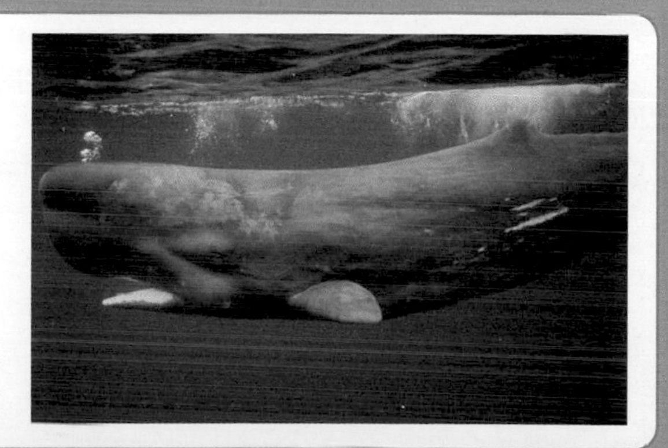

Classification

You will find out:

> how to classify animals and plants into different groups

> about the characteristics of the arthropods

Classify organisms

Look at the four pictures. We can group them in two ways. We could group all those animals that fly or we could group all those that have feathers. The group depends on the characteristics we choose.

Scientists put living organisms into groups because they are similar in some way. They put birds into a group because they all have feathers and a beak. Grouping organisms is called classification.

FIGURE 1: Three of these creatures can fly. Which is the odd one out? Three of them have feathers. Now which is the odd one out?

The five kingdoms

Organisms can be sorted into five different kingdoms.

Kingdom	Characteristics	Examples
plants	use light energy to produce food during photosynthesis; cells have a cell wall made from cellulose	roses, oak trees, wheat
animals	feed on other organisms; multicellular, which means their bodies are made up of lots of different cells; no cell walls	slugs, ladybirds, lions
fungi	make spores instead of seeds when they reproduce; cells have a cell wall made from chitin	yeasts, mushrooms, moulds
protoctista	most made up of just one cell	algae, *Euglena, Amoeba*
prokaryotes	have no nucleus; have a cell wall but not made from cellulose	bacteria

Arthropods

The different kingdoms can be split into smaller groups. Arthropods belong to the animal kingdom. They are invertebrates that have jointed legs, a segmented body and a hard exoskeleton made of chitin.

The arthropods are split into groups called classes:

> insects have three body sections and six legs, e.g. beetle

> arachnids have two body sections and eight legs, e.g. scorpion, spider

> crustaceans have two body sections and at least ten legs e.g. crab

> myriapods have two body sections and lots of legs on their body, e.g. millipede.

FIGURE 2: Why is the scorpion classified as an arthropod and an arachnid?

Questions

1 Sort the following organisms into their correct kingdom:

• wasp • rabbit • tulip • bread mould • *E. coli* bacteria

2 Name the four arthropod classes.

3 Which substance is found in the cell wall of a mushroom and the skeleton of a spider?

organism kingdoms phylum

Complete classification

All organisms are classified into a number of different groups, starting with their kingdom and finishing with their species.

For example, the black rat:

1. > Kingdom – animal
2. > Phylum – chordate
3. > Class – mammals
4. > Order – rodents
5. > Family – muroidea
6. > Genus – *Rattus*
7. > Species – *rattus*

Difficulties in classifying organisms

Different kinds of living things are on a continuous spectrum, showing different stages of evolution. Also, new organisms with unusual characteristics are always being discovered. This makes it difficult to place organisms into distinct groups.

The platypus is an animal that lives in Australia. It produces milk for its young, which suggests that the platypus belongs to a group of animals called mammals. However, the platypus represents a stage in evolution when mammal ancestors still laid eggs. This means that, unlike other mammals, it does not give birth to live young. For this reason the platypus is placed in its own subclass within the mammals.

The importance of classification

Classification can show both evolutionary and ecological relationships. The platypus is placed in the same class as other mammals because it shows a stage in the evolution of mammals. Whales and dolphins are also mammals but have been placed in the order Cetacea. This order identifies different species of mammals that all live and feed in water. By classifying them in this way we can understand something about their ecological relationships.

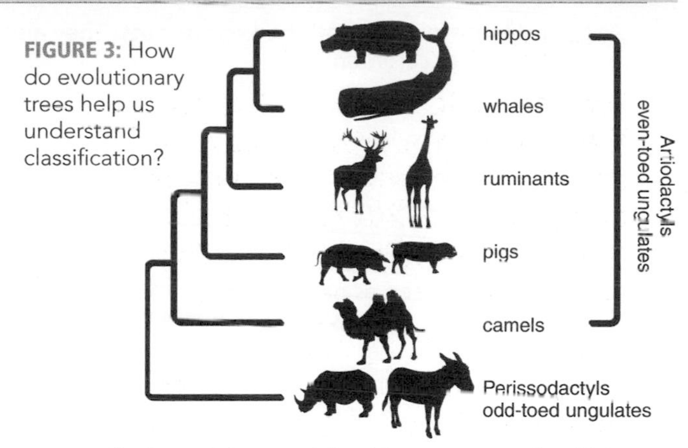

FIGURE 3: How do evolutionary trees help us understand classification?

Evolutionary trees

Evolutionary trees are diagrams that show close relationships between animals and their common ancestors. This one suggests that whales are closely related to hippos.

Questions

4 Which group would contain the most organisms, a kingdom or a genus?

5 Which class of organisms do humans belong to?

Artificial and natural classification

Organisms are often classified in two ways: artificially or naturally.

Artificial classification is based on one or two simple characteristics to make identification easier. For example, puffins, penguins and herring gulls are called seabirds because they live on or near the sea.

Natural classification is based on evolution. According to this, organisms are placed into smaller and smaller groups depending on recent common ancestors. For example, puffins and herring gulls are both birds, but puffins belong to the family Alcidae (auks) and herring gulls to the family Laridae (gulls and terns).

Changing classification

DNA sequencing has enabled scientists to become more certain of how closely related organisms are. The closest relation to an elephant was thought to be an animal called a hyrax. Evidence from DNA now suggests it is a manatee. The DNA evidence involved identifying the genes for similar characteristics. If these gene sequences are similar in both organisms then the species are related. The more sequences they have in common the more closely related they are. This process would not have been possible when animals were first classified because it requires computers to identify common genes.

FIGURE 4: What have scientists used to place the manatee and the elephant into similar groups?

Questions

6 Why might an organism be moved from one group of classification to another?

7 Which phylum do the elephant, manatee and hyrax all belong to?

Dogs and hamsters

Dogs and hamsters can both be kept as pets. They also have other characteristics in common:

> they have fur
> they give birth to live young.

Dogs and hamsters are in the same group, the mammals, because they share these characteristics.

But dogs and hamsters are also very different. They are different species.

All the dogs we keep as pets belong to the same species. But they can look very different. This is because humans have bred lots of different types of dogs.

You will find out:

> what the term 'species' means
> that members of the same species can be very different

Did you know?

There are more than 1000 different breeds of dogs in the world.

Labrador

Poodle

Labradoodle

FIGURE 5: How have humans made lots of different types of the dog species?

The dogs in the picture are very different, but they belong to the same species. They all:

> have fur
> eat meat
> have sharp claws and teeth
> can mate and give birth to a litter of puppies.

Species and habitats

The white whale and the narwhal both live in the same habitat: the cold Arctic Ocean. Similar species often live in similar habitats because they share features that allow them to survive there.

FIGURE 6: How are the narwhal and the white whale similar?

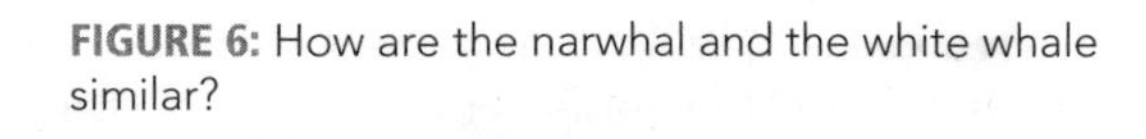

Questions

8 Describe two characteristics shared by hamsters and dogs.

9 Look at the pictures of the labrador and the poodle. Describe one way in which they look different.

What is a species?

A tiger and a lion are both cats. However, they are two different species. Lions look different from tigers; the most obvious difference being their lack of stripes.

> Members of the same species can breed. Lions breed with other lions to make young lions. These young lions will be fertile. This means they can grow up and have young of their own.

> Members of different species cannot naturally breed together. A lion will not naturally breed with a tiger. However, scientists have artificially mated a lion with a tiger to produce a tigon. The tigon is sterile. This means that it cannot reproduce.

All cats belong to the same family. The family is called Felidae. Each species of cat is given its own scientific name. Lions have the name *Panthera leo*. Tigers are called *Panthera tigris*. They both have the same first name, which means they are closely related. This close relationship is the reason why scientists can breed from them.

Panthera tigris is the bionomial name for the tiger, and is made up of the genus and species names. Binomial names are used so that people who speak different languages can use the same name for each species.

The lion and the tiger share a recent ancestor but have evolved to live in different habitats. Lions live in open grassland, whereas tigers prefer forest.

Their different coat patterns are part of their adaptation to their habitats, and act as camouflage. This means they blend in with their surroundings, making them difficult to see. Because of this, they are able to surprise their prey during a hunt.

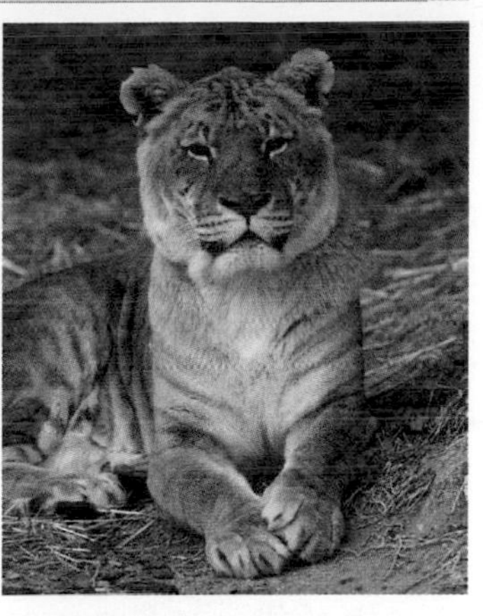

FIGURE 7: Will the tigon be able to have young?

Did you know?

What do you get if you cross a zebra with a horse?
A zorse.

What do you get if you cross a zebra with a donkey?
A zonkey.

It may sound like a joke, but these animals can exist.

Zorses and zonkeys are not classed as species because they cannot reproduce.

Questions

10 Explain why a lion and a tiger are different species.

11 Suggest one reason for the different coat patterns of a lion and a tiger.

Organisms that are difficult to classify

Organisms are often difficult to classify because they are constantly changing due to evolution. There are also many examples of organisms that don't quite fit the ideas of classification.

Bacteria

Members of the same species can interbreed to produce fertile offspring. Bacteria do not interbreed; they reproduce asexually. This means they simply split in half to produce new offspring. Therefore, bacteria cannot be classified into different species using the 'fertile offspring' idea.

Dolphins and whales

Dolphins and whales are special animals. They have many of the characteristics of mammals, yet they look and live like fish such as sharks.

Their ancestors lived on land, but gradually evolved to live in the sea. However, they are still classed as mammals because they give birth to live young and produce milk to feed them.

Mules

Hybrids are the result of breeding two animals from different species, such as the tiger and the lion to produce a tigon.

One of the first successful hybrids was produced from a male donkey and a female horse. This produced a mule. It was bred to provide a strong powerful animal.

However, because hybrids cannot breed (they are sterile), a mule is not a species and therefore is difficult to classify.

Questions

12 Suggest which part of bacteria scientists use to classify them into species.

13 Discuss the problems of classifying hybrids.

14 Would you expect dolphins to have lungs or gills? Give reasons for your answer.

Energy flow

You will find out:

> about energy transfers in food chains
> about pyramids of numbers and pyramids of biomass

Animals and plants

There are about 10 million different animals and plants in the world. The badger is just one of them. To survive the badger needs to find lots of different types of food. This includes earthworms that live in the soil. If there were no earthworms the badgers would have to eat other things like roots and seeds. If there were no badgers there would be more earthworms.

Animals and plants living in the same habitat depend on each other and on energy from the Sun.

FIGURE 1: Why is the Sun so important to the badgers and other animals in the world?

Food chains

Each time an organism is eaten, energy is transferred from that organism. This energy transfer is shown in a food chain.

> The energy at the start of a food chain comes from sunlight. Plants use this energy in photosynthesis. Plants are called **producers** because they produce food.

> All other organisms in a food chain are called **consumers** because they get their food from other organisms.

> Food chains can be linked to form food webs. The links in food chains and food webs show different **trophic** (feeding) **levels**.

> Energy is passed from one organism to another, from one trophic level to another.

Food chains usually start with green plants. However, algae are also producers, so they can start off a food chain. Consumers can often belong to more than one trophic level. When badgers are eating seeds they are primary consumers; when they eat earthworms they are secondary consumers.

Remember!
When drawing food chains the direction of the arrow is important. It shows the flow of energy.

Questions

1 Where does the energy in a food chain come from?

2 Why are plants called producers?

3 What is meant by the term 'trophic level'?

4 Why can badgers be primary and secondary consumers?

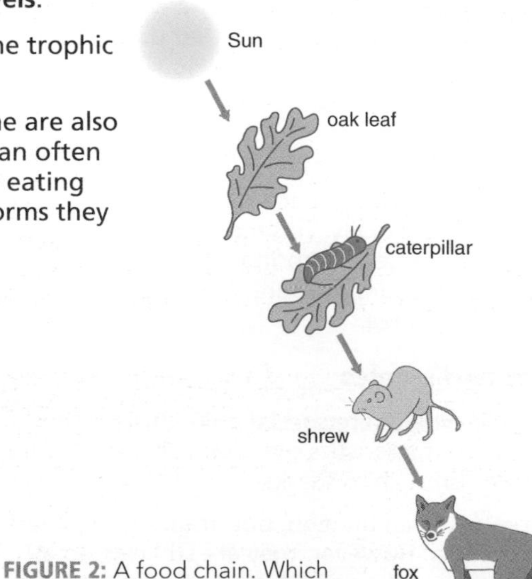

FIGURE 2: A food chain. Which organisms are the consumers?

food webs   food chains

Food pyramids

For a particular area or ecosystem the numbers of organisms at different trophic levels can be counted and the information shown in a pyramid of numbers. An animal that eats plants is called a **herbivore**. An animal that eats other animals is called a **carnivore**.

A pyramid of **biomass** is a better way of showing trophic levels. The mass of the organisms is used. However, there are changes in 'wet' mass over time so dry mass is usually used.

The pyramid of numbers can have an odd shape if the producers are very large or if a small parasite lives on a large animal. However, the large mass of the producer or the small mass of the parasite means the biomass always shows a pyramid shape.

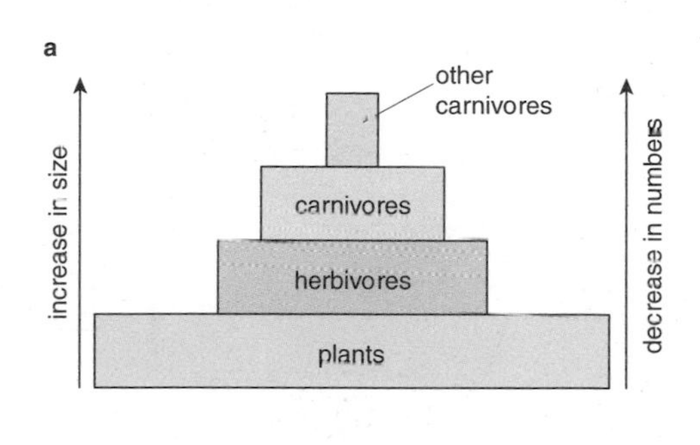

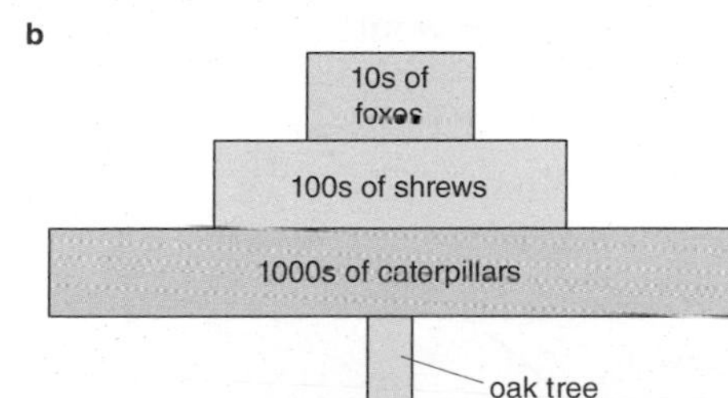

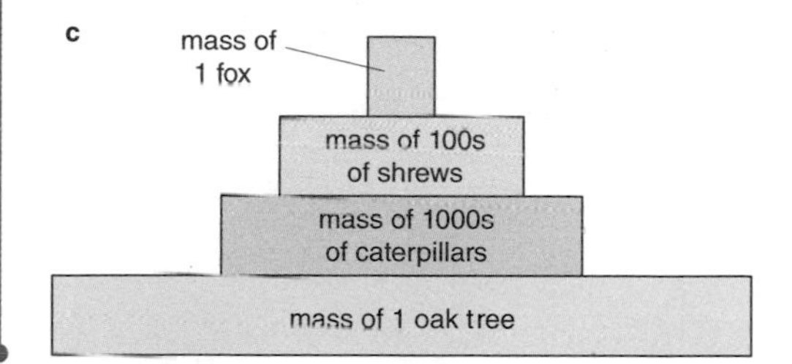

FIGURE 3: **a** A pyramid of numbers; **b** a pyramid of numbers for a woodland; **c** a pyramid of biomass for a woodland. Why is the pyramid of biomass for a woodland a different shape to the pyramid of biomass?

Questions

5 Ten rabbits eat 100 carrots and one fox eats the 10 rabbits. Draw the pyramid of biomass for this situation.

6 Describe the difference between a pyramid of numbers and a pyramid of biomass.

Pyramid problem

The pyramids in Figure 3 are very simple food chains. The oak tree is the first trophic level and the fox the last trophic level.

Some animals will eat both animals and plants. These animals are called **omnivores**. Omnivores can fit into more than one trophic level. It is difficult to construct a pyramid if there are omnivores in the food web.

To construct a pyramid of biomass, the dry mass of each organism needs to be found. To find the dry mass of an organism, the animal or plant needs to be placed in a warm oven until all the water has been lost from it. This is not possible with live organisms, which means that in many cases biomass has to be estimated.

Questions

7 Why are omnivores difficult animals to fit into a pyramid of numbers?

8 Explain why it would be difficult to calculate the biomass of all the organisms in a food chain.

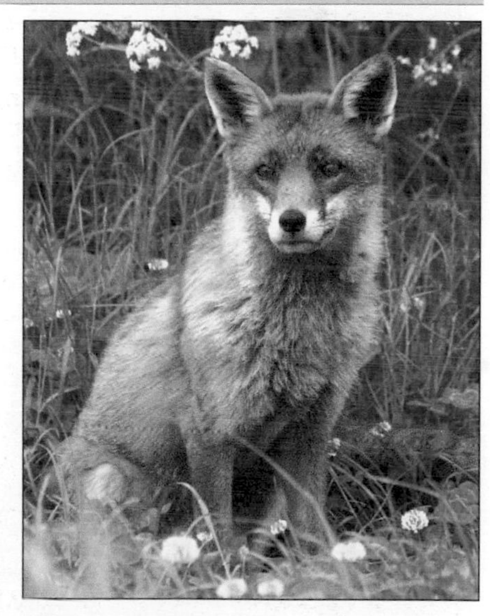

FIGURE 4: Foxes eat many different things, including worms and birds and fruit. Why would it be difficult to construct a food web that included a fox?

Food webs

> **You will find out:**
> - about why all organisms need plants to survive
> - about how energy is lost from a food chain
> - about efficiency of energy transfer

Animals in any habitat will usually eat more than one type of food.

The relationship between all the animals and plants in a habitat can be seen in a food web.

> All the animals in the food web rely on the plants. Some of the animals eat the plants and others eat the animals that eat the plants. If there were no plants nobody would have any food.

> The numbers of animals and plants in the food web are always changing. When the numbers of one species change it can affect another species.

Look at the food web in Figure 5. What would happen to the number of bees if there were no roses?

Energy enters the food chain from the Sun when plants **photosynthesise**. The energy flows through the food chain as animals feed.

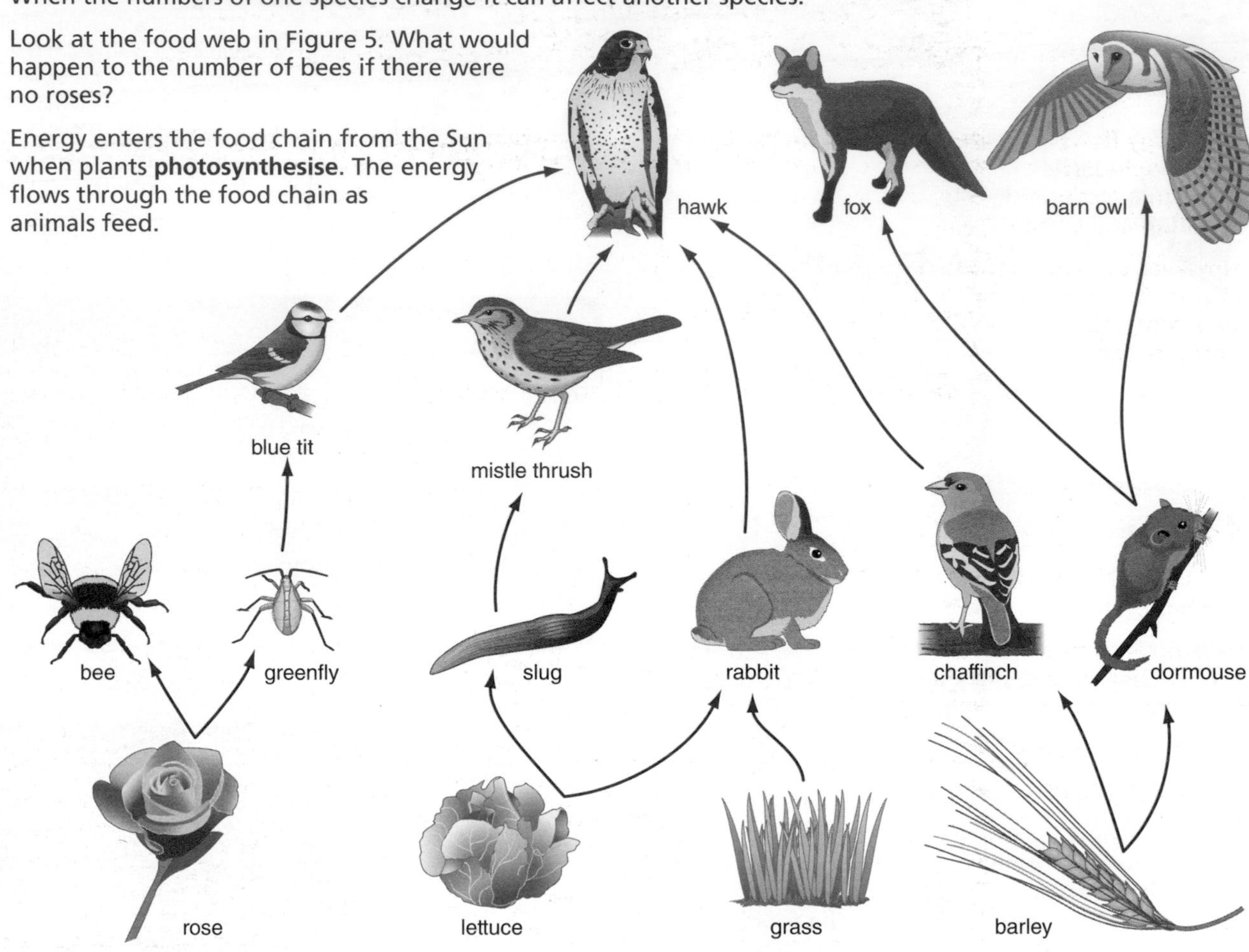

FIGURE 5: Blue tits feed on greenfly. Why do blue tits need there to be plenty of roses?

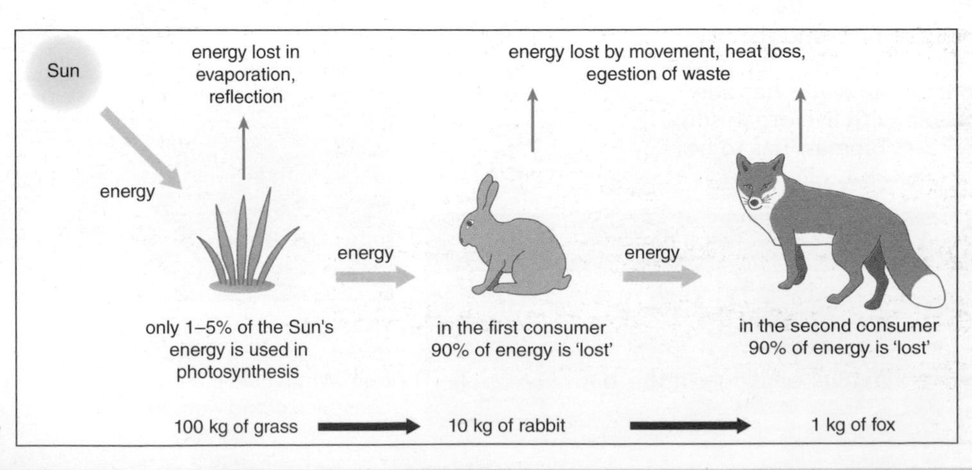

FIGURE 6: Energy flow in a food chain. In what ways is energy lost in a food chain?

energy in the food chain   decomposers

Questions

9 Use the food web in Figure 5 to answer these questions.

a Write down the name of one producer in the food web.

b What would happen to the number of dormice if there were more owls?

c One year there were more slugs than normal. Explain why that year there were fewer lettuces.

10 If a rabbit eats 200 g of grass, how much of the energy in the grass is lost from the food chain? Use Figure 6 to answer this question.

Energy flow

As energy flows along a food chain, some is used for growth. At each trophic level as much as 90% of the energy is transferred into other less useful forms such as heat from **respiration**, **egestion** and **excretion**.

However, some of the energy lost can be used in other food chains. The energy lost in the excretory products, faeces and uneaten parts can be used by decomposers to start another food chain.

Questions

11 Energy lost from a food chain can start another food chain, explain how?

12 Look at Figure 6. Energy is lost from the fox as heat. Which process in the fox makes heat?

Efficiency of energy transfer

The shape of a pyramid of biomass shows that the energy level decreases with increasing trophic level.

Because each trophic level 'loses' up to 90% of the available energy, the length of a food chain is limited to a small number of links. An animal at the end of a long food chain is vulnerable, as it depends on earlier food links not being interrupted.

The efficiency of energy transfer can be calculated as shown below.

There are 3056 kJ of energy in 1 m^2 of grass and only 125 kJ are used for a cow's growth. The efficiency of energy transfer can be calculated:

$$\text{efficiency} = \frac{\text{energy used for growth (output)}}{\text{energy supplied (input)}} \times 100$$

efficiency = 0.04 or 4%

This shows that the energy transfer from grass to cows is very inefficient.

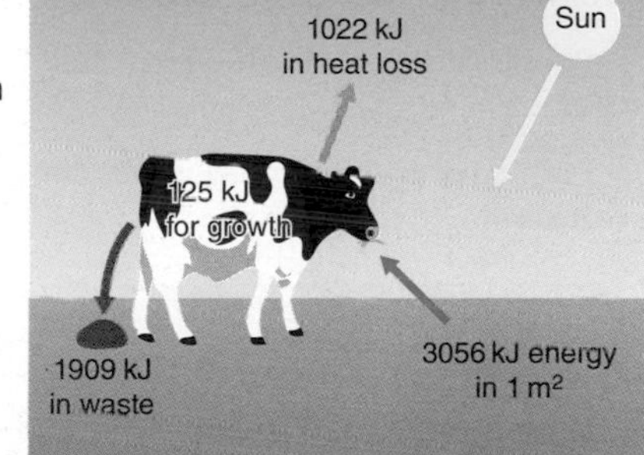

FIGURE 7: Calculating the efficiency of energy transfer.

Questions

13 Explain why food chains rarely have more than five links.

14 Explain why humans would have a better energy efficiency if they were vegetarians.

Recycling

You will find out:

> how elements are passed on to different organisms

- how carbon is recycled in nature
- how some carbon is locked up in limestone

Rubbish disposal

About 30 million tonnes of rubbish are thrown away annually in the UK. Scientists are always trying to solve the problems of what to do with it.

The chemicals in plants and animals are less of a problem because they are recycled naturally.

FIGURE 1: Why are landfill sites filling up?

Recycling naturally

Fish in a tank are fed every day. The fish eat the food and digest it. They use the digested food molecules to build up new chemicals for growth.

If some of the fish die, they can be buried and will **decay**. Chemical elements released during decay, such as carbon and nitrogen, are recycled.

The elements are recycled when plants take them in and use them for growth.

> Plants recycle carbon when they take in carbon dioxide.

> Plants recycle nitrogen when they take in nitrates.

FIGURE 2: What do fish use digested fish food for?

Questions

1 What does a fish do to its food?

2 What does a fish use the chemicals in its food for?

3 What happens to a dead fish when it is buried?

4 Write down the names of two elements that are recycled.

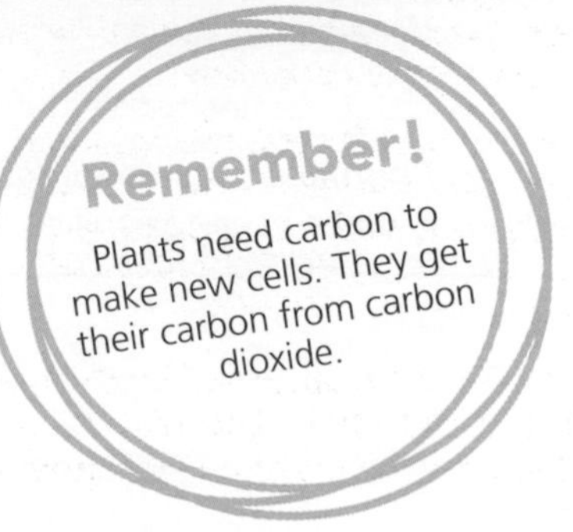

carbon cycle recycling rubbish

The carbon cycle

Carbon is one element that is recycled naturally. Carbon dioxide is a compound that contains carbon.

The atmosphere contains approximately only 0.04% carbon dioxide. However, this is enough to supply all the plants in the world. Plants use carbon dioxide to make food in photosynthesis.

As plants are eaten, the carbon in them passes on to animals. Carbon passes along food chains and webs, even when plants and animals die and **decompose**.

Carbon dioxide is put back into the atmosphere by:

> plants and animals respiring

> decomposers such as bacteria and fungi in the soil respiring

> burning (combustion) of fossil fuels, such as coal and oil.

Recycling of nutrients takes longer in acidic or waterlogged soils than in well-drained soils, because the lack of oxygen hinders microbial respiration.

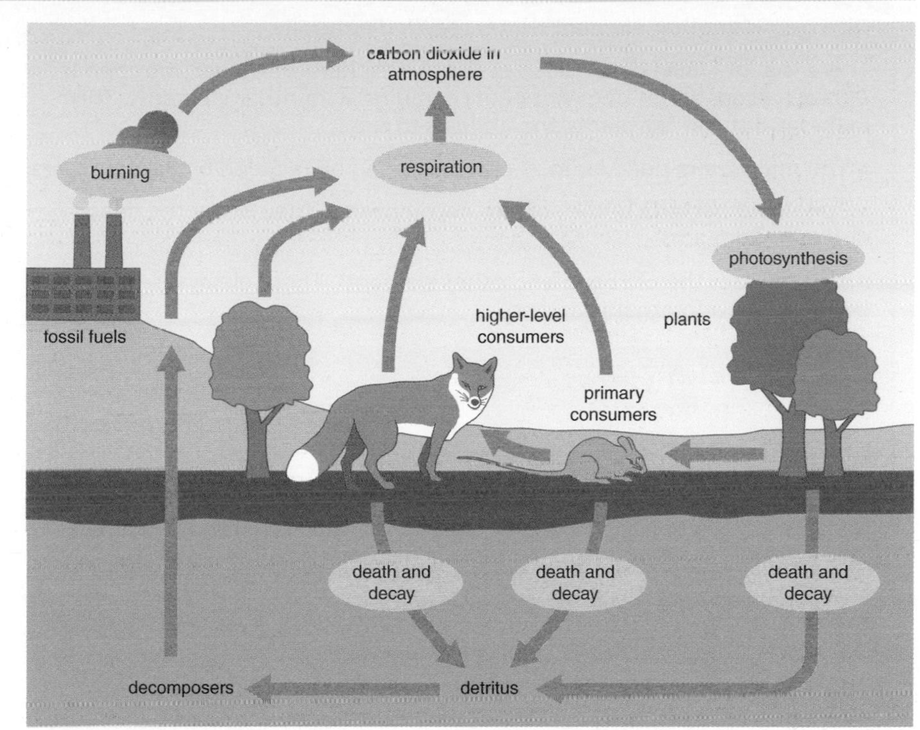

FIGURE 3: The carbon cycle.

Questions

5 Why do plants need carbon dioxide?

6 Suggest what would happen to life on Earth if there were no carbon dioxide.

7 Write down two processes that release carbon dioxide.

More on carbon cycling

Carbon is also recycled in the sea.

The shells of marine organisms such as molluscs contain carbonates. Other marine animals such as corals and microscopic algae also cover themselves with calcium carbonate.

When these organisms die, their shells collect on the sea floor. Over millions of years they form a type of sedimentary rock called limestone.

Limestone is a very soft rock. It is attacked by rain that is often acidic. Carbon dioxide reacts with the rain to form carbonic acid. The acid rain weathers the rock and carbon dioxide is released.

Carbon dioxide can be absorbed by the oceans and held in a 'carbon sink'. Lots of carbon dioxide are also released from erupting volcanoes and forest fires.

FIGURE 4: How does the carbon in dead coral become carbon dioxide?

Questions

8 Limestone is not a good choice to use in city buildings. Explain why.

9 Scientists describe limestone as a useful 'carbon sink'. Suggest why.

10 Why do forest fires release a lot of carbon dioxide?

limestone and weathering

Recycling nitrogen

Nitrogen is another element that is recycled naturally.

There is a lot (78%) of nitrogen in the atmosphere. However, nitrogen is an unreactive gas and it does not easily combine with other elements. This makes it difficult for plants and animals to use.

> The nitrogen in dead animals and plants can be recycled by **decomposers**.

> Fungi and bacteria in the soil are decomposers. They decay the dead animals and plants.

> This releases the nitrogen and other elements. These elements can then be used by living organisms.

Questions

11 Why is nitrogen gas difficult for plants to use directly?

12 Write down two types of decomposers.

You will find out:

- how nitrogen is recycled in nature
- why plants find nitrogen difficult to obtain

Remember!

Plants have to take nitrogen in from the soil. They can't use the nitrogen in the atmosphere directly.

The nitrogen cycle

Plants and animals need nitrogen to make proteins for growth and for making enzymes that control the reactions necessary for life.

Nitrogen passes along food chains and webs

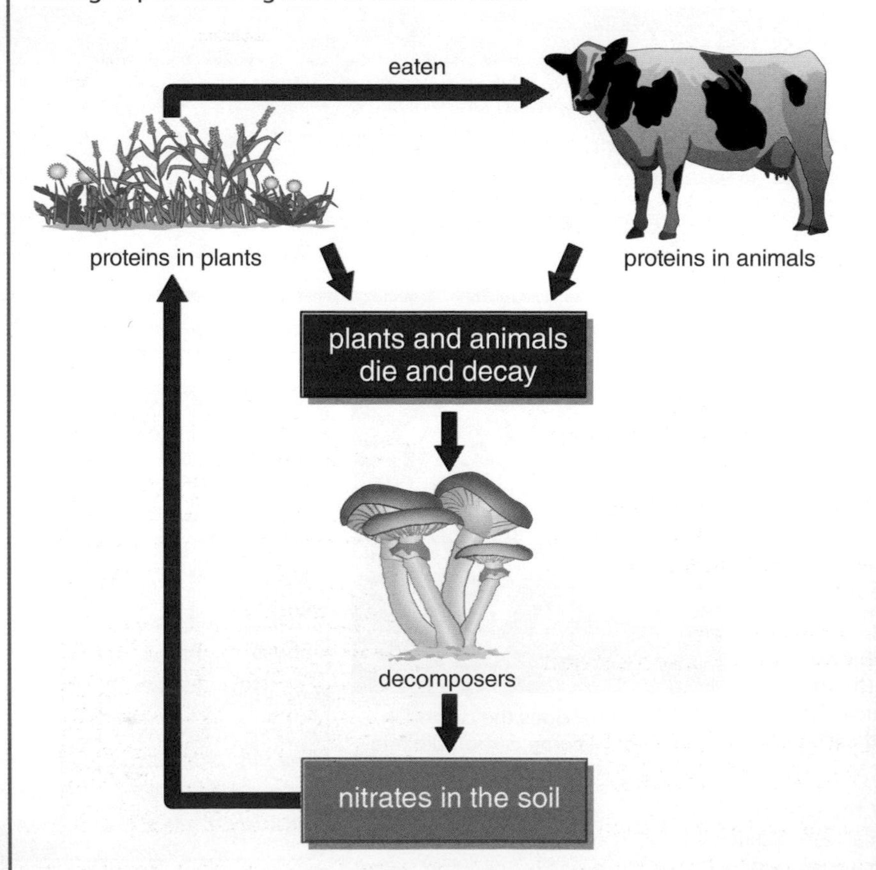

FIGURE 5: How do proteins in plants become nitrates in soil?

Did you know?

Each day, lightning produces about 250 000 tonnes of nitrogen fertiliser in the world. This is because the heat from the lightning makes the nitrogen react with oxygen. Nitrates form and are used by plants for growth.

Questions

13 Why do plants and animals need nitrogen?

14 Suggest how nitrates get into plants.

15 What happens to the proteins in dead plants and animals?

ammonia decomposer denitrifying bacteria insectivorous

More on nitrogen cycling

Decomposers such as soil bacteria and fungi are important links in the nitrogen cycle. They break down proteins in dead bodies and urea in liquid animal waste, releasing ammonia.

A baby's nappy smells of ammonia because bacteria convert the urea in urine into ammonia. This can cause 'nappy rash' on the baby's skin because ammonia is a strong alkali.

Although there is a lot of nitrogen in the atmosphere, plants rely on the following bacteria to make it available.

> **Nitrifying bacteria** convert ammonia into nitrates.

> **Nitrogen-fixing bacteria** in the soil or in the root nodules of peas, beans and clover 'fix' nitrogen gas and convert it into ammonia or nitrates and then into amino acids to form proteins.

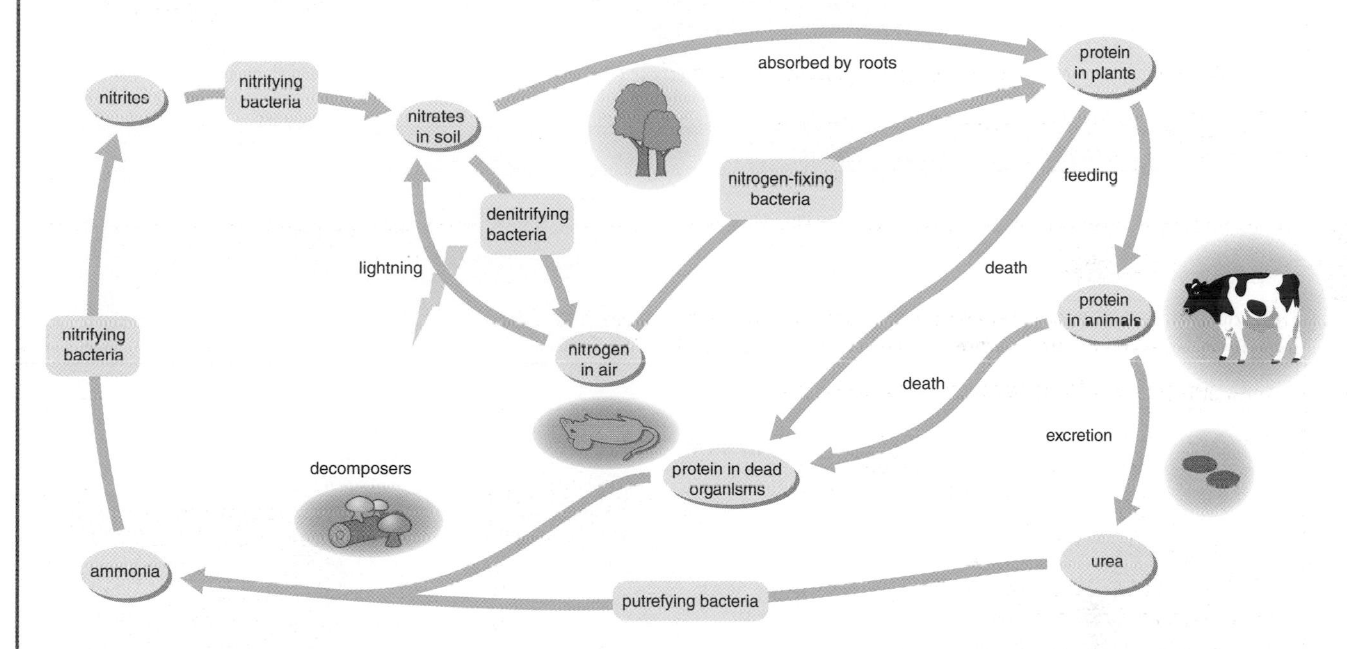

FIGURE 6: Why are denitrifying bacteria a problem for plants?

Some bacteria break down nitrates in the soil and release nitrogen back into the atmosphere. These are called **denitrifying bacteria**.

Lightning also fixes nitrogen gas in the atmosphere. During a thunderstorm, energy from the lightening is used to combine nitrogen and oxygen to form nitrogen oxides. These oxides dissolve in the rain and form nitrates in the soil.

A few plants obtain their nitrogen in a different way. Insectivorous plants such as the pitcher plant catch insects, digest their bodies and absorb the nitrates.

Questions

16 Suggest a suitable habitat for an insectivorous plant.

17 Explain why nitrifying bacteria are needed by animals.

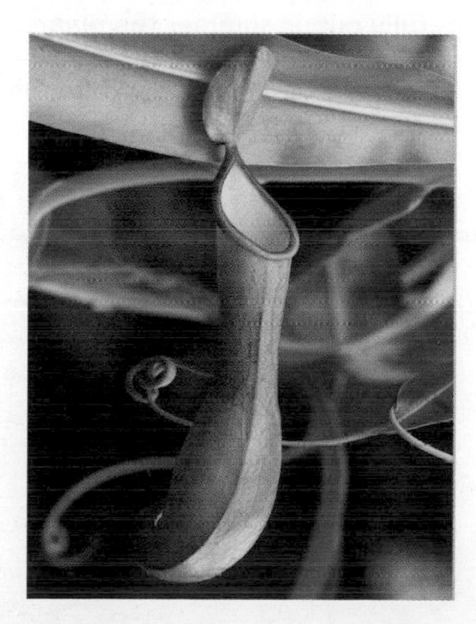

FIGURE 7: The pitcher plant. How does it obtain nitrogen?

Interdependence

You will find out:

- > what animals and plants compete for
- > how competition changes a population

Keeping a mate

Male elephant seals keep several females to mate with. They have to stop other males from trying to steal them.

They do this by fighting each other.

Did you know?

A male elephant seal can weigh 2700 kg.

That is about the same as 35 men.

FIGURE 1: These male elephant seals are confronting each other ready for a fight. What do they compete for?

Competition and population

Plants and animals compete for food, water, shelter, light and minerals to survive. If they do not, their species will die out.

Swallows and martins are two birds that compete for the same insect food.

The distribution and **population** size of these birds never stays the same. In summer, you will see lots of them because there is plenty of food. In winter, they must fly south because there is little food left in Britain.

Plant competition

If you go for a walk in a beech wood in spring, you will find lots of different plants growing. Yellow primroses and bluebells are two examples.

These plants grow and flower in spring to catch as much light as possible before the beech tree leaves are fully out. In summer the plants find it difficult to grow as the larger trees take most of the light, water and minerals. The plants and the beech trees are in competition with each other.

FIGURE 3: What plants are in competition in this picture?

FIGURE 2: House martins (right) and swallows (left) compete for the same food. Can you tell from their names where they build their nests so that they do not compete for the same nest sites?

Remember!

If you are asked about animals or plants you have never heard of, don't be put off. All animals and plants compete for similar things.

Questions

1 Why is it unusual to see a swallow during a British winter?

2 Explain why bluebells grow before the trees in the wood are in leaf.

3 Pandas are unusual animals because they eat only bamboo. Suggest one reason why a panda could not live in a beech wood.

ecological niche plant competition invasive species

Animal competition

Competition for mates

Bowerbirds live in Australia. The male bowerbird likes to collect brightly coloured objects. He then spends hours arranging them in his display area, called a bower.

He does this to attract a mate. Like other animals, his species will die out if its members do not breed. The bowerbird is competing with other males of his species.

The male with the best bower attracts the females first. Often a male will steal objects from another male when it is away from its bower.

Competition for food

Animals of different species often eat the same food. Lions and cheetahs live in the African grasslands and compete for similar prey. However, the smaller cheetah tends to keep to its own small area away from the lion, which would try to steal its food.

When cheetah cubs grow up they have to move away from their mother's territory so they are not competing with her for food. If they stayed, the population of the area would grow too large and there would not be enough food to go round.

FIGURE 4: Why does the male bowerbird collect brightly coloured objects?

Questions

4 Describe two ways in which males of a species can compete for a mate.

5 Explain why cheetah cubs have to move away when they become adults.

6 A red squirrel and a grey squirrel are in much closer competition than a badger and a red squirrel. Suggest reasons why.

Ecological niches

All organisms have a role to play in an ecosystem. For example, the 'role' of a squirrel is to live in woods and eat acorns. This role is called the squirrel's **ecological niche**.

It is important that different species have slightly different niches. The seven-spot and two-spot ladybirds compete with each other for greenfly. However, the seven-spot ladybird eats larger greenfly. This way the two species of ladybirds have a slightly different niche.

Close competition

In Britain, there are two types of squirrel:

> the native red squirrel
> the American grey squirrel.

At one time they both occupied a similar niche, but in different countries. Now they compete for a similar niche in Britain. This has resulted in a decline in the number of red squirrels.

Interspecific and intraspecific competition

The competition between the red squirrel and the grey squirrel is called interspecific competition, because it is between two different species.

Competition between different grey squirrels would be called intraspecific competition. This type of competition often has a greater effect on the population, because animals of the same species will have the same ecological niche. By contrast, different species have slightly different niches.

a

b

FIGURE 5: **a** Red squirrel. **b** Grey squirrel. What has happened to the size of the red squirrel population in recent years?

Questions

7 What do we mean by the term 'ecological niche'?

8 Explain why introducing the grey squirrel to Britain caused a fall in the red squirrel population.

Friend or foe?

You will find out:
- how predators and prey affect each other
- why some organisms cannot live without the help of others

In any habitat there are lots of different animals living together.

- Some of these animals help each other out.
- Other animals hunt other animals for food.

Friends

An oxpecker is an African bird that feeds on insects.

The buffalo is a large mammal that feeds on grass.

The two animals are not in competition. Instead, they help each other out.

Oxpeckers sit on buffalo. They feed on the insects that live on the buffalo's skin. The oxpeckers help to clean the buffalo by getting rid of ticks and fleas. In return the oxpeckers get plenty of food.

FIGURE 6: The oxpecker is sometimes called a 'cleaner species'. Why?

Foes

A food chain shows who eats who. In this food chain the badger eats the shrew. The badger is called a **predator** and the shrew its **prey**. These two animals are foes.

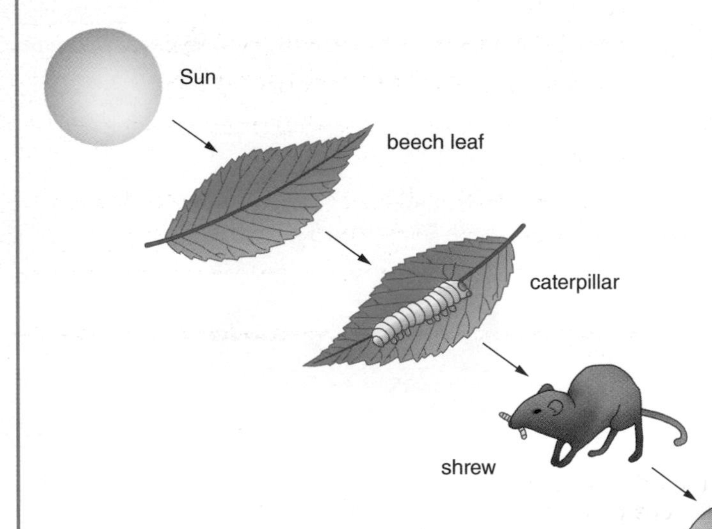

FIGURE 7: What is the badger's prey?

Questions

9 How is an oxpecker helpful to a buffalo?

10 Look at the food chain from a garden.

lettuce → snail → hedgehog

a One year the gardener decides not to grow lettuce. Why might the hedgehogs move to another garden?

b Slugs also feed on lettuce. If the slug population increases what might happen to the snail population?

c Explain why the snail population falls in winter, even though the hedgehogs hibernate.

11 Explain how a decrease in the shrew population results in a decrease in the badger population.

There are many ways in which animals of different species interact.

Predator–prey relationships

The badger is a predator. It hunts shrews as its prey. The number of shrews in a population will affect the number of badgers.

When there are more shrews, the badgers have more food so they can raise more young. This helps the badger population increase. The following year, the increased badger population eats more shrews, so the shrew population decreases.

Remember!

In the exam you may need to interpret data about population size and distribution to answer questions like 10 and 11 above.

cleaner species cyclic fluctuation food chain host legume mutualism

Parasitism and mutualism

Parasites

The tapeworm is a **parasite**. It lives in the digestive system of other animals, including humans. The tapeworm takes food away from its host so it can grow.

The tapeworm is unlikely to kill its host, as the tapeworm would also die. However, the host is likely to lose weight.

Cleaner species

The sharksucker is a fish that attaches itself to sharks. It cleans the shark's skin by eating its parasites. In return, the shark protects the sharksucker from predators. A relationship in which both animals benefit is called **mutualism.**

Populations and cyclic fluctuation

Predator–prey relationships play an important part in controlling populations.

The lemming is a small hamster-like animal that lives in the Arctic. Its population size can decrease and increase dramatically.

In years when there is a lot of food, lemmings reproduce rapidly. If the population grows too large, food becomes scarce and the female lemmings do not reproduce so rapidly.

The snowy owl hunts lemmings. In years when there are more lemming young, there are more prey and the owls successfully raise more young. The owl population increases, so the following year the lemming population decreases as there are more owls to hunt them.

This 'up and down' pattern or **cyclic fluctuation** can be shown on a graph. The numbers increase and decrease cyclically, but on average the populations of the lemmings and snowy owls stay the same.

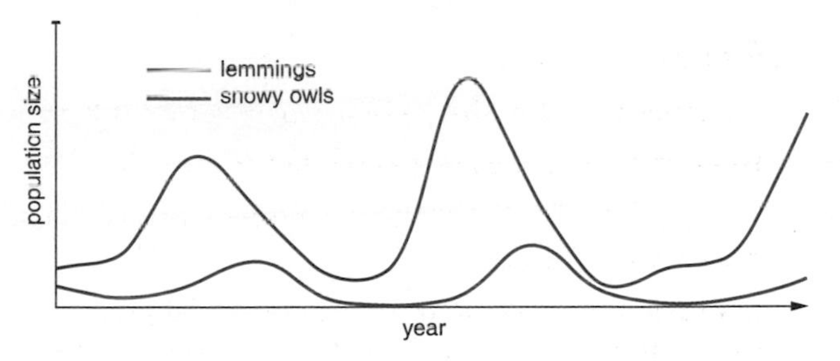

FIGURE 8: Why do the populations of lemmings and snowy owls have peaks and troughs?

Questions

12 Fleas live on the skin of dogs and feed on their blood.

a Explain how fleas benefit from dogs.

b Suggest how fleas may harm dogs.

c What type of relationship is this an example of?

13 The relationship between bees and flowers is an example of mutualism. Explain why.

More about relationships

Example of mutualism

Some species are totally dependent on others for their distribution and abundance. They cannot live without them. The pea plant is a **legume**. It has structures on its roots called root nodules. A type of bacteria lives inside the nodules that can convert nitrogen into nitrates. These bacteria are called nitrogen-fixing bacteria.

The bacteria give the pea plant extra nitrates to help it grow. The pea plant gives the bacteria sugar, which they turn into energy.

Out of phase fluctuation

The phases of predator fluctuation are always slightly behind the prey population. This is because the predator population can rise only if there is plenty of prey to feed the young. So while the prey numbers are low predator numbers will fall. When prey numbers are high again there is less competition for food, so more young survive to breed the following year.

Questions

14 Explain why the population of snowy owls falls soon after the population of lemmings falls.

15 Explain why it is important that the lemming population does not increase too much.

16 Describe two examples of mutualism.

Preparing for assessment: Applying your knowledge

To achieve a good grade in science, you not only have to know and understand scientific ideas, but you need to be able to apply them to other situations and investigations. These tasks will support you in developing these skills.

Food webs

Josh and Ben are down on the beach during their May half-term holiday. Their parents are having a barbecue and the food is taking longer to cook than they'd expected, so Josh and Ben go off to explore the rock pools at the water's edge.

One large rock pool that they spend quite a lot of time looking in has lots of living things in it.

By looking carefully and thinking about what they already know about some of the creatures, they sketch a simple food web in their notepad.

crab
sea anemone
shrimp
seaweed
mussel
barnacle
plankton

 Task 1

Read the information on which organisms Josh and Ben found.

> Which organisms in their food web are producers?

> Explain why the producers are not green plants. Which kingdom do they belong to?

> Sketch a pyramid of numbers for the food chain that includes plankton, shrimps and crabs.

> Explain why it would be difficult to construct a pyramid of biomass for this food chain.

 Task 2

Josh and Ben found two different species of crab. Crabs have a shield on their bodies called a carapace.

> Which class of arthropods do crabs belong to?

> Explain how scientists could prove the crabs were different species.

 Task 3

Josh tells Ben that the mussel shells contain carbonates.

> Explain how carbon in the atmosphere got into the shells of the mussels.

> Explain how mussels that lived millions of years ago are now contributing to carbon dioxide levels in the atmosphere.

 Task 4

Although the boys watch for some time, they never see any crabs being caught or eaten. Josh thinks that the crabs are in a good situation here and that their numbers will increase. Ben does not agree; he thinks the numbers will reach a maximum and then start to fall.

Who is correct? Explain your answer.

 Maximise your grade

Answer includes showing that you can...	
	Identify the producers and explain why they are not green plants.
	Identify which kingdom the producers belong to.
	Identify which class the crabs belong to.
	Explain how the population of predators and prey regulate one another.
	Draw a pyramid of biomass.
	Use ideas about the carbon cycle to describe how the carbon got into the shells.
	Use ideas about measuring biomass and trophic levels to explain why it may be difficult to construct a pyramid of biomass.
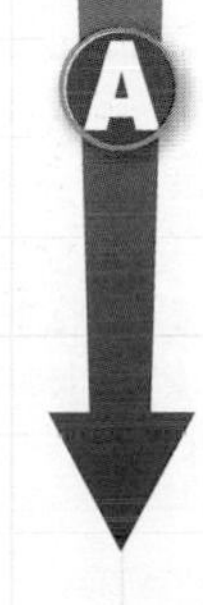	Explain why the population of predators may lag behind the prey population.
	Use ideas about DNA or fertile offspring to explain the term *species*.
	Explain how carbon is recycled, using ideas about marine organisms, limestone and weathering.

Adaptations

You will find out:

- > how animals and plants are adapted to their environment
- > how organisms are adapted to hot or cold habitats
- > how organisms are adapted to dry habitats

Adapted to feeding on dead bodies

Ever wondered why a vulture has a long neck and a bald head?

The answer is simple; a vulture feeds on dead bodies that predators leave. It needs a long neck to reach far inside the body to get at whatever meat is left.

If vultures had feathers on their head, the feathers would get covered in blood and flesh every time they ate. Head feathers are difficult for the bird to clean and would quickly become a home for parasites and bacteria. A bald head offers no hiding place for parasites and allows any blood to dry quickly in the hot sun, killing any parasites that do try to grow.

FIGURE 1: Vultures feeding. Why do vultures have a bald head?

Adapting to compete

Every living thing is adapted to live in its **habitat**.

> Polar bears are adapted to live in cold places.

> Camels are adapted to live in hot arid places.

Organisms with more successful adaptations are better at competing for food and other resources.

For example, koala bears have adapted to eat eucalyptus leaves. These leaves are poisonous to most other animals. Therefore, in a eucalyptus forest koalas are more likely to survive.

Questions

1 Match up the following living things with the habitat they are adapted to live in.

Living thing	**Habitat**
polar bear	soil
cod fish	Arctic
cactus	sea
earthworm	desert

2 Sharks compete with other sharks for food. The better-adapted sharks get more food. Write down one way in which a shark might be better adapted.

FIGURE 2: Why would a koala not survive in the Arctic?

Did you know?

Camels lose nearly 30% of their weight when they go without water for a long time.

To recover, they can drink quickly – 100 litres in just a few minutes.

Adapting to hot and dry environments

The desert is an example of a hot environment. Temperatures can reach as high as 40 °C.

To live in the desert, an animal has to be well adapted to survive the heat and scarcity of water and compete with other animals. Some of these adaptations are anatomical.

Camels and elephants are examples of an animal that lives in hot places.

The only body fat a camel has is in its hump. Fat all over its body would insulate the camel, causing it to overheat.

Elephants have large ears to increase surface area for heat loss from their body.

Camels do not need to sweat. Their body temperature can rise a bit above normal without harming them. This is a physiological adaptation for coping with lack of water.

Cacti are plants that have different anatomical adaptions to dry conditions.

> They have very long roots to reach as much water as possible.

> They are covered with a thick waterproof cuticle to reduce water loss.

> Water is stored in a fleshy stem.

> The leaves have become spines to reduce water loss and to stop animals getting at the water in the stem.

> The round shape reduces the plant's surface area. This cuts down water loss through evaporation.

Behavioural adaptations to hot and dry environments involve moving into shady areas to keep cool and drinking large amounts of water when it is available.

Questions

3 Deserts are very dry places. Why is the fact that camels do not need to sweat an advantage to them?

4 Explain how large feet help a camel to walk in the desert.

5 Camels are used for transport in many hot places. Explain why they are better adapted than horses for crossing the desert.

6 Explain three ways in which a cactus is adapted to reduce water loss.

7 Explain why there are lots of cacti in the desert but no roses.

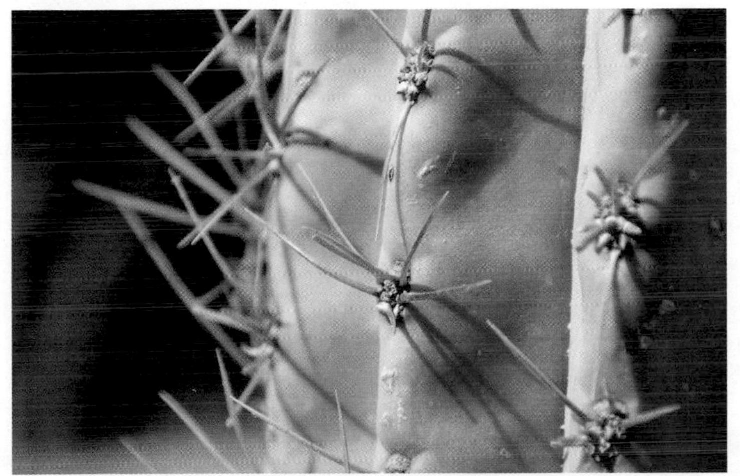

FIGURE 3: What functions do the cactus' spines have?

Remember!

Make sure you know the different meanings of 'describe' and 'explain'.
– The cactus has long roots – is to 'describe' an adaptation.
– The cactus has long roots to collect as much water as possible – is to 'explain' an adaptation.

The importance of surface area

Surface area to volume ratio of an organism is calculated by:

ratio = surface area / volume

Organisms that have a large surface area compared with their volume may lose more heat and water.

The shape of the cactus means it has increased volume and reduced surface area compared with other plants. Therefore less water is lost from cacti than from the other plants.

Question

8 Use ideas about surface area to volume ratio to explain why pine trees have needles instead of flat leaves.

surface area to volume ratio

Adapted to hunt

You will find out:
- > why predators are good at catching prey
- > why prey are good at getting away from predators
- > how polar bears cope with the cold

Alaskan brown bears like to catch salmon. They can be seen standing in rivers waiting for the salmon to swim past. To be good fishers they have to be well adapted.

> The bear has eyes at the front of its head giving it binocular vision. This helps it to judge distance and size. It knows how far to reach for the salmon to catch it.

> The bear has developed good hunting skills to catch the salmon. They know when the salmon will be in the river and lie in wait.

> They also time their breeding so that the cubs are born when the salmon numbers increase. This is called synchronous breeding.

Did you know?

A brown bear can weigh as much as 635 kg.

That is about the weight of seven men weighing in at 91 kg each.

FIGURE 4: How is the Alaskan brown bear adapted for fishing?

Adapted to escape

The salmon needs to swim past the bear to escape. It is adapted to do just that.

> It has a streamlined shape to help it swim fast.

> A lot of salmon swim up the river together in a shoal to breed at the same time. This reduces each individual's chance of getting caught.

> Salmon have eyes on the sides of their head. This gives them a wide field of view. They have more chance of seeing the bear.

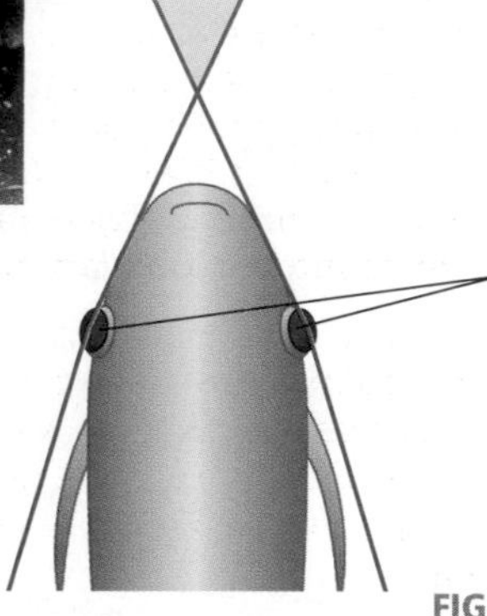

FIGURE 5: Salmon have a wide field of view. How does this help them escape the bears?

Other defence mechanisms

Some organisms don't try to outrun their predators. Instead they are poisonous or have stings. This stops the predators eating them. Some insects will mimic the warning colouration of wasps so animals think they have a sting. This also stops them getting eaten.

Some animals are coloured or patterned to blend in with their surroundings so they can hide from predators (cryptic colouration).

Questions

9 Copy and complete the sentences using the following words.

claws prey predators

Bears are________. They have sharp _______ to help them catch their _______.

10 What can judge distance better – a bear or a salmon?

11 Suggest how a wasp is adapted to avoid being eaten.

Adapting to the cold

Many organisms are adapted to live in cold habitats. For example, the polar bear has:

> thick fur and blubber for insulation

> a large body volume compared to its surface area, to stop it losing too much heat

> small ears, to reduce the surface area from which heat can be lost.

FIGURE 6: Why do polar bears have small ears?

Different bears for different habitats

Brown bears and polar bears live in different habitats. Brown bears would find it difficult to exist in the polar bear's habitat because of the way in which they are adapted.

> Polar bears hunt mainly seals. They wait at the water's edge for the seal to surface. When a seal appears the polar bear bites into the seal's head.

> Brown bears could not hunt in this way, as their colour would make them stand out in the snow. The seals would see them from under the water and would not surface.

Not all adaptations involve physical features. Some organisms have developed behaviour patterns to help them survive. Some animals will hibernate over the winter when food is scarce. This involves lowering their metabolic rate to reduce the need for food. Other animals will migrate to warmer countries that have more food.

Reptiles are unable to survive in cold temperatures, because they cannot regulate their body temperature like mammals can. They need to sit in the sun to warm up their body temperature or stay in the shade to prevent getting too warm. This is an example of a behavioural adaptation.

Questions

12 Give two reasons why a polar bear has fur-covered soles on its paws.

13 There is very little food in the Arctic. Explain how the polar bear is better adapted to catch seals than the brown bear.

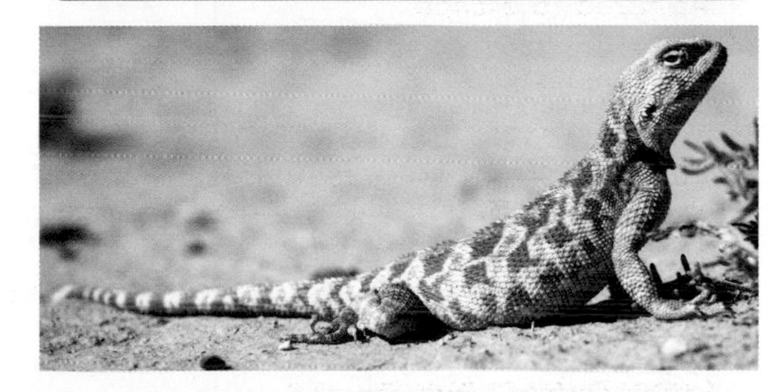

FIGURE 7: Why would lizards find it difficult to survive in cold climates?

Behavioural adaptations

Specialists or generalists

Some organisms are specialists: they are suited to only one type of habitat. A cactus cannot survive in a wet climate. Other plants such as grasses are generalists: they can survive different conditions. However, in the desert the cactus would be better able to compete for the limited water supply.

More adaptations

Penguins have a counter-current heat exchange mechanism. The warm blood entering their flippers flows past cold blood leaving. This warms the cold blood and cools the warm blood, so reducing heat loss from the flippers.

Some bacteria called extremophiles can live in hot springs. This is because they have enzymes that denature at temperatures much higher than 40 °C. Some plants have antifreeze proteins in their cells so they can live in extreme cold.

Question

14 Why can some plants survive in Arctic conditions without ice forming in their cells?

Natural selection

You will find out:

> how species can change over millions of years

> why animals become extinct

> about Darwin's theory of natural selection

Is Ardi one of your oldest ancestors?

The remains of 'Ardi' were found in Ethiopia. She is thought to have lived 4.4 million years ago.

The discovery of Ardi's skeleton has changed the way scientists believe humans evolved. They now think we no longer evolved from chimp-like animals. Instead, chimps and humans evolved from some long-ago common ancestor.

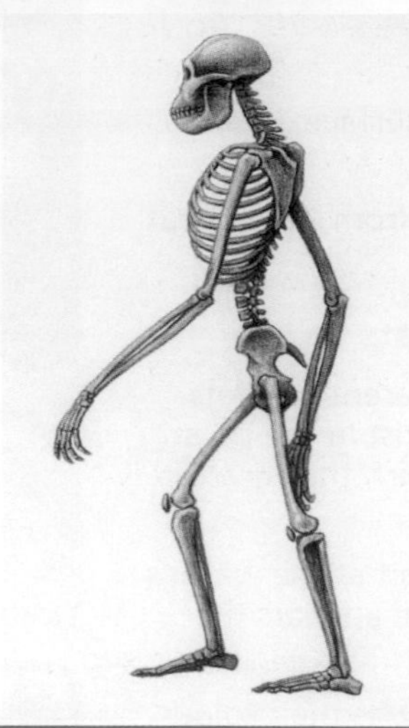

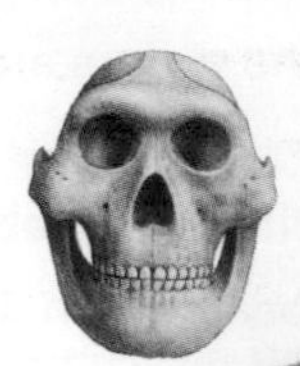

FIGURE 1: What word describes the way the ancestors of animals like Ardi changed into humans over millions of years?

Evolution of the elephant

There are two species of elephant living on Earth today. In the past there have been others.

> All of the elephant species had one common ancestor. This ancestor lived millions of years ago.

> Over time the ancestor changed into different elephant species. This change took place over millions of years and is called **evolution**.

The mammoth was one of the species of elephant. When the environment changed it did not evolve. Instead the mammoth became **extinct**.

Over millions of years, plants and animals have become adapted to their environment. Environments can change, leaving the animals and plants in the environment no longer suited to living there.

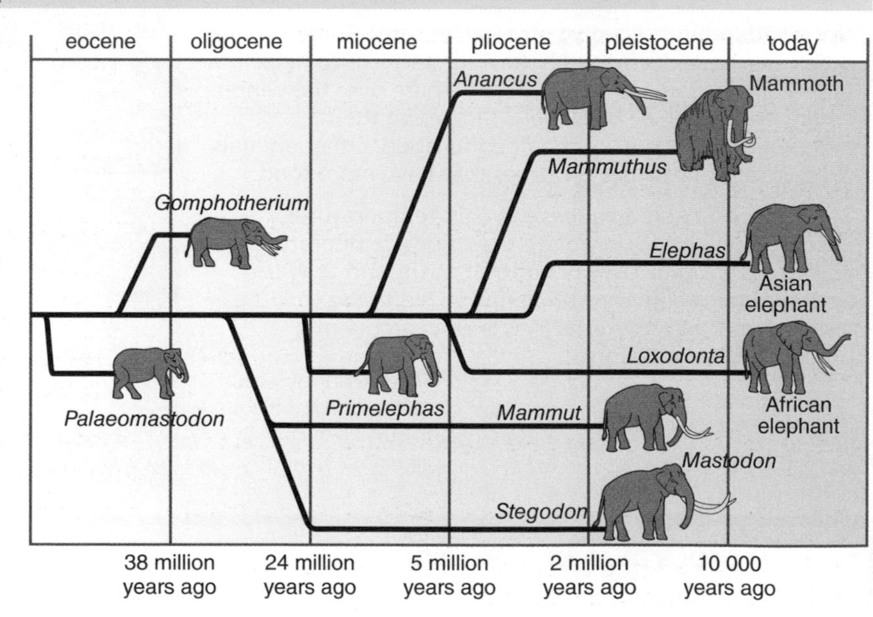

FIGURE 2: The diagram shows one possible way in which elephants evolved – why can't we be sure?

To survive, a species needs to adapt and evolve. If it does not evolve it will become extinct. In any species, it is only the best adapted or fittest that survive.

This survival of the fittest is called **natural selection**. The theory of natural selection was first put forward by Charles Darwin.

There have been many theories about why animals and plants evolve. Most scientists accept the theory of **natural selection** suggested by Charles Darwin in 1859.

Questions

1 Copy and complete the sentences using the words from this list: • changed • evolution

Animals and plants have _____ over time.

This change is called _______.

2 What happens to a species that cannot change over time?

Natural selection

Darwin's theory of evolution by natural selection

Charles Darwin sailed around the world from 1831 to 1836 on board the ship HMS *Beagle*. He developed the theory of natural selection.

> Within any species there is always natural variation.

> Organisms produce far more young than will survive.

> There will be competition for limited resources such as food.

> Only those best adapted will survive, called survival of the fittest.

> Those that survive pass on their successful adaptation to the next generation in their genes.

Adaptations and genes

Adaptations that help an animal or plant survive are passed on to the next generation.

Charles Darwin did not know exactly how these adaptations were passed on.

We now know that genes control adaptations. These genes are found inside cells.

When organisms reproduce, the genes are passed on to the next generation.

Problems with evolution

When Charles Darwin first suggested his theory of natural selection, many people objected for different reasons.

> Some people thought he did not have enough evidence to back up his theory.

> Many people disagreed because they thought God had created all species.

> Some people objected to the idea that humans may have evolved from apes.

Natural selection as a theory is now widely accepted. This is because many scientists have discussed and tested the theory. They have also used it to explain many observations, such as the change in the elephant fossils.

FIGURE 3: Which theory did Charles Darwin write about in his book *On the Origin of Species*?

Questions

3 Which part of the cell controls adaptations?

4 Darwin suggested humans evolved from apes. Why might some people object to this?

5 Use Darwin's theory of evolution to explain how the elephant developed a longer trunk.

New species

Evolution over a long period of time can lead to the formation of a new species. However, for this to happen, the process of natural selection needs to take place in geographic or reproductive isolation (without breeding between populations).

> Over time, the changes add up and may result in a new species.

> Where different species are competing, the less well adapted species may become extinct.

Although Darwin's theory has not been proven, new developments have helped to support it. For example, we now know more about inheritance and DNA. This has helped explain how the adaptations are passed from one generation to the next.

Question

6 A new species is more likely to evolve on an island than on the mainland. Explain why.

Adapted for survival

Large numbers of penguins live together in the Antarctic. The penguins lay their eggs in the same place.

FIGURE 4: Why do penguins live in large groups?

Not all the penguins are the same. Some of them will be better adapted to the cold.

> Some penguins will be able to live and breed closer to the South Pole and survive really bad weather.

> Others are better swimmers and can catch more fish.

When the penguins leave the land and swim in the sea to catch fish, predators such as killer whales and leopard seals try to catch them.

The faster swimmers have more chance of getting away. Being better adapted to swimming helps penguins to survive.

FIGURE 5: In what way is the penguin adapted to help it get away from leopard seals?

You will find out:

> why animals and plants need to adapt to a changing environment

> about modern-day examples of natural selection

> how the evolutionary theories of Darwin and Lamarck are different

Did you know?

Emperor penguins are the tallest species of penguin. They measure in at up to 130 cm tall.

Penguins may swim up to 15 to 20 km a day searching for small fish to eat.

One penguin travelled 100 km in a single day.

Questions

7 Give two reasons why a penguin that swims faster has more chance of surviving.

8 Suggest one way in which penguins are adapted to the cold.

Examples of natural selection

Peppered moths

> Genes control the colour of the peppered moth. Some moths are dark and some are pale.

> In parts of the country that have high pollution, lichen growing on the bark of trees is killed by the pollution and the bark becomes darkened. Dark moths are camouflaged on dark tree bark, so there are more dark moths in polluted areas.

> In cleaner regions, pale-coloured lichen grows. Dark moths stand out so are more likely to be eaten. It is the pale form of the moth that survives to pass on its genes to the next generation.

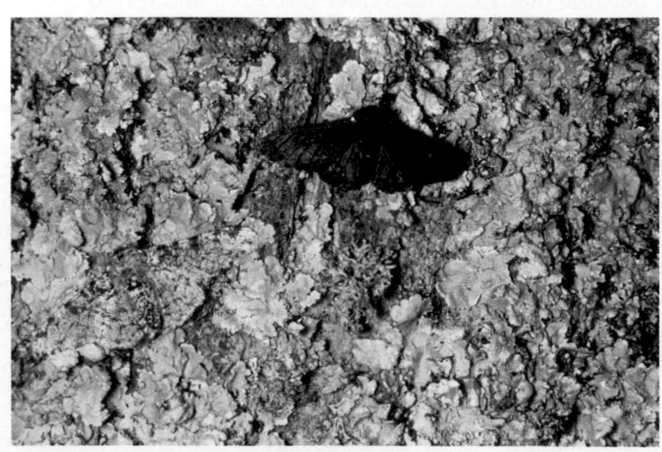

FIGURE 6: How many moths are there in this picture?

penguin adaptations   rats and warfarin   peppered moth

Rats

Rats have evolved to become resistant to the poison warfarin.

Bacteria

> To destroy harmful bacteria that invade our bodies, doctors give us antibiotics.

> More and more bacteria are developing resistance to antibiotics.

> If an antibiotic does not kill all the bacteria, the surviving bacteria reproduce and pass on their resistance to the next generation.

FIGURE 7: How will this rat pass on its resistance to warfarin to the next generation?

Questions

9 How is the colour of a peppered moth passed on to its offspring?

10 Suggest one reason why warfarin resistance in rats is a problem to humans.

11 Suggest why it is important to finish a course of antibiotics.

Lamarck's theory of evolution

Jean Baptiste de Lamarck had a different theory of evolution, called the law of acquired characteristics. According to this theory, giraffes acquired long necks to feed, and this characteristic was passed on.

Now that it is known that genes pass on characteristics, Lamarck's theory of acquired characteristics has been discredited.

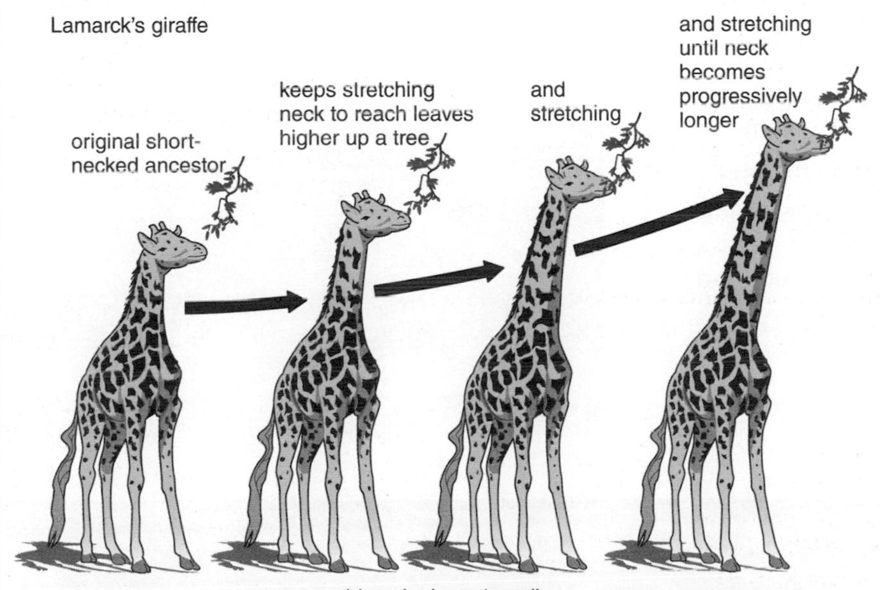

FIGURE 8: According to Lamarck, giraffes stretching their necks to reach leaves high in trees strengthen and gradually lengthen their necks. These giraffes then have offspring with slightly longer necks. Why has Lamarck's theory of evolution been discredited?

Questions

12 Explain how Darwin's theory of evolution differs from that of Lamarck.

13 Use Lamarck's theory to suggest how rats might have become resistant to warfarin.

Population and pollution

You will find out:

- > about increases in the human population
- > how the increase in population is causing more pollution

Running out of water

The world is running out of water. In Beijing in China more than one-third of the wells have run dry. The countries of the Middle East and North Africa face similar problems. In all these cases, the situation can only get worse because the human population continues to increase.

The situation is so bad, that the Chinese have thought about towing icebergs from the Arctic.

FIGURE 1: How might an iceberg help solve the world's water shortage?

More people = more resources

The world population is increasing. In 1800, there were about 1 billion (1000 million) people. Now there are about 6 billion. All these people are using more and more of the Earth's **finite** (non renewable) **resources**.

> Fossil fuel resources such as coal, oil and gas are burned for energy.

> Mineral resources, such as limestone, are used for building. Aluminium ore is mined to make foil, drinks cans and takeaway trays.

Using more and more resources

As more people use more resources, more waste is produced. All this waste is polluting the Earth.

> Household rubbish is piling up in landfill sites.

> Sewage can end up in rivers and oceans, killing fish.

> Burning fossil fuels releases carbon dioxide and sulfur dioxide.

Questions

1 Name one of the Earth's resources that is used for building.

2 Name two fossil fuels.

3 Why is there more household rubbish now than there was in 1800?

FIGURE 2: When minerals are taken from the ground, what is left behind?

The effects of pollution

Global warming

An increase in the use of fossil fuels has resulted in an increase in carbon dioxide in the atmosphere. The higher carbon dioxide levels are contributing to **global warming**. Scientists are concerned that global warming is changing the climate, which could lead to a rise in sea levels among other things.

Ozone layer depletion

The **ozone layer** is a layer of the Earth's atmosphere that protects us from the Sun's harmful ultraviolet (UV) rays.

An increase in use of chemicals called CFCs from car exhausts and aerosols has led to a depletion of this layer. More people are suffering from skin cancers as a result of more UV rays reaching Earth and damaging their skin.

exponential growth   finite resources   fossil fuel

Acid rain

Sulfur dioxide is produced when fossil fuels burn. Sulfur dioxide reacts with rainwater, making it more acidic.

The **acid rain** falls on the Earth, killing trees and making lakes acidic, causing fish to die.

Population and pollution

The graph shows the past, present and predicted future world human population.

The human population still seems to be in the growth phase. The population is growing at an ever-increasing rate. This is called **exponential growth**.

The growth is due to the birth rate being greater than the death rate.

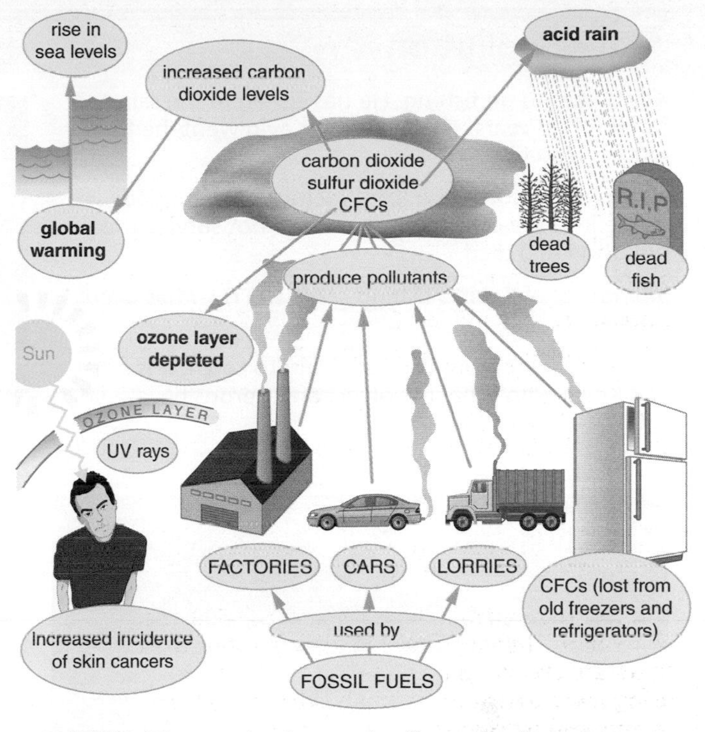

FIGURE 3: What are the three main effects of pollution?

Questions

4 Which gas contributes to global warming?

5 How has the use of CFCs led to an increase in skin cancer?

6 Name one gas that causes acid rain.

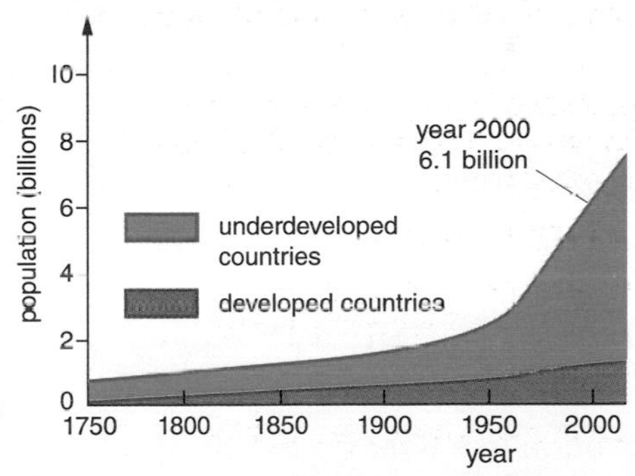

FIGURE 4: What type of population growth does this graph show?

Carbon footprint

The exponential growth of the world's human population has huge consequences for the planet. The Earth's resources such as oil are quickly being used up and pollution is increasing at a great rate, including the production of greenhouse gases.

A 'carbon footprint' measures the total greenhouse gas given off by a person or organisation, in a given time. Usually your carbon footprint is calculated over an entire year. Your carbon footprint would include the carbon dioxide given out when you use a car or heat your home.

Developed and underdeveloped countries

The world population figures show the greatest rise in population is occurring in underdeveloped countries, such as Africa and India.

However, if the countries that use the most fossil fuels are considered, it turns out to be the developed countries, such as the USA and Europe, that are causing the problem. America is the heaviest user of oil, using about 50 litres per person each day.

Questions

7a Suggest one reason why there is a greater demand for oil now than there was 200 years ago.

b Explain why this demand is likely to increase even more in the future.

8 Suggest one reason why a developed country such as America uses more oil than a poorer country such as Ethiopia.

9 When you buy a packet of sweets you increase your carbon footprint. Suggest three reasons why.

You will find out:
> about living things and pollution
> about indicator species

River pollution

Kevin likes to go fishing. He has fished in the same canal for 20 years. The last time Kevin went, he found lots of dead fish.

Kevin found out that a local factory had emptied waste into the canal. The fish could not survive the pollution and died.

In another of Kevin's fishing spots the river has been cleaned.

There are a lot more fish for Kevin to catch. It is not just Kevin who is happy: otters and herons have returned to eat the fish too.

FIGURE 5: Why has this fish died?

FIGURE 6: a Heron. **b** Otter. What food source do these animals need?

Questions

10 What can pollution in a river do to fish?

11 How did cleaning up the river help the otters?

Measuring pollution

The presence or absence of an **indicator species** is used to estimate levels of pollution.

> The stonefly larva is an insect that can live only in clean water.

> Mayfly larva can live in slighted polluted water.

> The bloodworm, waterlouse, sludgeworm, rat-tailed maggot and are animals that can live in polluted water.

Remember!

You only need to remember that indicator species are used.
You do not have to remember the level of pollution each species tolerates.

FIGURE 7: The waterlouse is an indicator species. Can it live in polluted water?

river pollution

> Lichen grows on trees and rocks, but only when the air is clean. It is unusual to find lichen growing in cities. This is because it is killed by the pollution from cars.

The amount of pollution can also be measured directly using different methods. Oxygen probes attached to computers can measure the exact levels of oxygen in a pond. Special chemicals can be used to indicate levels of nitrate pollution from fertilisers.

Questions

12 A water sample from a river contained bloodworms but no stonefly larvae. Is the water polluted? Explain your answer.

13 Explain why it is unusual to find lichen growing in cities.

14 Describe how you could measure the pH of a stream.

Using indicator species to test for water pollution

Animals have different sensitivities to environmental conditions. In rivers and ponds, different animals can tolerate different pollution levels.

> The sludge worm can live in polluted water. This is because it can cope with the low oxygen levels that occur.

> The alderfly cannot live in polluted water. It cannot tolerate low oxygen levels.

Water that contains lots of different species is usually a healthy environment.

Animal	Sensitivity to pollution
stonefly larva	sensitive
water snipe fly	sensitive
alderfly	sensitive
mayfly larva	semi-sensitive
freshwater mussel	semi-sensitive
damselfly larva	semi-sensitive
bloodworm	tolerates pollution
rat-tailed maggot	tolerates pollution
sludgeworm	tolerates pollution

FIGURE 8: Do polluted ponds support much life?

Questions

15 The animals in three different water samples were identified.

Sample	Animals found
A	stonefly larva, water snipe fly, alderfly, mayfly larva, damselfly larva
B	rat-tailed maggot, sludgeworm, bloodworm
C	sludgeworm, freshwater mussel, damselfly larva

a Which sample had the largest variety of animals?

b Place the samples in order from cleanest to most polluted.

c Which water sample contained little dissolved oxygen?

d Suggest which sample could be from a fast-running stream. Explain your answer.

e Suggest which sample could be from a polluted pond. Explain your answer.

f Suggest one advantage and one disadvantage of using chemical tests instead of indicator species to test for pollution.

Sustainability

You will find out:

> why species become endangered

> how species become extinct

> how species can be saved from extinction

Whaling in Whitby

In the 18th and 19th centuries, Whitby in North Yorkshire was the biggest whaling port in England.

The Volunteer was a typical Whitby whaling ship. In the year of 1811, it caught 23 whales.

On its return to port, the barrels of blubber were used to produce many tonnes of oil. The horrible stench of the oil could be smelled all over the town.

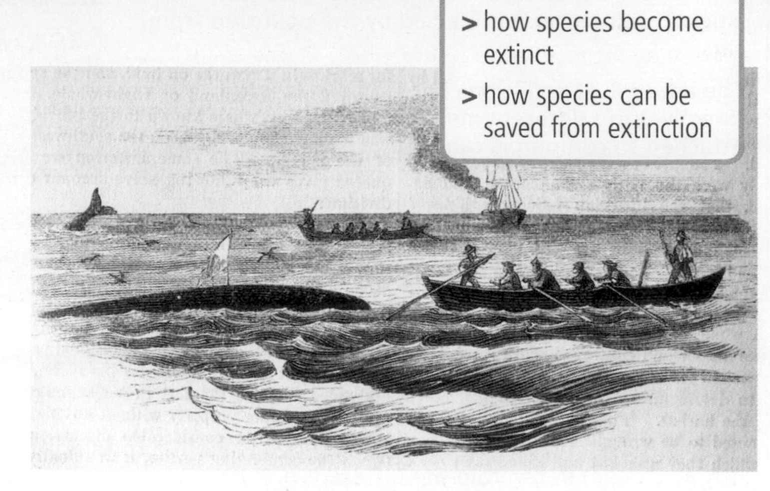

FIGURE 1: When was Whitby the biggest whaling port in England?

Why animals become extinct

If an environment changes, an animal may not be as well adapted. It will be forced to compete with other species. This may result in species becoming endangered or extinct.

The environment can change in two ways.

> Naturally, such as an ice age.

> Artificially, normally as a result of human activities, such as destruction of habitats for farming or building and pollution of rivers or lakes by industry or sewage.

Other animals, such as the great auk, have become extinct because humans have hunted them. The great auk was hunted for food. As the birds became scarce, they were also hunted for a well-paid trade in skins and eggs. The last known living pair and one egg were stolen in Iceland in 1844.

Saving animals from extinction

The giant panda is one of the best-known endangered animals. It has been the symbol for the WWF (formerly World Wildlife Fund) since 1961.

The panda lives in the forests of China. It eats bamboo. To help with its conservation, large areas of forest are now protected. Hunting the giant panda is illegal.

Only 61% of the panda population is protected. Outside of the reserves, panda habitat is still being destroyed and poaching remains a problem. To stop this, people need to be educated as to the importance of saving the panda.

Many pandas live in captivity. Zoos around the world help to increase the panda population by breeding them.

Zoos often exchange pandas for **captive breeding** programmes.

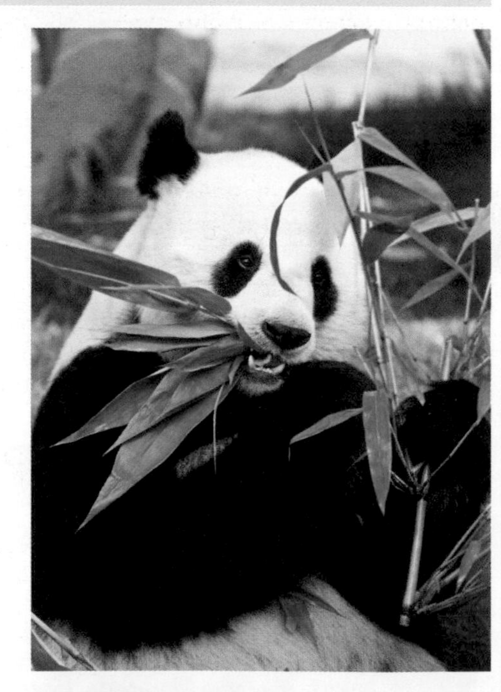

FIGURE 2: How is the giant panda being protected?

Did you know?

Every year between 20 000 and 100 000 species reach extinction.

panda conservation

It is also possible to set up artificial ecosystems where pandas can live safely.

Plants are also being saved. Places like Kew Gardens store seeds from all around the world. These **seed banks** keep rare plants safe from extinction.

Questions

1 Describe three ways in which an animal could become extinct.

2 How are zoos helping to increase panda numbers?

3 What problems do pandas living outside the reserves face?

Why save the panda?

Pandas have a highly specialised diet – they eat only fresh green bamboo shoots, so they have a limited habitat in which to live.

The pandas became endangered because their habitat area and their numbers fell below a critical level. This means there aren't enough breeding pairs to keep the population growing, and to provide genetic variation.

Panda habitat is found at the top of the Yangtze Basin in China, in an area populated by millions of people. The area contains many rare plants and animal species. Some of the plant species could be of important medicinal value. Protecting the panda will ensure its habitat remains, which also benefits the local people.

Tourism could become an important economic benefit to local people and lead to improved transport and water resources. By protecting the environment the local people will still be able to collect food and other resources as their ancestors have done for hundreds of years.

Questions

4 Suggest three reasons why tourists visiting pandas in the wild would benefit the local people.

5 How will protecting the panda help:

a medical research?

b preserve traditions of the local people?

The story of the Chatham Island black robin

The Chatham Island black robin lives on isolated islands near New Zealand. In 1976 there were only seven birds left. Such a low number is not a viable population in the wild, which means that these birds would soon become extinct.

The Chatham Islands had been taken over by cats. The isolation of the islands meant the birds could not escape these new predators.

A conservation programme was set up to save the black robin. This used other birds as foster parents for the chicks. This was because there was only one breeding pair left. These two birds did not have the time to hatch and feed their young.

There are now 250 birds, all descendants of the last breeding pair. This has led to further problems:

> the island habitat is too small, so some birds are being moved to other islands

> there is very little genetic variation in the population, so it would be vulnerable to environmental changes

> there is increased competition with other birds on the islands.

FIGURE 3: Why was the black robin population no longer viable?

Questions

6 Explain why the black robin was at risk of extinction.

7 The robins were raised by tomtit birds. How might this affect the population of tomtits?

8 Explain why there is little genetic variation in the population of black robins.

Chatham Island black robin   seedbanks

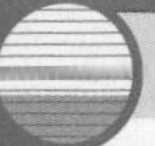

Different types of whales

You will find out:
- > why some whales are close to extinction
- > how fish populations can be protected
- > how woodland can be sustained

Whales are mammals that live in saltwater. There are two types of whales.

> The baleen whale eats small animals called krill. The whale sieves the water through baleen plates in its mouth, trapping the krill.

> Toothed whales, such as the sperm whale, have lots of teeth to trap food such as squid.

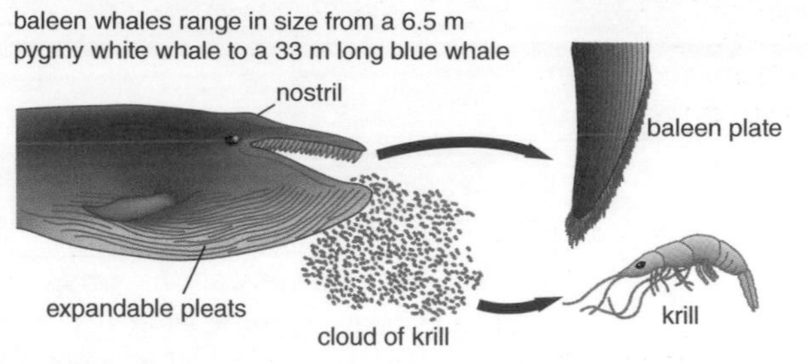

FIGURE 4: How are baleen whales adapted to eat tiny krill?

FIGURE 5: Why do some countries still hunt whales?

Where whales live

Different whale species are found in different seas. This is because they eat different food. Some whales even move from place to place following their food. This is called migration.

Minke whales are close to extinction

Whales have been hunted for many years for oil and meat. Minke whales are still hunted. If too many more minke whales are caught they could become extinct.

Sustainable resources

It is important not to take too many things from the environment.

> Whales, fish and trees are all **sustainable resources**. If only a few of the resources are taken they will never die out. This is because the ones left behind will be able to reproduce enough to replace them so that the species do not become extinct.

> Oil is not a sustainable resource. It takes millions of years to form, yet we have used up most of it in only 150 years.

Questions

9 Why do different whale species live in different places?

10 Why may minke whales become extinct?

11 Why are fish a sustainable resource?

Uses of whales

Whales have been hunted for hundreds of years. Their parts have many uses. Selling the whale parts earns the hunters money.

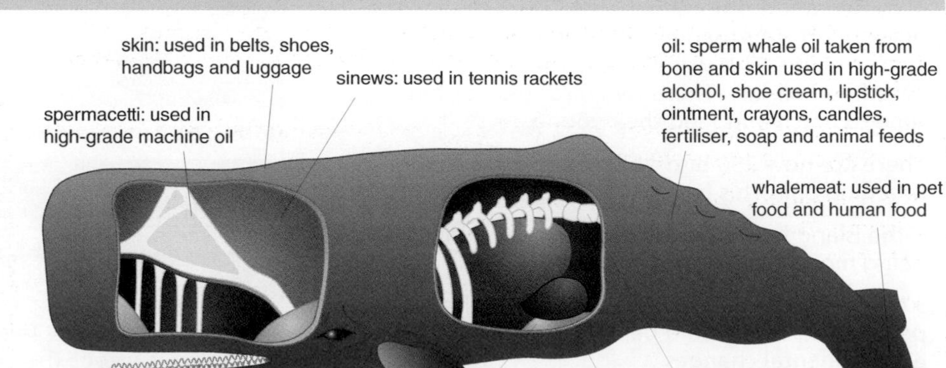

FIGURE 6: Do you think we could do without or use different materials to make all these things?

When whaling was first stopped in 1986 (it has now started again), many owners of whaling ships had to find new employment. Tourist whale-watching trips have given whales new commercial value.

Whales in captivity

Some whales are kept in captivity. This can be for important reasons such as research or captive breeding programmes, or just for our entertainment. However, many people object when whales lose their freedom.

Sustainable development

To ensure we do not run out of important resources we need to plan for the future. **Sustainable development** is a way of taking things from the environment but leaving enough behind to ensure a supply for the future and prevent permanent damage to populations and the environment.

Fishing quotas

Scientists have worked out how many fish can be taken from the sea and leave enough to reproduce and maintain the population. Fishermen have then been set quotas.

Many fishermen see the quotas as a threat to their livelihood. They need to be educated as to the importance of quotas. Without them there will not be enough fish for future generations.

Managed woodland

In managed woodland, trees are only cut down if other trees are planted to replace them. The number of trees cut down will also depend on how long it takes the new trees to grow. Pine trees grow faster than oak trees, so it is easier to maintain their numbers.

Questions

12 What commercial value do live whales have?

13 Suggest arguments for and against keeping whales in captivity.

14 Why are fishermen set quotas?

Reasons for whaling

In 1986, a halt to whaling was agreed by the International Whaling Commission. This caused problems for countries such as Norway, Iceland and Japan. Many small communities in these countries rely on whaling for food and income. In the 1990s, they started whaling again.

Some scientists also see a need to kill some whales. They think it will help them to understand how whales can survive at extreme depths. However, they could study the whales without killing them. Migration patterns and whale communication can only be investigated if the animal is alive.

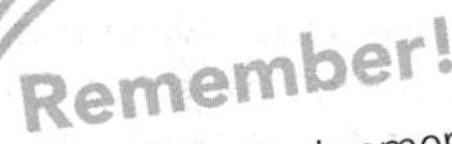

Sustainable development is a way of providing food for an increasing world population without harming the environment.

Sustainable development and world population

As the world population increases it is even more important to carry out sustainable development to protect endangered species. This requires planning and co-operation at a number of levels, including local, national and international.

Fossil fuels will run out, yet there is an increase in demand for fuel. Therefore we must manage alternative fuels such as wood.

The demand for food and other resources could lead to an increase in whaling. The whaling nations will need to work together to prevent extinction.

When whaling quotas are set, other factors will need to be taken into account. These include:

> pollution levels

> overfishing of the whales' food source

> food webs in the ecosystem.

Questions

15 Suggest why it is difficult to police whaling.

16 Waste products of industry are often dumped at sea. Suggest why this could make maintaining fish numbers difficult.

sustainable development whaling

Preparing for assessment: Analysis and evaluation

To achieve a good grade in science, you not only have to know and understand scientific ideas, but you need to be able to apply them to other situations and investigations. These tasks will support you in developing these skills.

Task

- Find out if there is a relationship between surface area to volume ratio of an organism and the rate of heat loss from the organism.

Context

Organisms lose and gain heat from their surface area. Organisms that live in cold habitats need to reduce heat loss. The Arctic fox has small ears and short legs to reduce its surface area to volume ratio.

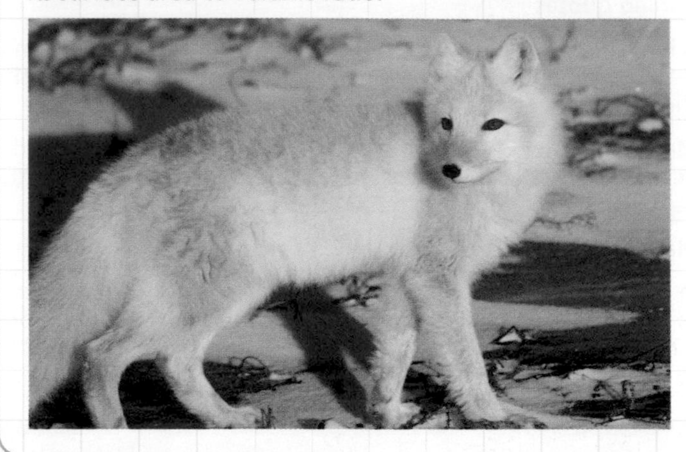

How to do this experiment

Flasks of hot water can be used to represent organisms with different surface area to volume ratios.

1. Measure the volume and surface area of each flask.
2. Calculate the surface area to volume ratio.
3. Fill each flask with boiling water.
4. Place a bung containing a thermometer inside each flask.
5. Once the water is at 90 °C start the stopclock.
6. Record the temperature every minute for 20 min.

flask	volume in cm^3	SA / V ratio
1	100	1.15
2	500	1.66

Results, analysis and evaluation

A group of students found these results when they recorded the temperature of water in two flasks.

time in s	temperature of water in °C	
	SA / V ratio of 1 to1.15	SA / V ratio of 1 to 0.66
0	90	90
1	88	89
2	86	87
3	84	87
4	82	86
5	80	85
6	79	84
7	78	83
8	77	84
9	75	82
10	74	81
11	73	80
12	72	79
13	71	79
14	70	78
15	68	77
16	67	77
17	66	76
18	65	76
19	64	75
20	63	74

Think about the scale for the axes. Use sensible divisions but make the graph as large as possible.

Should these be straight lines or smooth curves? Never draw dot to dot.

1. Plot graphs of temperature (y-axis) against time (x-axis).

2. Draw lines of best fit through the points.

3. Is there a relationship between surface area to volume ratio of an organism and the rate of heat loss from the organism? Explain your answer.

Look at the gradient of each line of best fit.

4. How could the students who found the results improve their technique?

How many flasks did they use?

5. How can the students improve the precision of their results?

Do not confuse precision with accuracy; precision is about the repeatability of results.

6. The body of an elephant has a small SA / V ratio and this animal lives in hot climates. They have evolved to have large ears. The ears increase the surface area for heat loss. Use the data and your scientific knowledge to explain why it is important that elephants have large ears.

B2 Checklist

To achieve your forecast grade in the exam you'll need to revise

Use this checklist to see what you can do now. It gives you many of the important points you will need to know. Refer back to the relevant pages in this book if you're not sure and to see if there is anything else you need to know. Look across the three columns to see how you can progress.

Remember you'll need to be able to use these ideas in various ways, such as:

> interpreting pictures, diagrams and graphs
> applying ideas to new situations
> explaining ethical implications
> suggesting some benefits and risks to society
> drawing conclusions from evidence you've been given.

Look at pages 278–299 for more information about exams and how you'll be assessed.

To aim for a grade E	To aim for a grade C	To aim for a grade A
describe the characteristics used to place organisms into the five kingdoms and the different classes of arthropods **recognise** that organisms of the same species may show great variation	**classify** organisms into kingdom, phylum, class, order, family, genus and species **explain** the importance of classification of species in terms of evolutionary and ecological relationships **define** the term 'species' and use the binomial system to name a species	**understand** that classification systems can be natural or artificial **explain** how use of DNA sequencing has led to changes in understanding of classification **explain** how similarities and differences between species can be explained in terms of both evolutionary relationships and ecological relationships
explain the term 'trophic level' **describe** how changes in one organism in a food chain or web may affect the other organisms **explain** how energy from the Sun flows through food chains **interpret** data on energy flow in food chains and webs	**explain** why pyramids of numbers and pyramids of biomass for the same food chain can be different shapes **explain** how some energy is transferred to less useful forms at each stage (trophic level) in the food chain	**explain** the difficulties in constructing pyramids of biomass **explain** how the efficiency of energy transfer explains the length of food chains and the shape of a pyramid of biomass **calculate** the efficiency of energy transfer
recall that when animals and plants die and decay the elements in their bodies are recycled **recognise** that bacteria and fungi are decomposers **recall** that nitrogen is taken up by plants as nitrates **explain** why nitrogen gas cannot be used directly by animals or plants in terms of its reactivity	**explain** how carbon is recycled in nature to include ideas about: • photosynthesis, respiration, combustion and decay • marine organisms • weathering of limestone • carbon sinks • volcanoes **explain** how nitrogen is recycled in nature to include ideas about: • nitrates • decomposers • feeding • the bacteria involved	

To aim for a grade E	To aim for a grade C	To aim for a grade A
interpret data which show that animals and plants can be affected by competition for resources **explain** how the size of a predator population will affect the numbers of prey and vice versa **recall** that some organisms benefit from the presence of organisms of a different species	**describe** how organisms within a species compete in order to survive and breed **explain** how the populations of some predators and their prey show cyclical fluctuations in numbers **describe** parasitism and mutualism	**use** the terms 'interspecific' and 'intraspecific' **explain** the term 'ecological niche' **explain** why the predator and prey cycles are out of phase with each other **explain** why nitrogen-fixing bacteria in the root nodules of leguminous plants are an example of mutualism
explain how some animals are adapted to be successful predators or avoid being caught as prey **recall** that animals and plants that are adapted to their habitats are better able to compete for limited resources	**explain** how adaptations to cold environments help organisms survive **explain** how adaptations to hot environments help organisms survive **explain** how adaptations to dry environments help organisms survive	**understand** that some organisms are biochemically adapted to extreme conditions
understand how when environments change, some species survive or evolve but many become extinct **recall** that many theories have been put forward to explain how evolution may occur **recall** that most scientists accept the theory of natural selection first put forward by Darwin	**recall** that adaptations are controlled by genes **explain** the main steps in Darwin's theory of natural selection **describe** examples of change by natural selection **recognise** that the theory of natural selection is now widely accepted	**explain** how Lamarck's ideas of evolution are different from Darwin's theory **recognise** that the theory of natural selection has developed **understand** why speciation requires isolation of a population
explain how as the human population increases more resources are used, which means more pollution **understand** that pollution can affect the number and type of organisms that can survive in a particular place	**understand** why the human population is increasing **explain** the causes and consequences of global warming, ozone depletion and acid rain **explain** how indicator species help indicate the level of pollution **describe** how pollution can be measured	**explain** how developed countries impact on the use of resources and the creation of pollution **explain** the term 'carbon footprint' **discuss** the consequences of exponential growth **interpret** data on indicator species **describe** the advantages and disadvantages of different methods to measure pollution
explain why organisms become extinct or endangered **describe** how endangered species can be helped **discuss** the reasons why certain whale species are close to extinction **recognise** that a sustainable resource can be removed from the environment without it running out	**explain** reasons for conservation programmes **recognise** that both living and dead whales have commercial value **explain** the term 'sustainable development'	**explain** why species are at risk of extinction if there is a lack of genetic variability **evaluate** a given example of a conservation programme **recognise** that some aspects of whale biology are still not fully understood; describe issues concerning whaling **understand** that sustainability requires planning and co-operation at all levels

B2 Exam-style questions

Foundation Tier

AO2 **1 (a)** Rachael investigates leaf litter. She finds five different animals in some leaf litter and describes their features. The table shows her observations.

animal	description
A	has six legs, wings and antennae
B	has black spots on red wings
C	has eight legs
D	has a long body with different segments but no legs
E	has a long body with lots of legs

(i) Which of the five animals is not an arthropod? [1]
(ii) There are three different arthropod groups described. Write down the names of these three arthropod groups. [1]

AO1 **(b)** Rachael finds out that leaf litter decays releasing the elements inside their cells.
One element that is recycled when leaf litter decays is nitrogen.
Explain how nitrogen is recycled in nature. [3]
[Total: 5]

2 The graph shows the relative population sizes of a predator and its prey over several breeding seasons.

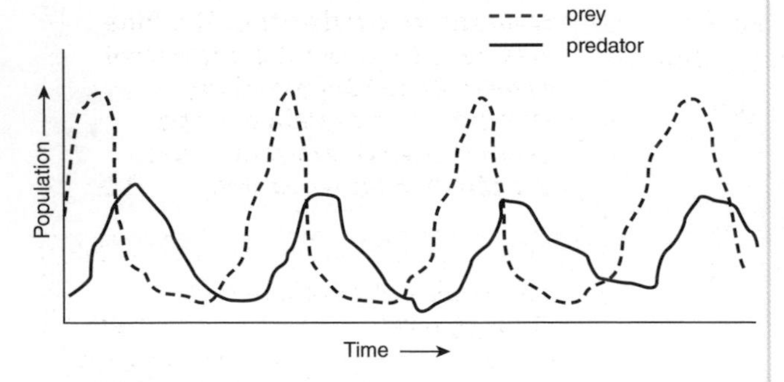

AO2 **(a)** Describe and explain the effect of the prey species on the predator population. [2]

AO1 **(b)** Deer are prey that are adapted to escape predators.
Explain **ways** in which prey are adapted to escape predators. [2]
[Total: 4]

3 Lewis investigates different water samples to find out how clean they are.

Look at the table; it names the animals he looks for and the type of water they can live in.

clean water	some pollution	very polluted
alderfly stone fly lava	flatworm mayfly larva	rat-tailed maggot sludgeworm

Lewis takes water samples from a stream. He looks for the presence of each animal. The table shows his results:

animal	present
alderfly	no
flatworm	no
mayfly larva	no
rat-tailed maggot	yes
stone fly lava	no
sludgeworm	yes

AO3 **(a)** Lewis writes a statement about his results.

The stream is polluted.

Is Lewis' statement correct?
Explain your answer. [2]

AO1 **(b)** Air can also become polluted by gases such as sulfur dioxide.
How does sulfur dioxide get into the air? [1]
[Total: 3]

AO1 **4** Black rhinoceroses are endangered animals.

(a) Suggest reasons why the black rhinoceros became endangered. [4]

(b) Black rhinoceroses can be helped to stop them becoming extinct. Describe how. [4]
[Total: 8]

AO1 recall the science AO2 apply your knowledge AO3 evaluate and analyse the evidence

Worked Example – Foundation Tier

Read this information about Arctic foxes.

- Arctic foxes are predators that are adapted to live in cold conditions.
- In winter Arctic foxes have white fur. In summer the fur is brown.
- Arctic foxes have short legs, small ears and fur on the underside of their paws.

1 Explain how Arctic foxes are adapted to live in cold conditions and hunt for their food. [6]

The Arctic fox can live in cold conditions because they have thick fur to keep them warm.

*They may also have a layer of fat that helps **insulate** the body.*

*The fur on the underside of their feet helps them to grip the ice, so they can run after their prey. The fur could also reduce **heat loss** from conduction on the cold ice.*

*The short legs and small ears reduce the **surface area** for heat loss.*

*The fox is a predator; it hunts its prey. The white fur provides **camouflage** in winter when there is snow. The brown fur provides camouflage when the snow melts in summer.*

*Arctic foxes have eyes at the front of their heads to help them **judge distance**. This means they can judge how far they will need to jump to catch the prey.*

How to raise your grade!

Take note of these comments – they will help you to raise your grade.

Students are asked to *explain* adaptations so just saying they have thick fur is not enough at C grade. Students need to explain the following: The fur traps air, which is an insulator reducing heat loss. 0/1

This is correct. The idea of heat loss by conduction is not in B2. However, it explains why the fur is needed on the feet and will be credited. This is why this question is partly AO2; students can apply their knowledge of physics into their biology questions. 1/1

This is correctly explained at Grade C level. 1/1

This part of the question is targeted at low demand. This idea is correct. However it is a low level answer. To reach E standard students must explain why the camouflage is important to the predator's hunting strategy. 1/2

This is correctly explained at Grade E level. 1/1

This student has scored 4 marks out of a possible 6. This is below the standard of Grade C. With a little more care the student could have achieved a Grade C.

These longer 6 mark answers usually have marks awarded for the quality of written communication shown by this symbol ✏ so answers need planning, and care is needed with spelling, punctuation and grammar.

Remember!

In the exam you may be asked about an animal you know nothing about. Just explain the adaptations you know as they may apply to the animal in the question.

B2 Exam-style questions

Higher Tier

1 Look at the table; it shows the classification of two similar animals.

taxon (group)	animal X	animal Y
kingdom	Animalia	Animalia
phylum	Arthropoda	Arthropoda
	Insecta	Insecta
order	Coleoptera	Coleoptera
family	Coccinellidae	Coccinellidae
genus	*Adalia*	*Coccinella*
species	*bipunctata*	*septempunctata*
common name	2 spot ladybird	7 spot ladybird

AO1 **(a) (i)** Write down the missing taxon (group). [1]

AO2 **(ii)** Write down the binomial name for the 7 spot ladybirds. [1]

AO2 **(iii)** The two ladybirds had a recent common ancestor. What evidence is there in the table to support this? [1]

AO1 **(b)** The two ladybirds compete for the same food. Write down the term used to describe competition between two species. [1]

[Total: 4]

2 Look at the diagram. It shows energy transfer in an ecosystem.

producer (87 000 kJ) → primary consumer (14 000 kJ) → secondary consumer (1600 kJ) → tertiary consumer (90 kJ)

AO2 **(a) (i)** The percentage energy transfer from producer to primary consumer is 16.09%. Calculate the percentage energy transfer from secondary consumer to tertiary consumer. [1]

AO2 **(ii)** Explain the differences between the two values. [3]

AO1 **(b)** Some of the energy not used by the tertiary consumers is used to start other food chains. Explain how. [2]

[Total: 6]

3 Look at the table; it shows the volume and surface area of some different animals.

animal	surface area in m^2	volume in m^3	surface area to volume ratio
manatee	9.4	1.0	9.4
walrus	11.0	1.7	

AO2 **(a)** Calculate the surface area to volume ratio of the walrus. [2]

AO3 **(b)** The walrus lives in cold water; the manatee lives in warm water. Using the information in the table, say whether or not the two animals could live in the same habitat. Explain your answer. [3]

AO2 **(c)** Manatees are large grey animals that swim in warm shallow seas. They graze on plants found on the sea bed. One of their close relations is the elephant. To breathe they need to come to the surface.

Use Darwin's theory of natural selection to explain how the ancestor of the manatee and the elephant might have evolved differently to produce two separate species. [4]

[Total: 9]

AO1 recall the science | AO2 apply your knowledge | AO3 evaluate and analyse the evidence

Worked Example – Higher Tier

Read this information about a beech wood.

- The wood contains lots of beech trees growing close together.
- Every autumn the amount of leaf litter increases.
- By spring the amount of leaf litter has been reduced.
- The leaf litter has not blown away or been removed by humans.

1 Use ideas about recycling elements to explain what happens to the leaf litter over winter.
The quality of written communication ✎ will be assessed in your answer to this question. [6]

*The leaf litter **decomposes** over the winter or it is eaten by animals like earthworms. Bacteria and fungi are decomposers. They decompose the leaf litter.*

*After the decomposers break down the leaves **bacteria** then make **nitrates**.*

Some of the nitrates end up back in the air as nitrogen.

Plants take in the nitrates to make proteins. Animals will then eat the plants.

Carbon in the plants is also recycled. The animals that eat the leaf litter take in the carbon and use it for growth. Later this carbon is put back into the air as carbon dioxide.

The bacteria also recycle some of the carbon.

How to raise your grade!

Take note of these comments – they will help you to raise your grade.

More detail is needed on exactly what the decomposers do. Students need to say that decomposers convert protein in the leaves into ammonia. 1/2

Students *need to identify these bacteria as nitrifying bacteria. Also make it clear that they convert ammonia into nitrates.*0/1

This is correct, but students should mention this is because of denitrifying bacteria. 1/1

This is a valid standard demand point and should be included.1/1

When describing the recycling of carbon it is important to explain that it is respiration in all organisms that returns carbon to the air as carbon dioxide.

Students should also explain that the carbon returns to the plants during photosynthesis.1/1

For the most part the information is relevant and presented in a structured and coherent format. Specialist terms (shown in bold) are used for the most part appropriately. There are occasional errors in grammar, punctuation and spelling.

This student has scored 4 marks out of a possible 6. This is below the standard of Grade A. With a little more care the student could have achieved a Grade A.

Remember!

Learn the different types of bacteria and their role in the nitrogen cycle. Try using cards with the bacteria, the chemical they change and the chemical they make. Mix the cards up and then sort them into the correct groups.

C1 Carbon chemistry

Ideas you've met before

Using crude oil

Crude oil is a substance that needs separating.

Distillation, filtering and chromatography are ways to separate mixtures.

Crude oil and coal are fossil fuels which have been made over millions of years.

Fossil fuels will run out and we need to look at alternative sources for our needs.

 **How can alcohol and water be separated?**

Materials

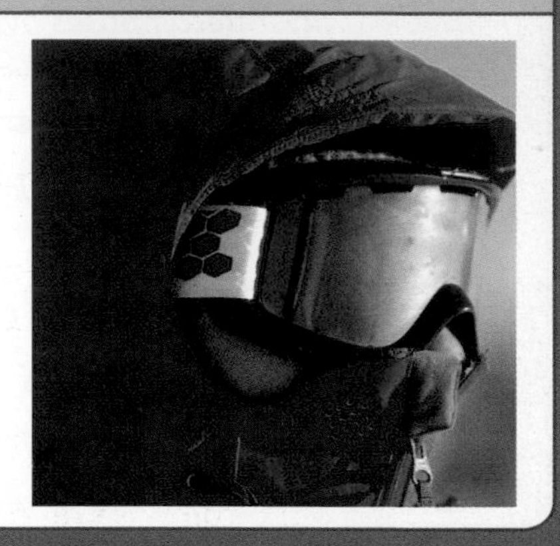

New synthetic materials are very important to our lifestyle changes.

Plastics have helped advances in machinery and communications to happen.

There are problems with plastics being non-biodegradable.

Choices of materials with different properties can be made for different objects.

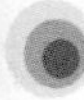 **What two properties would be needed for a fishing rod?**

Gases of the atmosphere

Air is a mixture of gases.

We need oxygen for breathing.

Particles of a gas move rapidly and randomly.

Acid rain causes damage to buildings and to the natural environment.

 **How are the particles of a solid arranged?**

Food and cooking

Ripening fruit and cooking foods involve chemical reactions.

A chemical reaction is irreversible.

When a chemical reaction takes place a new product is made.

When food is cooked it changes texture and appearance.

 **Is making cheese from milk a chemical change or not?**

In C1 you will find out about...

- > how and why fractional distillation separates crude oil
- > how cracking keeps the supply of petrol up with demand
- > complete and incomplete combustion

- > how small monomers make long polymers
- > how to draw a displayed formula of a polymer
- > how waterproof but breathable fabrics are made
- > the problems of the disposal of polymers

- > the composition of the air and how it remains constant
- > how the levels of atmospheric pollution can be controlled
- > the forces of attraction between particles vary

- > the main types of food additives and why they are used
- > how emulsifiers help water and oil to stay mixed
- > how pigments, binding media and solvents make paints
- > thermochromic and phosphorescent pigments

Making crude oil useful

You will find out:

> how crude oil can be separated into fractions

> how and why fractional distillation works

> how the forces between molecules affect their boiling points

Precious oil

Crude oil is one of our most important natural resources. It is a finite resource and will eventually run out. It is non-renewable because it cannot be made again.

Different parts of crude oil are used to make transport fuels, plastics, medicines, fabrics and dyes.

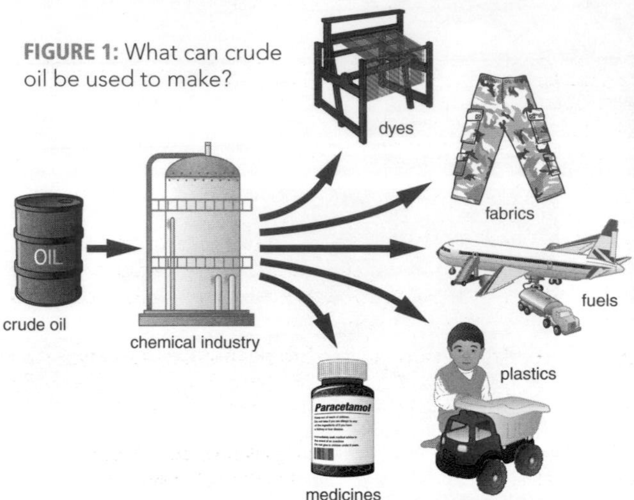

FIGURE 1: What can crude oil be used to make?

FIGURE 2: How is crude oil taken from below the surface of Earth?

Fossil fuels

Coal, gas and crude oil are **fossil fuels**. These fuels are non-renewable fuels. They take a very long time to make and are used up faster than they are formed.

Separating crude oil

Crude oil is separated by **fractional distillation**:

> the oil is heated up until parts of it have boiled

> the parts are cooled down

> the liquids are collected.

The crude oil is separated into different fractions (parts). The process works because each fraction has a different **boiling point**.

The fractions from crude oil include: • LPG • petrol • diesel • paraffin • heating oil • fuel oils • bitumen.

LPG is **liquid petroleum gas** that contains propane and butane gases.

Questions

1 What kind of fuel resource are crude oil and coal?

2 How is crude oil separated?

3 Look at Figure 3. How many fractions is crude oil separated into?

More on fossil fuels and the separation of crude oil

Fossil fuels are finite resources because they are no longer being made, or are being made extremely slowly. The conditions on Earth are not the same as they were millions of years ago. When these fossil fuels are used up there will be no more. They are called a **non-renewable** source as they cannot be made again, or are being used up faster than they are formed.

Crude oil was made from the bodies of plankton compressed over 260–650 million years. It is made up of a mixture of many types of oils. All these oils are **hydrocarbons**. A hydrocarbon is made up of molecules containing carbon and hydrogen only.

The oils are separated by fractional distillation.

fossil fuels boiling point liquids fractional distillation liquid petroleum gas

- Crude oil is heated at the bottom of a tower.
- Fractions with a high boiling point 'exit' at the bottom of the tower.
- Other fractions boil and their gases rise up the tower.
- The tower gets colder the higher up it is. The column has a **temperature gradient**. Fractions with lower boiling points, such as petrol and LPG, 'exit' or condense at the top of the tower, where it is colder.

Fractions containing mixtures of hydrocarbons are obtained. Fractions contain many substances with similar boiling points.

Oil that does not boil sinks as a thick liquid to the bottom of the tower. This fraction is called bitumen. It is used to make tar for road surfaces. Bitumen has a very high boiling point.

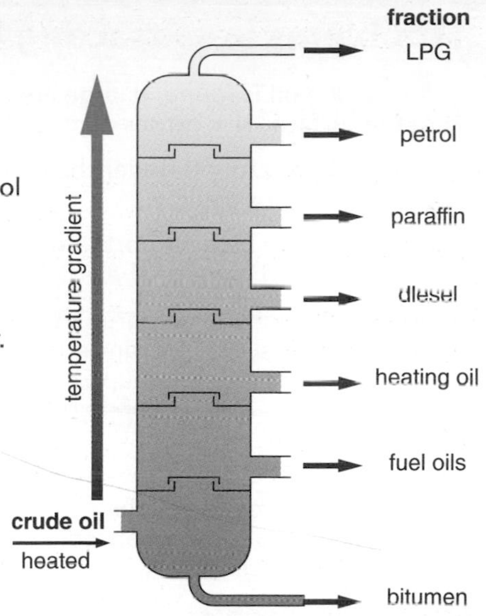

FIGURE 3: A fractional distillation column. What characteristic of the fractions in crude oil means that they can be separated using this method?

Questions

4 Why is crude oil a non-renewable resource?

5 Where do fractions with high boiling points exit the tower?

6 Where is the tower coldest?

More on fractional distillation

Crude oil can be separated by fractional distillation because the molecules in different fractions have different length chains. This means that the forces between the molecules are different.

The forces between molecules are called **intermolecular forces**. These forces are broken during boiling. The molecules of a liquid separate from each other as molecules of gas.

Large molecules, such as those that make up bitumen and heavy oil, have very long chains so there are strong forces of attraction between the molecules. This means that they are difficult to separate. A lot of energy is needed to pull each molecule away from another. They have high boiling points.

Small molecules, such as petrol have short chains. The molecules do not have very strong attractive forces between them and are easily separated. This means that less energy is needed to pull the molecules apart. They have very low boiling points.

strong intermolecular forces between large molecules

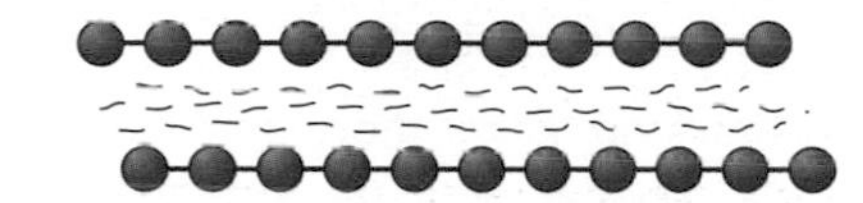

FIGURE 4: Larger molecules such as bitumen have strong intermolecular forces of attraction (shown by continuous wavy lines). Where in the fractionating column do these molecules exist?

weak intermolecular forces between small molecules

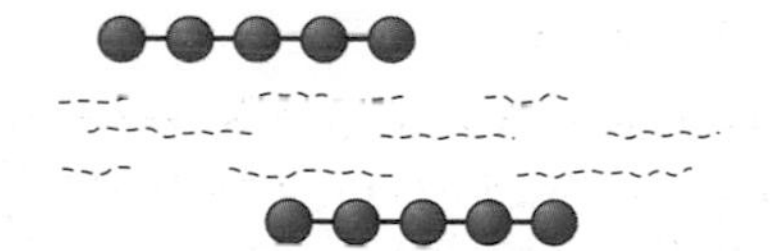

FIGURE 5: Shorter molecules have weaker forces of attraction (shown by the gaps in the wavy lines).

Questions

7 Why can the molecules of crude oil be separated by fractional distillation?

8 What are the forces between molecules called?

9 What is the relationship between the size of a molecule and the strength of the forces of attraction between molecules?

10 What is the relationship between the force of attraction between molecules in a fraction and the boiling point of the hydrocarbons in the fraction?

Remember!

Intermolecular forces between hydrocarbon molecules break during boiling, but covalent bonds within the molecule do not.

hydrocarbons intermolecular forces non-renewable temperature gradient

Problems in extracting crude oil

You will find out:

- about environmental problems in extracting crude oil
- about political problems associated with the distribution of crude oil
- how the supply of petrol keeps pace with demand

> When crude oil is found, a large area is taken over for drilling and pumping oil to the surface. This causes damage to the environment.

> Crude oil is also found under the sea. It has to be pumped out using oil rigs. This is a dangerous activity.

> Crude oil has to be transported through pipelines or by tanker. Any oil spills from these would damage the environment.

> If sea-going tankers run aground and are damaged by accident, the oil spills and forms an oil slick. This can cause enormous harm to wildlife and habitats. Beaches have to be cleaned up which is expensive and can take years.

However, not many people want to give up their car, their heating or their plastic goods.

The need for petrol and other products from crude oil has to be balanced with the cost to the environment.

More petrol needed

More petrol is needed, but less paraffin is needed. Paraffin can be broken down or 'cracked' into petrol. This can be done on an industrial scale or in the laboratory.

> To crack liquid paraffin a high temperature and a catalyst are used.

> Large molecules are not so useful. Liquid paraffin is cracked into smaller, more useful molecules.

> **Cracking** makes more petrol.

liquid paraffin (large molecules) on mineral fibre
aluminium oxide
hydrocarbon gas (small molecules)
very strong heat
water

FIGURE 6: What does cracking produce from large molecules?

Did you know?

Petrol, diesel and paraffin are hydrocarbon molecules.

Diesel and paraffin contain large hydrocarbon molecules.

Petrol contains small hydrocarbon molecules.

Cracking turns large hydrocarbon molecules into smaller, more useful hydrocarbon molecules.

Questions

11 Name three uses of crude oil.

12 Name one disadvantage of getting petrol from crude oil.

13 How is petrol made from fractions that contain large molecules of oil?

Environmental damage

Extracting crude oil causes damage to the environment. Oil fields are usually found over very large areas. Once extraction has ceased it is very expensive to return the landscape to its former state.

Extracting oil from under the sea is a dangerous and skilled activity. Oil slicks caused by leaks from tankers can harm animals, pollute beaches and destroy unique habitats for long periods of time. Damage to birds' feathers can cause death. When a bird is covered with oil it is no longer buoyant and often ingests the oil. Clean-up operations to save the animals and clean beaches are extremely expensive. The detergents and barrages used in clean-ups pose their own environmental problems.

FIGURE 7: What do you suggest would be the best substance to clean oil off the bird's feathers?

cracking petroleum   oil slick   paraffin uses   industrial cracking

Supply and demand

There is over-supply of long-chain hydrocarbon fractions, such as heavy oil, and a shortage of short-chain hydrocarbons, such as petrol.

Cracking

Cracking turns the surplus long-chain hydrocarbons into useful short-chain hydrocarbons.

> The long-chain hydrocarbons are heated at high temperature over a catalyst.

> They split into smaller molecules such as petrol.

Hydrocarbons from crude oil are called alkanes. When a large alkane is cracked it becomes a smaller alkane and an alkene.

An alkene has a double bond. This makes it useful for making polymers.

Product	Supply in tonnes	Demand in tonnes
petrol	100	300
diesel	200	100
heating oil	250	50

FIGURE 8: What is used in these cracking towers to make more petrol on an industrial scale?

Questions

14 What type of fractions are cracked?

15 What conditions are needed for cracking to take place?

16 Look at the table. How much more petrol is needed than is supplied?

17 Explain the difference between the supply of heating oil and its demand, using the figures in the table. What could be done with the excess?

Political problems

Oil is a finite resource which may run out in the next 100 years.

There has started to be more research into finding replacements.

However, demand for oil and its products is enormous and increases each year. Not only is it used for fuels for transport and heating, but there is also a large demand for the fraction called naphtha. Medicines, plastics and dyes are made from this fraction. There is a conflict between the needs for making petrochemicals and for making fuels.

Extracting crude oil can cause political problems. The UK is dependent on oil and gas from politically unstable countries.

Oil-producing nations can set prices high and cause problems to nations that do not produce oil. The supply of oil in the future is uncertain and depends on finding deposits and political stability.

It is difficult for industry to match supply with demand for petrol.

The solution to this problem of supply and demand is a process called cracking.

Expensive industrial cracking plants have to be built near to refineries so that large hydrocarbon molecules can be cracked to make smaller, more useful molecules.

The supply of petrol can then be matched more closely with demand as more petrol is produced.

FIGURE 9: A cracking plant built beside an oil refinery. Why are cracking plants needed?

Questions

18 How does industry match the demand for petrol with the supply from crude oil?

19 Explain why the naphtha fraction is important.

Using carbon fuels

You will find out:

> what makes a good fuel
> how to use a word equation to describe a fuel burning

Choosing a fuel

Have you ever considered how suitable the fuel is that is put into a car? It will need to flow easily.

It is important to choose the right fuel for the job.

Fuels are used for cooking, transport and heat.

Heat is released when a fuel burns. This is useful energy.

Lots of oxygen is needed for complete combustion.

Complete combustion of a hydrocarbon fuel produces carbon dioxide and water only.

FIGURE 1: Which fuel is suitable for a car?

Choosing a fuel

A fuel is chosen because of its characteristics:

> energy value
> ease of use
> cost
> pollution caused.
> availability
> storage method
> toxicity

FIGURE 2: Which fuel is best in a quad bike engine? Explain why.

Fuel	Able to flow?	How it burns
petrol	yes	instantly
coal	no	glows for a time

How fuels burn

When a fuel burns it gives off gases.

Waste gases can be seen in the exhaust of a car.

Water is given off in the form of steam. Other gases depend on how well the fuel burns and what else is in the fuel.

Oxygen is needed for fuels to burn.

The word equation for a fuel burning in air is:

fuel + oxygen → carbon dioxide + water

The two word equations for incomplete combustion are:

fuel + oxygen → carbon monoxide + water

fuel + oxygen → carbon + water

This happens when there is not enough oxygen.

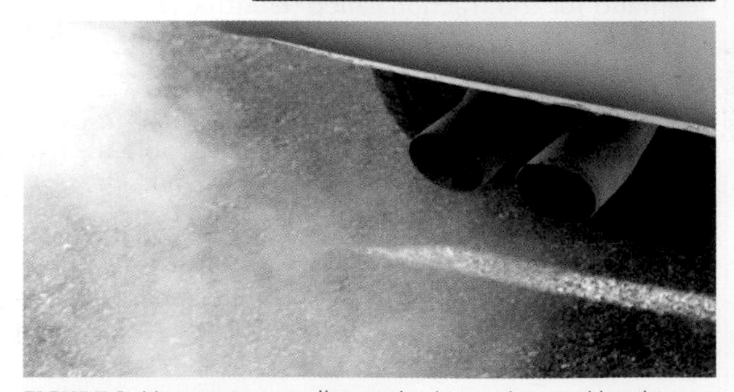

FIGURE 3: How can you tell petrol is being burned by this car's engine?

Questions

1 What characteristic must a fuel used in a car have?

2 What gas is always given off when a fuel burns?

oxygen gas carbon dioxide exhaust fumes vehicles

Comparing fuels

Coal and petrol are fuels with different characteristics.

Which could be used in a power station and which in a car engine?

Characteristic	Coal	Petrol
energy value	high	high
availability	good	good
storage	bulky and dirty	volatile
cost	high	high
toxicity	produces acid fumes	produces less acid fumes
pollution caused	acid rain, carbon dioxide, soot	carbon dioxide, carbon monoxide
ease of use	easier to store	flows easily and easy to ignite

When coal and petrol are compared, coal produces more pollution than petrol.

However, coal is not as dangerous to store as it does not catch fire so readily. Coal is not volatile (able to turn into a gas easily) but it is dirty and takes up a lot of storage space.

Burning hydrocarbon fuels produces energy.

fuel + oxygen ⟶ carbon dioxide + water

It can be shown in the laboratory that a hydrocarbon fuel burns in oxygen to give carbon dioxide and water. Lime water is used to test for carbon dioxide. Cobalt chloride paper is used to test for water, which is produced as steam.

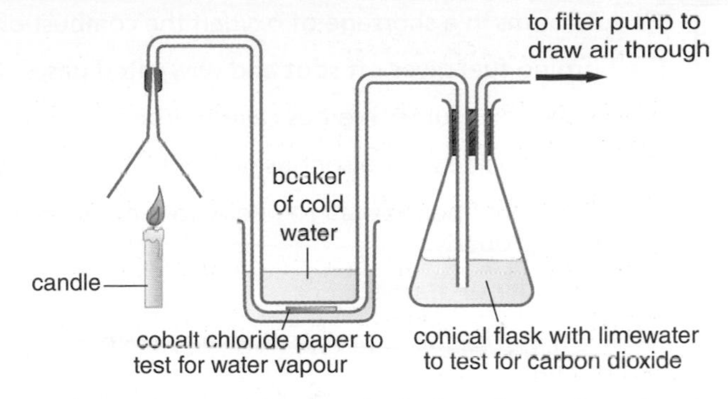

FIGURE 4: What is made when the hydrocarbons of candle wax burn?

Questions

3 Explain why you would use coal as the fuel for an open fire rather than petrol.

4 How can you test for the products of combustion of fuels?

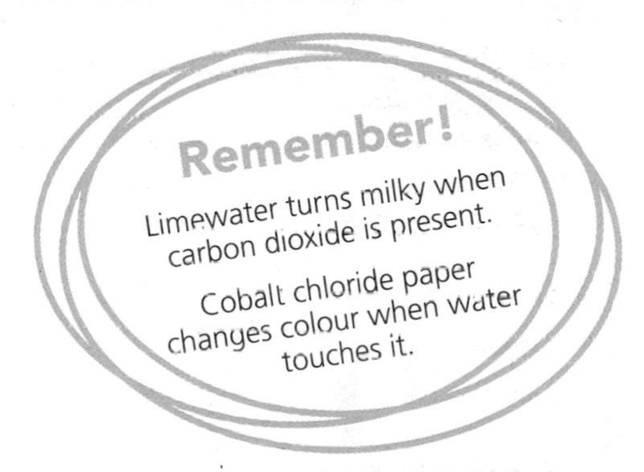

More on choosing fuels

Coal would not be easy to use in an engine for a car as it does not ignite easily. Look at the table above. Why else would coal not be good to use?

You would choose petrol or diesel as a fuel for a car. Petrol and diesel are liquids so they can circulate easily in the engine. Use the table above to give another reason why petrol or diesel would be better to use than coal would be.

These fuels are so easy to use and, as the population increases, more and more fossil fuels are being consumed, which has a big impact. Governments are concerned because of the increasing carbon dioxide emissions that result when fossil fuels are burned. Countries with huge populations, such as India or China, are using more fuel, which adds further to gas emissions. Many governments have pledged to try to cut carbon dioxide emissions over the next 15 to 20 years. It is a global problem that cannot be solved by one country alone.

Questions

5 What concerns governments about the increasing population?

6 Suggest what individuals could do to reduce carbon dioxide emissions.

volatile liquid   chemical toxicity   products of combustion   population growth

Incomplete combustion

If a fuel burns in a shortage of oxygen the combustion is incomplete.

The burning fuel gives off soot and unwanted gases. These gases are toxic.

One of the toxic fumes is a gas called carbon monoxide.

Carbon monoxide is a poisonous gas and is very dangerous if it is breathed in.

A heater should be checked regularly to make sure it is burning properly.

Bunsen burner flame

A Bunsen burner flame produces energy from burning gas.

> If the air hole is open the flame burns in plenty of oxygen. Combustion is complete and a blue flame is seen. This means there is more energy released.

> If the air hole is closed there is a shortage of oxygen. Combustion is incomplete and less energy is transferred.

A blue flame from a Bunsen burner transfers more energy than a yellow Bunsen flame as complete combustion gives a blue flame. Incomplete combustion gives a yellow flame and so less energy is released.

When combustion is incomplete, a yellow flame is seen. This is because a yellow flame produces a lot of soot. Carbon monoxide, soot and water vapour are produced as well as carbon dioxide.

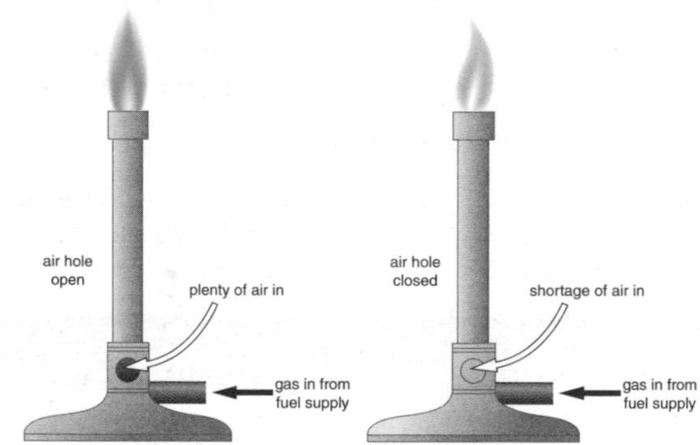

FIGURE 6: What colour is the flame from a Bunsen burner when the air hole is open?

You will find out:
> about complete combustion and incomplete combustion
> about pollution caused by burning fossil fuels
> how to write a balanced symbol equation to describe a fuel burning

FIGURE 5: Why should heaters in people's homes be checked regularly?

Questions

7 What is given off when fuels burn in a shortage of oxygen?

8 Why is this a problem?

9 Which Bunsen burner flame transfers the most energy?

10 What are the products of incomplete combustion?

Complete and incomplete combustion

Complete combustion occurs when a fuel burns completely in air.

A fuel, such as methane, uses oxygen in the air to produce products.

When there is a plentiful supply of oxygen the products are carbon dioxide and water.

methane + oxygen ⟶ carbon dioxide + water

The advantage of complete combustion is that more energy is released and no toxic gases or soot are produced. More energy is released during complete combustion than during incomplete combustion.

FIGURE 7: What has caused this stone gargoyle to erode?

carbon monoxide incomplete combustion

Incomplete combustion

Incomplete combustion occurs when a fuel burns in limited oxygen. The products are carbon monoxide and water. Complete combustion is better than incomplete combustion because:

> less soot is made

> more heat is released

> toxic carbon monoxide gas is not produced.

FIGURE 8: Trees killed by acid rain in Scandinavia. Suggest why forests are particularly badly hit in these countries.

Pollution problems

A fossil fuel contributes to the greenhouse effect when it burns because it produces carbon dioxide. This is believed to contribute to global warming.

Fuels also cause other problems. Coal has sulfur in it. Burning coal gives off sulfur dioxide. Sulfur dioxide dissolves in rainwater to make acid rain, which damages stone buildings and statues and kills fish and trees.

Questions

11 If carbon monoxide was the product made during incomplete combustion of methane, how would you construct the word equation for the reaction?

12 What are the products of complete combustion?

13 Which type of combustion releases more useful energy?

14 What effect does carbon dioxide contribute to?

15 What problems do fuels containing sulfur cause?

More on combustion

Complete combustion releases useful energy.

The formulae for the products of complete combustion are:

carbon dioxide CO_2

water H_2O

Methane is a common hydrocarbon fuel. The formula for methane is CH_4.

Complete combustion can be shown by the equation:

$CH_4 + O_2 \rightarrow CO_2 + H_2O$

The equation must be made to balance.

There are two oxygen atoms in the reactants and three oxygen atoms in the products.

The balanced equation is:

$CH_4 + 2O_2 \rightarrow CO_2 + 2H_2O$

There are now four oxygen atoms in both reactants and products.

Incomplete combustion happens when less oxygen is available. Carbon monoxide, CO, and carbon, C, can be made.

$CH_4 + O_2 \rightarrow C + 2H_2O$

Equations need to be balanced.

First, count up the numbers of atoms in each molecule (shown by the subscript numbers).

Don't change these numbers.

Then, if necessary, add to the molecule number (large number in front of formula) in order to balance the numbers on either side of the arrow.

Question

16 Write a balanced equation for the complete combustion of propane, C_3H_8.

Remember!

Complete combustion gives carbon dioxide and water. Incomplete combustion gives carbon monoxide, carbon and water.

You need to remember the formulae of carbon dioxide, CO_2, carbon monoxide, CO, water, H_2O, and oxygen, O_2.

greenhouse effect   acid rain   complete combustion   sulfur dioxide

Clean air

You will find out:

- > about the composition of air and how it remains constant
- > about a possible theory on the evolution of the atmosphere

Pollution of our air

Photochemical smog happens more often in large cities than it did 30 years ago. This is because of the increases in population and the use of motor vehicles. The gases from the vehicles react together in bright sunlight to produce smog. Photochemical smog irritates the eyes and the lungs.

Both Athens in Greece and Beijing in China have particular photochemical smog problems. The 2004 and 2008 Olympic Games took place in Athens and Beijing, respectively. Athletes were worried that photochemical smog would mean their performances may be below their best.

In the United Kingdom the number of people suffering from asthma is steadily increasing. Many people think this is because of an increase in the amount of pollution in the atmosphere.

FIGURE 1: What other problems are caused by photochemical smog in Athens?

Clean air

Air is a mixture of different gases. Clean air is air that has no pollutants caused by human activity. The main gases in clean air are:

> nitrogen > oxygen > carbon dioxide

> water vapour > noble gases (argon, neon, krypton, xenon).

The amount of water vapour in the air changes. The amounts of oxygen, nitrogen and carbon dioxide in the air remain almost constant.

The levels of gases in the air depend on:

> combustion of fossil fuels, which increases the level of carbon dioxide and decreases the level of oxygen

> respiration by plants and animals, which increases the level of carbon dioxide and decreases the level of oxygen

> photosynthesis by plants, which decreases the level of carbon dioxide and increases the level of oxygen.

Remember!

Do not confuse breathing and respiration. Breathing describes the way air enters and leaves the lungs. Respiration takes place in cells.

FIGURE 2: In what way does photosynthesis alter the levels of gases in air?

Questions

1 Suggest why the amount of water vapour in the atmosphere can change.

2 Respiration uses up oxygen and makes carbon dioxide. Write down the name of one other process that uses up oxygen and makes carbon dioxide.

nitrogen photochemical smog noble gases oxygen pollutants in the air

What is in clean air?

About 78% of clean air is nitrogen and about 21% is oxygen. Of the remaining 1%, only 0.035% is carbon dioxide.

These percentages do not change very much. This is because there is a balance between processes that use up carbon dioxide and make oxygen and processes that use up oxygen and make carbon dioxide. Some of these processes are shown in a carbon cycle.

Respiration: glucose + oxygen → carbon dioxide + water

Photosynthesis: carbon dioxide + water → glucose + oxygen

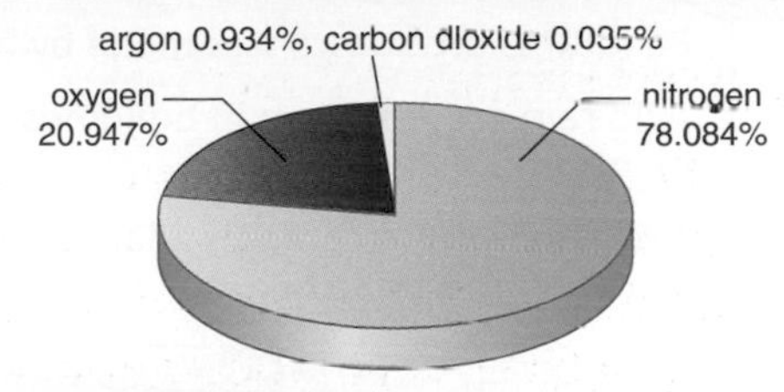

FIGURE 3: Why do the proportions of gases in clean air stay almost constant?

Evolution of the atmosphere

Scientists know that gases trapped in liquid rock under the surface of the Earth are always escaping. This happens in volcanoes.

Scientists speculate about the original atmosphere of the Earth. It is known that at some point in the Earth's history, microbes developed that could photosynthesise. These organisms could remove carbon dioxide from the atmosphere and add oxygen. Eventually the level of oxygen reached what it is today and it now stays fairly constant.

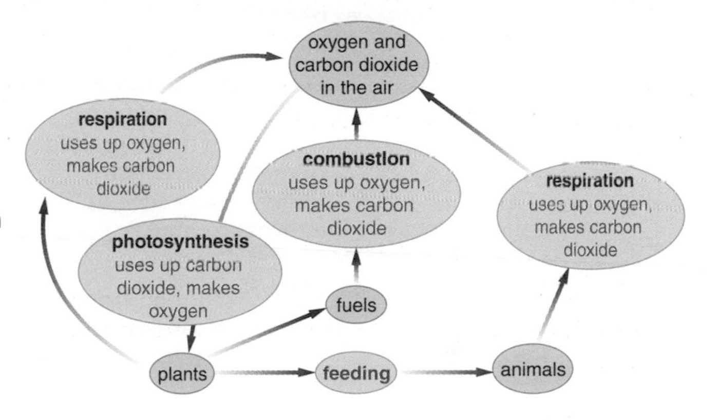

FIGURE 4: Carbon cycle. The arrows show the direction of movement of carbon compounds. What is the only process that produces oxygen?

Questions

3 How does photosynthesis change the level of oxygen?

4 Write the word equation for respiration.

Changing atmospheric composition

Although the composition of the air remains fairly constant there are changes in the percentages of some gases.

Over the past few centuries the percentage of carbon dioxide in air has increased slightly. This is due to a number of factors, for example:

> Deforestation – as more rainforests are cut down less photosynthesis takes place.

> Increased population – as the population increases, the world's energy requirements increase.

Origin of the Earth's atmosphere

The Earth is over four and a half thousand million years old. During this time the Earth's atmosphere has evolved into its present day composition. Scientists can only guess about this evolution over millions of years. One thing they do know is that the gases came from the centre of the Earth in a process called degassing. One theory (based on the composition of gases vented by present-day volcanic activity) is as follows.

> The original atmosphere contained ammonia and, later, carbon dioxide.

> Degassing of early volcanoes produced an atmosphere rich in water and carbon dioxide.

> Chemical reaction between ammonia and rocks produced nitrogen and water. The condensing water vapour formed the oceans.

> Carbon dioxide dissolved in the ocean waters.

> Nitrogen levels increased because of its lack of reactivity.

> Much later, organisms that could photosynthesise evolved. These organisms converted carbon dioxide and water into oxygen. Oxygen levels increased.

> As the percentage of oxygen in the atmosphere increased the percentage of carbon dioxide decreased until today's levels were reached.

> The percentage of nitrogen slowly increased. Since nitrogen is very unreactive, very little nitrogen was removed from the atmosphere.

Questions

5 Explain why it is impossible for scientists to be certain of the answer to the question, 'How did the Earth's atmosphere evolve?'

6 Describe how an increase in the use of motor vehicles may affect the composition of the atmosphere.

Atmospheric pollutants

You will find out:

- > how to evaluate the influence of human activity on the composition of the atmosphere
- > about atmospheric pollutants and their origins
- > how the levels of atmospheric pollutants can be controlled

Pollutants are substances made by human activity that harm the environment.

Acid rain erodes stone in buildings and statues, corrodes metals and kills plants and fish.

The atmosphere contains a large number of pollutants. The main ones are shown in the table below.

Pollutant	Environmental effects
carbon monoxide	poisonous gas formed by incomplete combustion of petrol- or diesel-powered motor vehicles
oxides of nitrogen	photochemical smog acid rain formed by reaction of nitrogen and oxygen at very high temperatures such as in an internal combustion engine
sulfur dioxide	acid rain formed from sulfur impurities when fossil fuels burn

FIGURE 5: This statue of a lion is made from limestone. Suggest why its features are eroded.

Reducing pollution

Scientists are finding ways to reduce the amount of these pollutants.

Most cars are now fitted with **catalytic converters**. A catalytic converter reduces the levels of carbon monoxide and oxides of nitrogen.

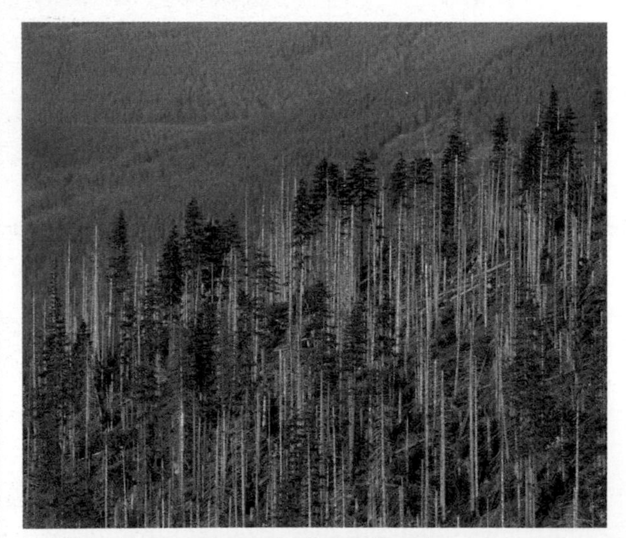

FIGURE 6: These trees in the Czech Republic have died from the effects of acid rain.

Question

7 Look at the table below. How does a catalytic converter help to reduce atmospheric pollution?

Catalytic converter	Emissions in grams/kilometre		
	of carbon monoxide	of hydrocarbons	of oxides of nitrogen
not fitted	5.59	1.67	1.04
fitted	0.61	0.07	0.04

oxides of nitrogen how catalytic converters work erosion by acid rain

Pollution control

Atmospheric pollution affects the environment and people's health. These effects will get worse unless atmospheric pollution is controlled. The European Union and the United Kingdom government have introduced many laws regarding pollution control. These have had some effect, but still more controls are needed. There have been a number of initiatives to reduce road congestion and to encourage car users to reduce emissions, and you can research these.

A car fitted with a catalytic converter produces exhaust fumes that contain only a very small amount of these pollutants.

Catalytic converters

Exhaust gases from a car contain several pollutants, including nitric oxide (an oxide of nitrogen) and carbon monoxide.

A catalytic converter changes carbon monoxide into carbon dioxide. At the same time oxides of nitrogen are converted into nitrogen.

Carbon dioxide is a greenhouse gas, but it is less dangerous to the environment than carbon monoxide.

The reduction in carbon monoxide means less toxic gas. Carbon monoxide prevents oxygen being taken up by blood cells.

Less oxides of nitrogen means less risk of photochemical smog.

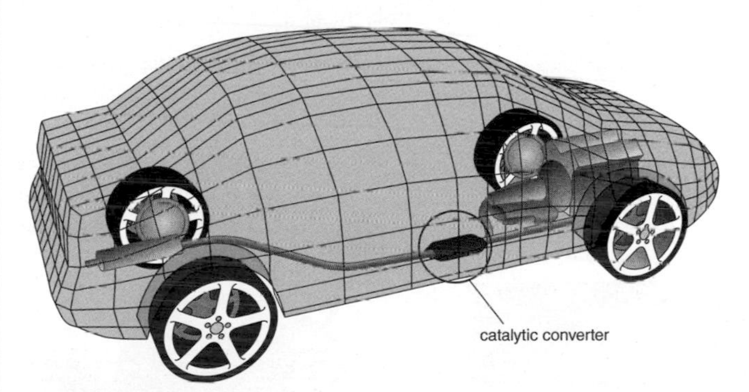

FIGURE 7: Why are catalytic converters in motor cars good for the atmosphere?

Did you know?

Solar-powered cars do not produce greenhouse gases or acid-rain emissions.

Did you know?

The pH of acid rain in some parts of Central Europe is so low that it's able to attack church roofs that contain gold.

Question

8 In a catalytic converter nitric oxide reacts with carbon monoxide to make carbon dioxide and nitrogen. Write the word equation for this reaction.

More on catalytic converters

High temperatures inside an internal combustion engine allow nitrogen from the air to react with oxygen to make oxides of nitrogen, because there is enough energy to break the strong bonds between nitrogen atoms.

Catalytic converters contain a rhodium catalyst.

The converter removes carbon monoxide from exhaust fumes by converting it into carbon dioxide.

A reaction between nitrogen monoxide and carbon monoxide takes place on the surface of the catalyst. The reaction forms nitrogen and carbon dioxide.

carbon monoxide + nitrogen monoxide → nitrogen + carbon dioxide

$$2CO + 2NO \rightarrow N_2 + 2CO_2$$

Questions

9 Cars that use a hydrogen–oxygen fuel cell instead of petrol are being developed. In these cars hydrogen and oxygen react to make water.

a Write the balanced equation for the reaction between hydrogen and oxygen.

b Suggest why a car using a hydrogen–oxygen fuel cell produces very little atmospheric pollutants.

Making polymers

You will find out:

- that polymers are large molecules made up of many small molecules
- about polymerisation
- how to draw displayed formulae for polymers

Out with the old, in with the new

Compare the two pictures. Our lifestyles are very different to those led by people in the Victorian age. You can see how plastics have changed our lives.

Plastics are made from **polymers**.

The molecules in plastics are called polymer molecules.

FIGURE 1: Were there positive benefits to a life without plastics?

What is a polymer?

A polymer is a:

- very big molecule
- very long chain molecule
- molecule made from many small molecules called **monomers**.

When lots and lots of monomers are joined to make a polymer the reaction is called **polymerisation**.

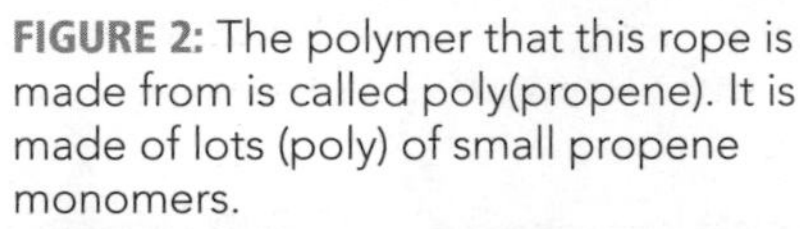

FIGURE 2: The polymer that this rope is made from is called poly(propene). It is made of lots (poly) of small propene monomers.

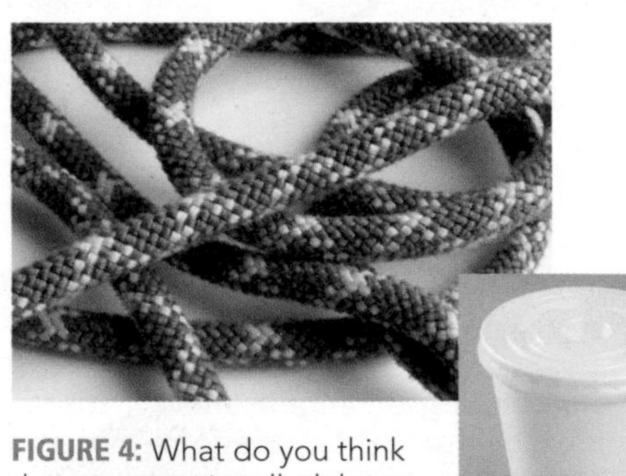

FIGURE 3: This shopping bag is made from a plastic polymer called poly(ethene). What are the monomers called?

FIGURE 4: What do you think the monomer is called that makes poly(styrene), used to produce this cup?

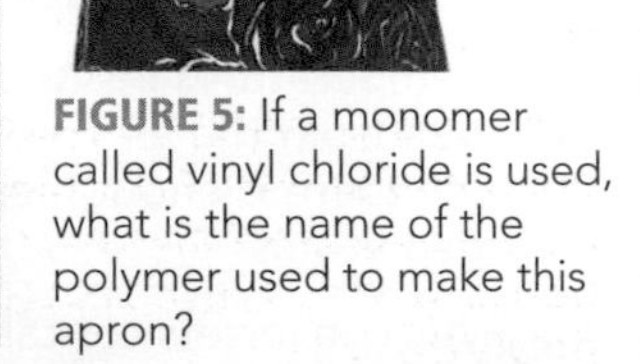

FIGURE 5: If a monomer called vinyl chloride is used, what is the name of the polymer used to make this apron?

Did you know?

A polymer always has a bracket around the name of its monomer.

What would be the 'monomer' of poly(paperclip)? Try making a paperclip chain yourself.

Questions

1. What is a polymer?
2. Write down the name of one polymer.
3. Write down the name of the monomer that makes poly(propene).
4. Name one use of a polymer.

monomers and polymers polyvinyl chloride polymerisation uses of polythene

Polymerisation

A polymer can often be made from many monomers of the same type joined together. This process is called addition polymerisation. It requires **high pressure** and a **catalyst**.

A small letter '*n*' is used in science to mean 'lots of'.

So a polymer made from the monomer ■ is written $(■)_n$ and can be represented like this.

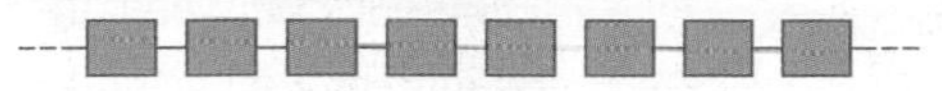

The dashed line means 'and longer'.

poly(ethene)

This is the displayed formula for the addition polymer of ethene, which is poly(ethene). The displayed formula of polymers are written as repeating units with a square bracket at each end. A small *n* follows the last bracket to show that there could be more repeating units and the chain could go on.

A polymerisation reaction needs:

> high pressure

> a catalyst.

FIGURE 6: A polymerisation plant. What conditions does this plant need to provide in order for a polymerisation reaction to take place?

Questions

5 What is the process called in which lots of monomers join to make long chain molecules?

6 What two conditions are needed for this process to take place?

Addition polymerisation

Addition polymerisation is the reaction of many monomers that have carbon-to-carbon **double bonds** to form a polymer that has single bonds.

The displayed formula of an addition polymer can be constructed when the displayed formula of its monomer is known.

Example

The displayed formula for the ethene monomer is:

ethene

During a polymerisation reaction, the high pressure and catalyst cause the double bond in the ethene monomer to break.

Each of the two carbon atoms needs to form a new bond and join with another ethene molecule, which also has two carbon atoms needing to form two new bonds, and so the reaction continues to give poly(ethene).

If the displayed formula of an addition polymer is known, the displayed formula of its monomer can be worked out by looking at its repeated units.

Example

This addition polymer:

has a repeated unit of two carbon atoms, three hydrogen atoms and one chlorine atom. The bond between the two carbon atoms would have been a double bond originally.

Therefore the monomer's displayed formula is:

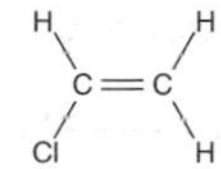

In addition polymerisation:

A monomer that makes a polymer contains at least **one double bond between carbon atoms.**	→ A polymer contains only single covalent bonds between carbon atoms.
When a molecule has one or more double covalent bonds between carbon atoms it is called an unsaturated compound. Many unsaturated monomer molecules (alkenes) form a saturated polymer during addition polymerisation.	→ When a molecule has only single covalent bonds between carbon atoms it is called a saturated compound.

Questions

7 Copy the monomer. Draw three repeat units of its polymer.

8 Look at the polymer. Draw its monomer.

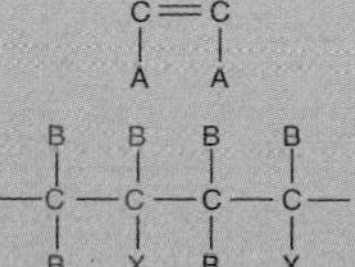

Alkanes and alkenes

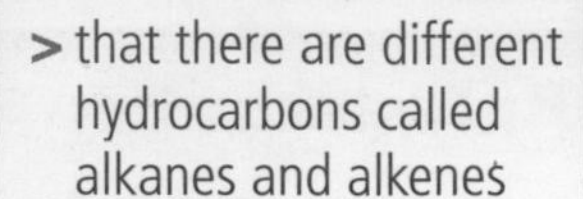
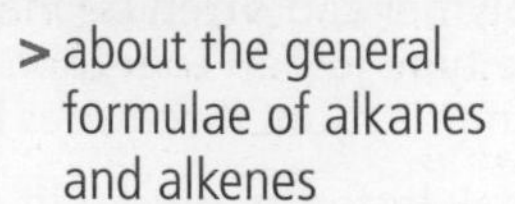

You will find out:
- > that there are different hydrocarbons called alkanes and alkenes
- > about the general formulae of alkanes and alkenes
- > how bromine solution can be used to tell the difference between an alkane and an alkene

Hydrocarbons are molecules made from carbon and hydrogen only.

Alkanes are hydrocarbons. **Alkenes** are hydrocarbons.

One atom of carbon and four atoms of hydrogen chemically combine to make a hydrocarbon called methane. Methane is an alkane.

CH_4

A hydrocarbon is named according to:

> the number of carbon atoms it has

> whether it has a double bond or single bonds.

If the name of the alkane is known it is easy to predict the name of the alkene.

C_5H_{12} is pentane. The alkene with five carbon atoms is pentene.

Look at these displayed formula. All the alkanes are hydrocarbons. All the alkenes are hydrocarbons.

Alkane	Displayed formula	Alkene	Displayed formula
methane	CH_4	–	–
ethane	$H_3C{-}CH_3$	ethene	$H_2C{=}CH_2$
propane	$H_3C{-}CH_2{-}CH_3$	propene	$H_3C{-}CH{=}CH_2$
butane	$H_3C{-}CH_2{-}CH_2{-}CH_3$	butene	$H_3C{-}CH{=}CH{-}CH_3$

Questions

9 Only two elements chemically combine to make a hydrocarbon. Which two?

10 Butene has the molecular formula C_4H_8. Is this hydrocarbon an alkane or an alkene?

Remember!

You need to recognise that propane has only single bonds so it is an alkane. You need to recognise that propene has a double bond between carbon atoms, so it is an alkene.

alkanes alkenes molecular formula covalent bonds

Recognising hydrocarbons

Hydrocarbons are made from hydrogen and carbon only.

Propane, C_3H_8, is a hydrocarbon.

```
    H   H   H
    |   |   |
H — C — C — C — H
    |   |   |
    H   H   H
```

propane

Propanol, C_3H_7OH, is not a hydrocarbon. It contains an oxygen atom.

```
    H   H   H
    |   |   |
H — C — C — C — OH
    |   |   |
    H   H   H
```

propanol

Hydrocarbons that contain single bonds only are called alkanes.

Look back at the displayed formula of butane in the table. How can you tell it is an alkane?

Hydrocarbons that have at least two carbon atoms joined together by a double bond are called alkenes. Look at the displayed formula of butene in the table. How can you tell it is an alkene?

The double bond between two carbon atoms in an alkene involves **two** shared pairs of electrons.

Bromine solution is used to test for a double bond in alkenes. It is an orange solution. When it is added to an alkene the orange solution turns colourless. This does not happen with an alkane.

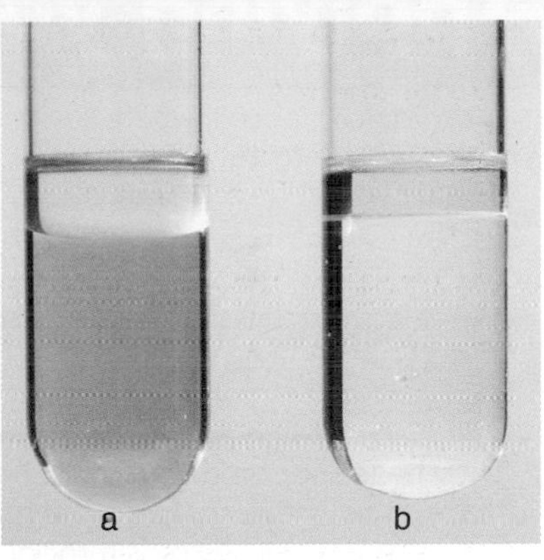

FIGURE 7: Using bromine solution to test for alkenes. Two test tubes of bromine solution were set up and labelled a and b. An alkane was added to one tube and an alkene to the other. Which tube, a or b, had the alkene added to it?

Questions

11 What is a hydrocarbon?

12 What is the difference between an alkane and an alkene?

More on hydrocarbons

The building blocks of many polymers are hydrocarbons.

Polymers are made from hydrocarbons that have at least one double covalent bond between two carbon atoms.

```
H       H
 \     /
  C = C      ethene
 /     \
H       H
```

These hydrocarbons are called alkenes and they are **unsaturated**.

Did you know?

The general formula of an alkene is C_nH_{2n}.

The general formula of an alkane is C_nH_{2n+2}.

Unsaturated molecules add together to make addition polymers in a process called addition polymerisation.

Ethane is **saturated.** It has no double bonds. It only has single covalent bonds between carbon atoms.

```
    H   H
    |   |
H — C — C — H
    |   |
    H   H
```

ethane

Why does bromine decolourise?

Bromine solution is used to test for alkenes. It is an orange solution. When an alkene is added the orange solution turns colourless. This is because the bromine solution has reacted with the alkene and formed a new compound. This is an **addition reaction**. The new compound is a **dibromo compound,** which is colourless.

```
H       H                    Br  H
 \     /                     |   |
  C = C    + Br2  ——>    H — C — C — H
 /     \                     |   |
H       H                    H   Br
```

When an alkane is added to the bromine solution it remains orange because an alkane does not react with bromine.

Questions

13 What is the difference between a saturated molecule and an unsaturated molecule?

14 Explain why there is a difference between the general formulae for an alkane and an alkene.

Preparing for assessment: Applying your knowledge

To achieve a good grade in science, you not only have to know and understand scientific ideas, but you need to be able to apply them to other situations and investigations. These tasks will support you in developing these skills.

The Dartmoor Challenge

Will, Sacha, Mike and Nisha are on their school summer camp in Dartmoor, in south-west England. They have spent a day in school at the start of the week with the rest of the group going over their plans. Now they are up on the moorland for 4 days. The group is planning to walk to 20 tors by the end of the week. A tor is a rocky outcrop, and some of them are quite high.

The weather on Dartmoor can change very quickly and in the summer it is often sunny and warm. However, clouds can gather quickly and rain can fall and make the unsuspecting walker very wet very quickly.

Staying dry is very important. If you get wet you soon start to feel cold and miserable and that makes it much harder to carry on walking. However, climbing the tors is hard work and you soon start to sweat on a warm day.

This is a problem when selecting suitable clothing. Outer clothes that are not waterproof would soon get wet through, but ordinary waterproofs like Nylon, though they keep rain out, keep perspiration in.

A special material called GORE-TEX® provides an answer, and many moorland walkers now use hats, coats, trousers and boots made from this material. GORE-TEX® is a membrane made from a substance called PTFE, combined with Nylon. It keeps external moisture from rain out, but lets water vapour from the body escape.

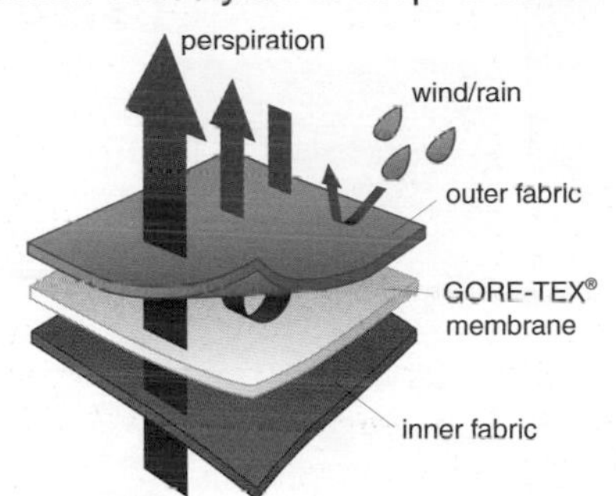

GORE-TEX® works because PTFE is hydrophobic ('hates' water) and causes water to gather in beads on its outer surface that are then too large to pass through to the inside (the water drops are 20,000 times larger than the pores). Water vapour (perspiration) from the body can still pass through to the outside as the molecules are around 700 times smaller than the pores in the membrane.

Task 1

Why do Will and his friends need coats that are 'waterproof and breathable'?

Task 2

Why is GORE-TEX® successful at keeping them dry but not getting them sticky with sweat?

Task 3

Draw and label diagrams to show the difference between beads of rainwater and molecules of water vapour.

Maximise your grade

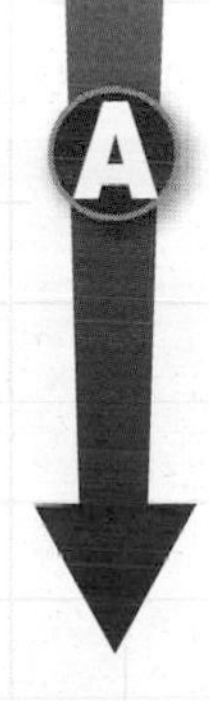

Grade	Answer includes showing that you can...
	Know which material can be used as waterproof clothing even if it does not let perspiration out.
F	Interpret information about materials that are • breathable **but not** waterproof • waterproof **but not** breathable • breathable **and** waterproof.
C	Explain why Gore-Tex® type materials are of great help to active outdoor people to cope with both rain and perspiration.
	Compare Gore-Tex® properties with other materials that are also waterproof but not breathable.
A	Explain how Gore-Tex® is laminated.
	Explain how liquid water cannot pass **or** how water vapour can.
	Explain why Gore-Tex® is laminated.
	As above, but with particular clarity and detail.

Designer polymers

You will find out:
- about polymers used for packaging and clothing
- how nylon is useful but is not breathable
- how GORE-TEX® has all the properties of nylon and is breathable

Polymers all around

Modern cars, communication equipment and building materials are all made from polymers in some way.

FIGURE 1: What are the attributes of polymers that make them so versatile?

Uses of polymers

Fabrics for clothes, paint for cars and cases for computers are all made from different polymers.

Polymers are chemicals such as nylon, poly(vinylchloride) (also called PVC) and polyester. Each polymer is chosen carefully for the job that it does best.

For example, nylon is used for clothing as it can make flexible fibres.

Look at the table. One property of PVC is that it is waterproof. This is an advantage when it is used to make raincoats. What properties would be needed to make a gutter and drainpipe. Would any of these polymers be useful?

Polymer	Property 1	Property 2	Use
PVC poly-(vinylchloride)	waterproof	flexible	raincoats
poly(ethene)	waterproof	flexible	plastic bags
poly(styrene)	insulates	absorbs shock	packaging
poly(propene)	strong	flexible	ropes

Did you know?

These boots are tested for toughness. They have a GORE-TEX® lining and are tested by walking through water for 400 km on a simulator.

Questions

1 Write down the names of three polymers.

2 What two properties make PVC a good polymer for raincoats?

3 Look at the properties of poly(propene) in the table. What other use could it have apart from making rope?

polystyrene properties poly(propene) what makes something waterproof

Breathable polymers

Nylon is tough, lightweight and keeps rainwater and ultraviolet light out. However, it also keeps in water vapour from body sweat. The water vapour from the sweat condenses and makes wearers wet and cold inside their raincoats. If nylon is laminated with a PTFE/polyurethane membrane, clothing can be made that is waterproof and breathable (PTFE is polytetrafluoroethene).

GORE-TEX® has all the properties of nylon and is breathable.

The discovery of GORE-TEX® material has helped active, outdoor people to cope with wetness from sweat. Water vapour from sweat can pass through the membrane, but rainwater cannot.

The table compares the properties of nylon and GORE-TEX®.

Nylon	GORE-TEX®
waterproof	waterproof
flexible	flexible
non-breathable	breathable

Questions

4 What is the disadvantage of using nylon to make outdoor clothes?

5 What is the advantage of the membrane used in GORE-TEX® material?

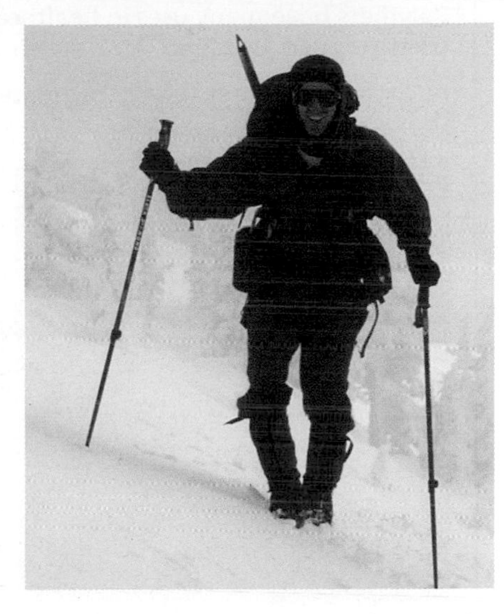

FIGURE 2: Suggest what type of material this hiker's clothes are made from.

More on breathable polymers

GORE-TEX® material is used to make waterproof and breathable clothing. GORE-TEX® is made from nylon laminated with PTFE/polyurethane membrane. The inner layer of the clothing is made from expanded PTFE, which is **hydrophobic**. Expanded PTFE is too fragile on its own and so is combined with nylon.

Scientist Bob Gore discovered a special way of processing PTFE. The PTFE is expanded to form a microporous membrane. Only small amounts of the polymer are needed to create this airy, lattice-like structure. This is why wind does not pass through the membrane. The holes in PTFE are too small for liquid water to pass through, but big enough for water vapour to pass through. This is why it is waterproof and yet breathable.

In expanded PTFE a membrane pore is 700 times larger than a water vapour molecule and therefore moisture from sweat passes through.

Questions

6 Explain why expanded PTFE repels water.

7 Explain why GORE-TEX® does not let the wind through but lets sweat out.

8 Explain why the size of the pores in the expanded PTFE membrane is significant.

water droplets from rainwater do not pass through

hydrophobic PTFE is expanded to form a microporous membrane

FIGURE 3: Why does water not soak through expanded PTFE?

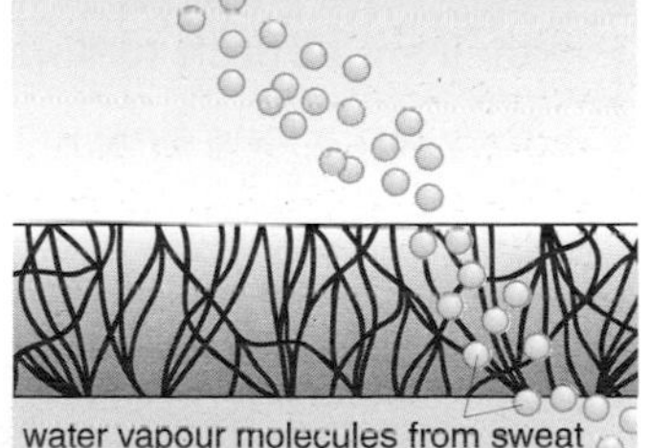

FIGURE 4: Suggest why sweat can pass through the membrane.

FIGURE 5: Suggest what happens to wind as it meets the microporous membrane.

Uses of polymers in healthcare

You will find out:

> about the uses of polymers in healthcare
> about problems with the disposal of waste polymers
> about new polymers being developed that are easy to dispose of

Polymers have many uses in healthcare. Figures 6–8 show some of the things they are used to make.

Polymers are better than other materials for some uses as the table below shows.

FIGURE 6: Contact lenses.

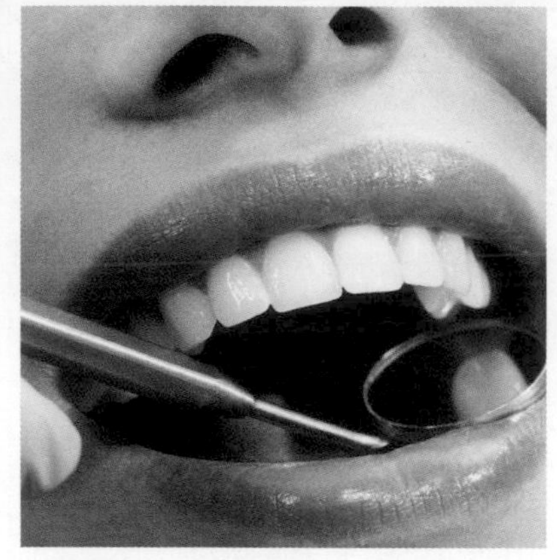

FIGURE 7: White fillings to fill cavities in teeth.

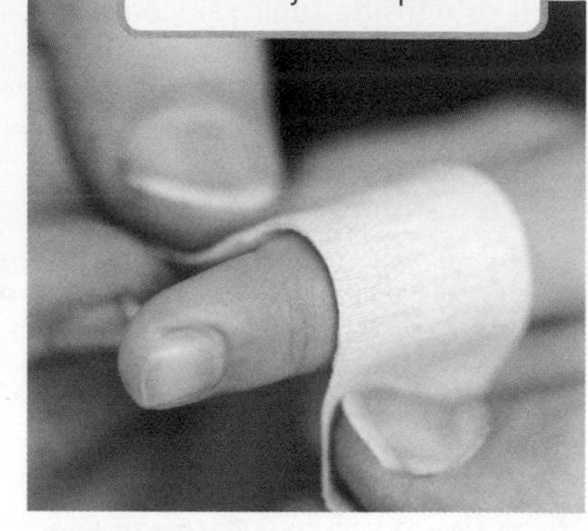

FIGURE 8: Wound-dressing materials.

Most polymers are non-biodegradable which means that they do not decay and are not decomposed by bacteria. This can cause problems.

Some ways that waste polymers can be disposed of are:

> in landfill sites

> by burning

> by recycling.

Some of the problems of using non-biodegradable polymers are:

> they are difficult to dispose of

> they last a long time in landfill sites.

The table shows how the use of a polymer is better than the use of other materials for making healthcare products.

Use	Polymer	Other material
contact lens	wet on the eye	dry on the eye
teeth filling	attractive	looks metallic
wound dressing	waterproof	gets wet

Questions

9 Write down two uses of polymers in healthcare.

10 What is the advantage of a wound dressing made from a polymer over a different dressing material?

11 What does non-biodegradable mean?

12 Write down one way to dispose of polymers.

FIGURE 9: What are the three main ways that polymers can be disposed of?

Disposing of polymers

Plastic laundry bags in hospitals dissolve in the hot water when they are put in the washing machine.

Dishwasher detergent tablets can be wrapped in new types of plastics that dissolve in the dishwasher.

Scientists are developing addition polymers that are biodegradable. These are disposed of easily when put in landfill sites. Shampoo bottles are sometimes made from plastics of these polymers.

Disposal problems for non-biodegradable polymers

Landfill sites get filled quickly and waste valuable land. Burning waste plastics produces toxic gases.

Disposal by burning or in landfill sites wastes a valuable resource. The crude oil used to make polymers is wasted and the land could be used for other purposes.

Difficulty in sorting different polymers makes recycling difficult.

Did you know?

There are now polymers made from starch that decompose and dissolve.

Biopol is a biodegradable plastic.

Question

13 Why is disposing of plastics by burning a problem?

Stretchy polymers and rigid polymers

The atoms of the monomers in each of the chains in a polymer are held together by strong covalent bonds.

The chains of the polymer are held together by weak intermolecular forces of attraction.

Plastics that have weak intermolecular forces of attraction between polymer molecules have low melting points and can be stretched easily as the polymer molecules can slide over one another.

Plastics that have strong cross-links between polymer molecules cannot slide over one another.

Remember, covalent bonds are bonds between atoms within a molecule. Intermolecular forces are forces between chains of molecules.

FIGURE 10: Which are stronger, covalent bonds or intermolecular forces of attraction?

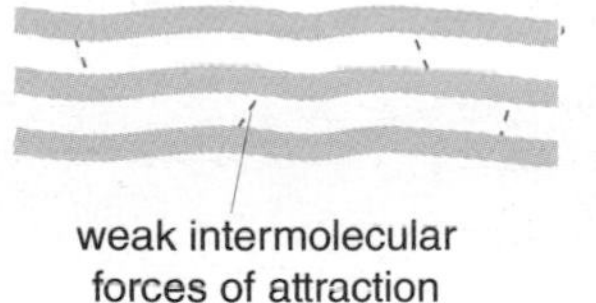

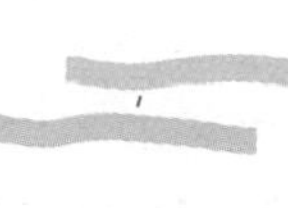

FIGURE 11: Why can polymers that have weak intermolecular forces of attraction stretch easily?

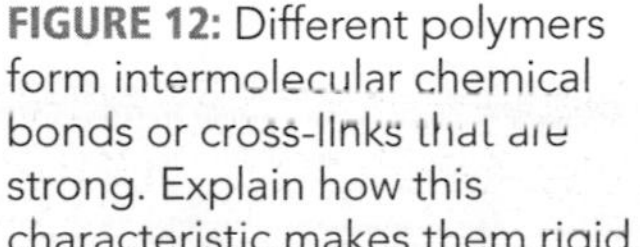

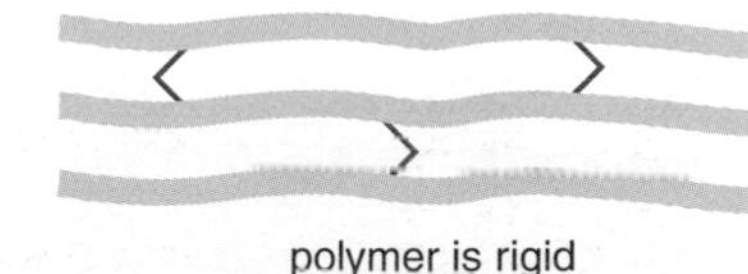

FIGURE 12: Different polymers form intermolecular chemical bonds or cross-links that are strong. Explain how this characteristic makes them rigid.

Type of polymer	Property 1	Property 2
weak intermolecular forces of attraction	stretches easily	low melting point
strong intermolecular chemical bonds or cross links	rigid	high melting point

Questions

14 Explain how the two different types of bonds or forces affect the structure of a polymer.

15 Some polymers stretch easily, some are rigid. Explain why.

intramolecular bonds   polymer cross-linking

Cooking and food additives

You will find out:

> how to recognise when a chemical change takes place

> about what happens to protein when food is cooked

> about potatoes and what happens when they are cooked

The discovery of fire

The discovery of how to make fire was enormously important.

Not only could people now keep warm, but they could also cook food.

FIGURE 1: Apart from taste what other advantages are there to cooking food?

The changing of food

A food changes when it is cooked. It cannot go back to its raw state. Cooking is a chemical change because a new substance is made. The process cannot be reversed, a new substance is made and the process cannot be reversed.

> The change is irreversible.

> An energy change takes place in the food.

These chemical changes happen when the chemicals in food are heated.

When a potato is cooked its taste and texture change irreversibly. The texture of the potato becomes softer and more fluffy.

FIGURE 2: Can the process of cooking potatoes be reversed?

Questions

1 What kind of change is there when a potato is cooked?

2 Write down the word that means 'it cannot change back'.

3 Which diagram, A, B, C or D, shows food that is not the result of a chemical change?

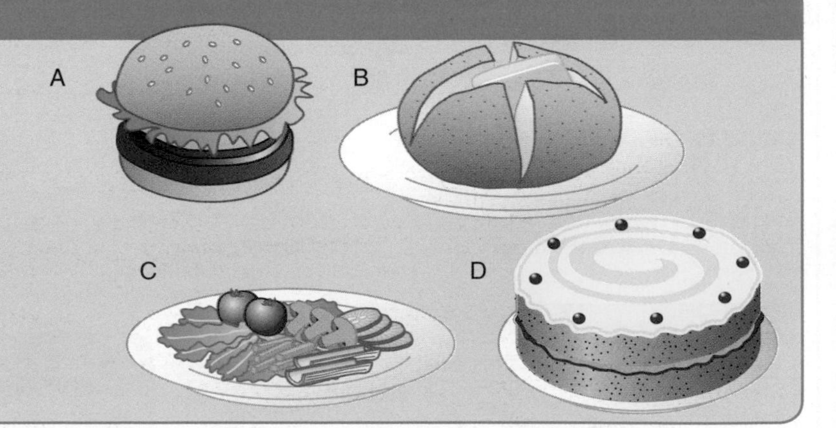

chemical change irreversible chemical change energy changes

How do proteins change during cooking?

Many of the chemicals in food are proteins.

Proteins are large molecules that have definite shapes.

The white of the egg is made of a protein called albumin. Albumin molecules change shape when the egg is cooked.

The protein molecules of meat also change shape when it is cooked. When a protein molecule changes shape it is called denaturing.

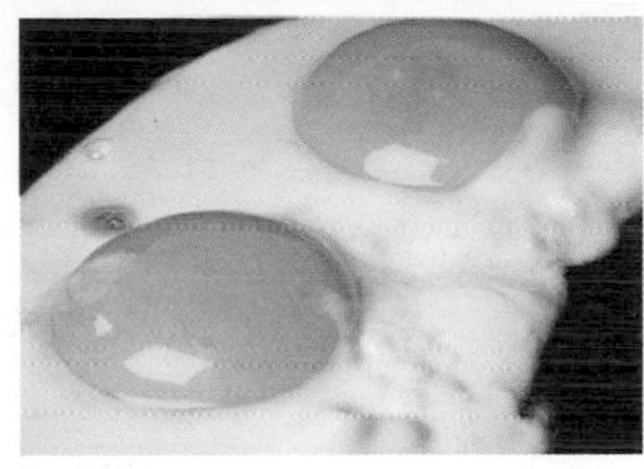

FIGURE 3: Eggs are a good source of protein.

FIGURE 4: Meat is a good source of protein.

Questions

4 What evidence is there that cooking produces a chemical change?

5 What happens to the protein molecules in an egg when it is cooked?

6 What is the name of the process that happens to proteins when they are heated?

More on proteins and carbohydrates

When a protein molecule in egg or meat is heated during cooking it changes shape. The shape change is permanent and irreversible. This causes the texture of meat or egg to change.

When a potato is cooked its cell walls break down or rupture resulting in a loss of rigid structure and a softer texture.

Starch grains burst, swell up and spread. This makes the potato easier to digest.

Questions

7 Explain how heating a protein may change it.

8 Why does cooking make potatoes more digestible?

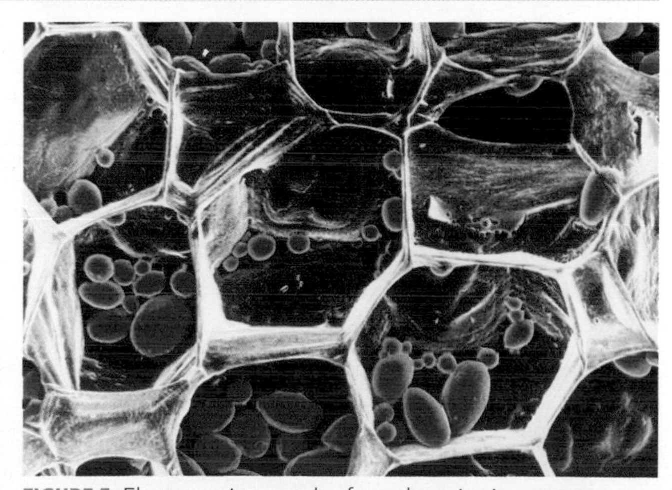

FIGURE 5: Electron micrograph of starch grains in a potato. Magnification ×700.

Cooking different foods

You will find out:

- > how carbon dioxide is used in cooking
- > which type of food additives are used
- > what emulsifiers are used for

Bread and cakes are made from flour.

Flour is made from wheat and must be cooked.

Baking

Baking powder is added to flour to make cakes rise.

Carbon dioxide is made when baking powder is heated in an oven. The carbon dioxide makes the cake rise.

Food additives

When foods are processed additives are often put in for different purposes.

The main types of food additive are:

> Antioxidants are used to stop food from reacting with oxygen. Ascorbic acid (vitamin C) is used in tinned fruit and wine as an antioxidant.

> Food colourants are used to give food an improved colour.

> Flavour enhancers are used to improve the flavour of food.

> Emulsifiers are additives that help oil and water mix in foods. They stop the oil and water from separating.

FIGURE 6: What is added to bread dough to make the mixture rise?

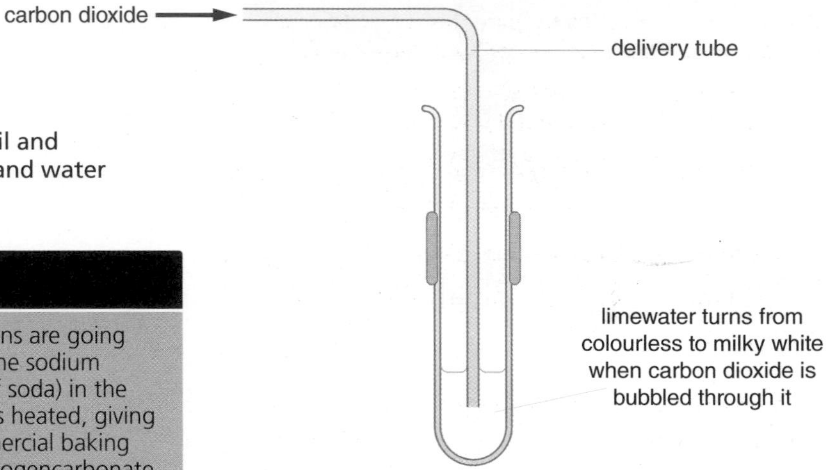

carbon dioxide

delivery tube

limewater turns from colourless to milky white when carbon dioxide is bubbled through it

FIGURE 7: The test for carbon dioxide is to use limewater.

Did you know?

When a cake is baking, two reactions are going on to make the cake rise. First, the sodium hydrogencarbonate (bicarbonate of soda) in the baking powder decomposes when it is heated, giving off carbon dioxide. However, commercial baking powder contains not only sodium hydrogencarbonate, but tartaric acid as well. When these two chemicals become moist and warm they react together to give off more carbon dioxide.

Questions

9 Why are emulsifiers sometimes added to foods?

10 What gas makes cakes rise during baking?

11 What is the test for carbon dioxide? Draw a diagram of the apparatus you would use

Remember!

Make sure you know the test for carbon dioxide. It turns limewater from colourless to a milky white.

antioxidant food colourants food emulsifier flavour enhancer

Baking powder and emulsifiers

To make a cake, baking powder is added to flour to make the cake rise. Baking powder is a chemical called sodium hydrogencarbonate. When it is heated it breaks down (decomposes) to give sodium carbonate, carbon dioxide and water.

The reactant is sodium hydrogencarbonate and the products of the reaction are sodium carbonate, carbon dioxide and water.

The balanced equation for this reaction is:

$$2NaHCO_3 \longrightarrow Na_2CO_3 + CO_2 + H_2O$$

The word equation for the decomposition reaction is:

$$\text{sodium hydrogencarbonate} \xrightarrow{\text{heat}} \text{sodium carbonate} + \text{carbon dioxide} + \text{water}$$

FIGURE 8: What is added to cake mixture to make it rise during baking?

Emulsifiers

Foods such as mayonnaise need emulsifiers.

Mayonnaise has egg as its emulsifier. An emulsifier has two parts to the molecule.

One part is a water-loving part that attracts vinegar to it. This is called the **hydrophilic** head.

The other part is a water hating part that attracts oil to it. This is called the **hydrophobic** tail.

FIGURE 9: An emulsifying molecule showing the hydrophilic head and hydrophobic tail.

Questions

12 What does hydrophilic mean?

13 What does hydrophobic mean?

More on emulsfiers

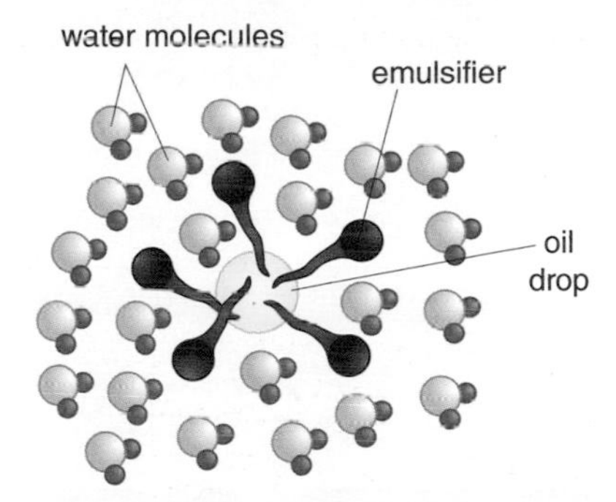

FIGURE 10: The hydrophobic tail is attracted into the oil but the head is not. The hydrophilic head is attracted to water and 'pulls' the oil on the tail into the water.

Emulsifiers help to keep oil and water from separating. They can do this because they have two parts to the molecule. The hydrophobic end bonds with the fat or oil, but the hydrophilic head does not. The hydrophilic head is attracted to the water molecules and keeps the water near to the oil.

Question

14 Explain how the two parts of the molecule in egg yolk act as an emulsifier.

Smells

You will find out:
- > that natural sources can be used to make cosmetics and perfumes
- > how perfumes are made synthetically
- > how we can smell perfumes

Face painting

Since Egyptian times, people have put coloured materials and creams on their bodies and faces to protect and attract.

FIGURE 1: Are there other reasons why humans decorate their faces and bodies?

A sense of smell

Onions and garlic both have a very strong smell. They are pungent.

Rose and honeysuckle flowers have very strong, pleasant smells. They have sweet scents.

The natural substance found in pine trees is used to give a pleasant smell to disinfectant.

Perfumes and cosmetics can be made from natural sources. Others can be made from a synthetic source.

Oil from roses can be distilled to make perfume. Lavender oil is made from lavender grown in Norfolk. These are perfumes from natural sources.

A similar perfume can be made synthetically. Chemicals are boiled to make an ester that has the smell needed.

FIGURE 2: Onions and garlic smell very strongly. Can you name any other natural substances that have a strong smell?

What makes a good perfume?

Not only does a good perfume have a pleasant smell, it also needs the properties shown in Figure 3.

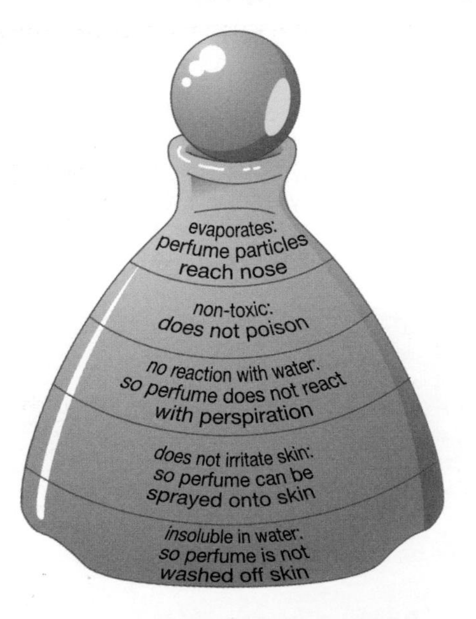

FIGURE 3: Suggest why a good perfume needs to evaporate easily.

Did you know?

If you stand by a scented flower, a gas goes into your nose.

Sensors in the lining of your nose pick up the scent and send a signal to your brain.

People who taste wine are said to have 'a good nose' as they smell many different wines each day.

Questions

1 What is the word used to describe a material that has a strong smell?

2 What is the method used for taking the perfume substance out of a natural material?

what is in perfume   cosmetic industry   disinfectant industry

Making a perfume

To make a perfume a chemical called an **ester** is made. An alcohol mixed with an acid makes an ester.

alcohol + acid ⟶ ester + water

Some nail varnish removers contain an ester.

The reasons why a good perfume needs certain properties are shown in Figure 3.

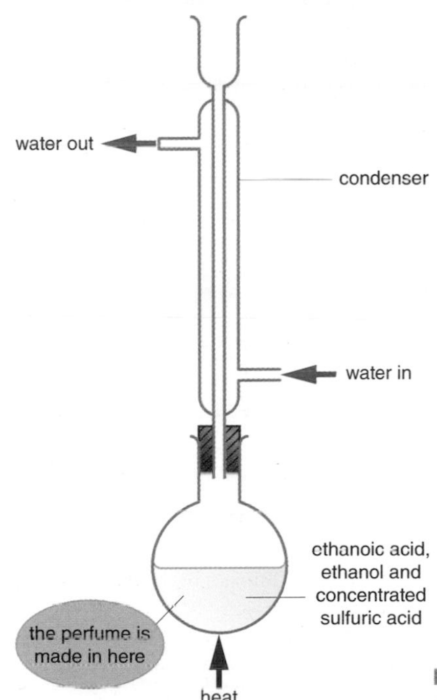

FIGURE 4: Suggest what would happen to the experiment if the flask was not heated.

The table gives some esters and the alcohol and acid they are made from.

Alcohol	Acid	Ester
ethanol	ethanoic	ethyl ethanoate
butanol	ethanoic	butyl ethanoate
propanol	butanoic	propyl butanoate

Questions

3 Name two uses of esters.

4 Describe how to make an ester.

5 Describe and explain three properties of perfumes.

Volatile liquids

If a liquid **evaporates** easily then the substance is **volatile**.

Volatile perfumes are liquids that have energetic, fast-moving particles at room temperature. There is only a weak attraction between particles in the liquid perfume so the **forces of attraction** between the molecules are easy to overcome. This means that particles with lots of energy can escape the attraction to other molecules in the liquid. They can escape from the surface of the liquid, becoming gas particles.

These gaseous particles move through the air until they reach the sensors of the nose.

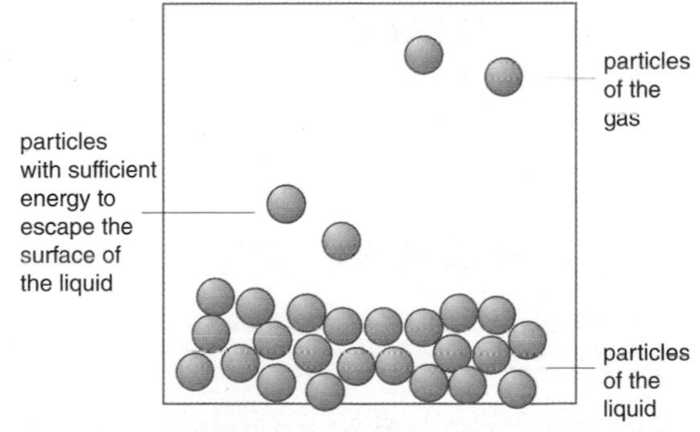

FIGURE 5: The movement of particles during evaporation.

Question

6 Explain how the smell of an open bottle of perfume can be detected on the other side of the room. Use the word 'volatile' in your answer.

Solubility

You will find out:

> that some substances are soluble and some insoluble

> about evaporation

> about attraction between particles

A substance that dissolves in a liquid is **soluble**. The substance dissolved in a **solution** is called the **solute**. The liquid that it dissolves in is called the **solvent**.

A substance that does not dissolve in a liquid is **insoluble**.

Water does not dissolve some substances. In Figure 6, oil does not mix with water.

Water does not dissolve nail varnish so it cannot be used to remove varnish from nails.

Nail varnish remover does dissolve nail varnish.

Cosmetics testing

All cosmetics must be thoroughly tested to make sure they are safe for human use.

> They must not cause rashes or itchiness.

> They must not cause skin damage or lead to cancer or other life-threatening conditions in long-term use.

The EU has banned all testing of cosmetics on animals.

FIGURE 7: Is nail varnish soluble in nail varnish remover?

FIGURE 6: Why do the oil and water stay separate and do not mix?

Questions

7 What does insoluble mean?

8 Using the table opposite what solvent would you use to remove a bloodstain from a shirt?

9 Why do cosmetics need to be tested?

Did you know?

To clean your clothes you have to choose the solvent you need carefully. You need different solvents to remove each type of stain.

Stain	Solvent
blood	water
oil	methylated spirit
nail varnish	propanone
wine	salt water

Solutions

A solution is a solute and a solvent that do not separate. Esters can be used as solvents.

Solvents can be used as cleaners, as shown in the table.

Solvent	Solute that is cleaned
oil	grease
ester	nail varnish
thinners	paint

cosmetic testing soluble and insoluble solvent, solute and solution

Cosmetics testing

Testing takes many years.

Scientists need to test cosmetics to make sure they are safe. Animals are no longer used to test cosmetics in the UK. The testing of cosmetics on animals has been banned by the EU because public opinion is strongly against it. People believe they have no right to cause animals unnecessary suffering.

Questions

10 Suggest why animals are used in cosmetics testing.

11 Suggest why some people feel animals should not be used in cosmetics testing.

Opinions on cosmetic testing

Some cosmetics are tested on animals. Some people do not agree with animals being used. What do you think?

Some people object as the animals may be harmed. The animals also have no control over what happens to them. They say that there are other ways of testing products that are less damaging to living things.

Other people say they feel safer if the cosmetics have been tested on animals. They say that an animal's reaction to a chemical closely mimics the reaction in a human. If there is no danger to the animal then it is likely that the cosmetic or perfume will be safe for use by people.

People have different opinions according to the differences in values that they hold about human needs and animal suffering. They are arguing on the issue of whether testing cosmetics on animals is ever justified.

Attractions between particles

Kinetic energy of particles

If you stand behind a car you can see and smell exhaust fumes. They are collections of tiny molecules of different gases that move through air molecules.

Gas particles have a high kinetic energy and move rapidly and randomly.

Evaporation

Particles of a liquid have less kinetic energy and are weakly attracted to each other. When some particles of a liquid increase their kinetic energy, the force of attraction between the particles is overcome and the particles escape from the surface of the liquid into the surroundings.

Attractions between particles are important

Water does not dissolve nail varnish. This is because the force of attraction between two water molecules is stronger than that between a water molecule and a molecule of nail varnish.

Also the force of attraction between two varnish molecules is stronger than that between a varnish molecule and a water molecule.

Questions

12 Do particles in a gas or a liquid have higher kinetic energy?

13 Explain evaporation. Use ideas about particles.

14 Explain why Karen cannot remove her nail varnish using water. Include the words 'will not dissolve' and 'forces of attraction' in your answer.

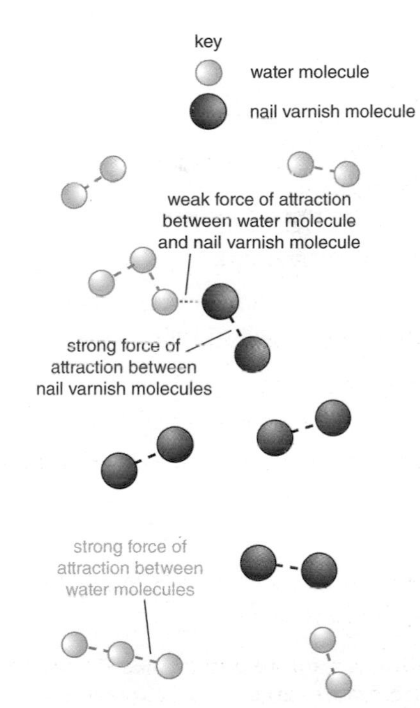

FIGURE 8: What would happen if nail varnish remover was used instead of water? What does this tell you about the strength of the force of attraction between a molecule of nail varnish remover and a molecule of nail varnish?

forces of attraction between particles   kinetic energy gas molecules

Paints and pigments

You will find out:

- about pigments, binding agents, solvents and colloids
- how oil paints and emulsion paints are similar
- how paint dries

Ancient paintings

Some paintings in Egyptian tombs are 4000 years old, and they look as if they were painted last week.

The paintings shown in the photograph are in caves in France.

They are the oldest cave paintings we know about. They are about 30 000 years old.

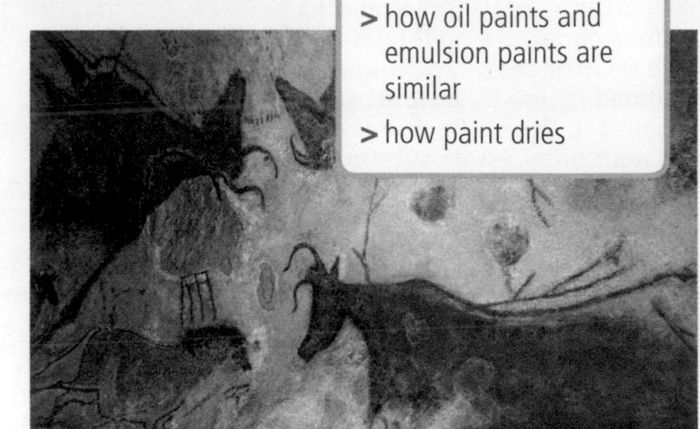

FIGURE 1: Many of the oldest paintings have been found in caves; why is this not surprising?

What is paint?

Painting a wall is simple. All you do is take coloured pigment, grind it into a powder, add a glue and stick it to the wall.

> A **pigment** gives paint its colour.

> The **binding** medium sticks the pigment in the paint to the surface.

> The **solvent** thins the paint making it easier to spread.

In oil paint the binding medium is oil. It is specially chosen to stick the pigment to surfaces. The tiny particles of pigment powder are dispersed (spread) through the oil.

If the oil is too thick it is dissolved in a solvent.

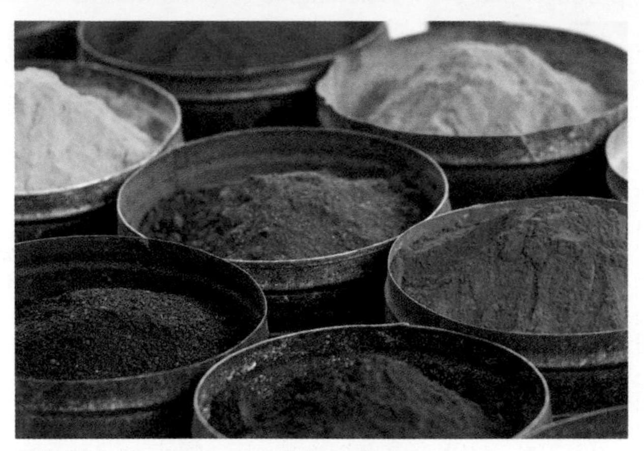

FIGURE 2: Pigments are solid powders.

Why do we use paint?

> For protection – woodwork outside is painted to protect the wood against rain. The oil sticks to the wood and forms a skin.

> To look attractive – when pictures or walls inside a house are painted, it is the pigment part of the paint we enjoy. The binder is used to stick pigment to the canvas or walls.

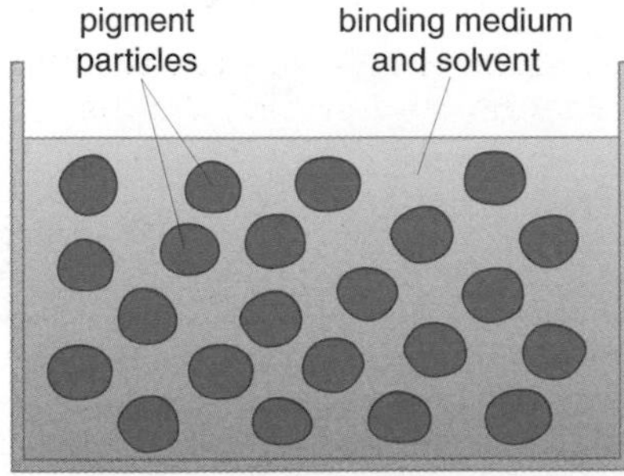

FIGURE 3: What is a mixture of pigment and oil dissolved in to make it easier to use?

Did you know?

A good glue for sticking pigment on to walls is egg white.

It takes a lot of eggs to paint a wall.

Questions

1 White paint is made of a white powder mixed into oil.

a Which is the pigment?

b Which is the binding medium?

2 What is a mixture of a solid powder in a liquid called?

pigments in paint   paint binding medium

Pigments

In the past the range of different coloured rocks used to make pigments was limited. Chemists now make more solids that are unusual pigments.

A good pigment does not fade. It hides the colour of whatever is underneath the paint and stays mixed throughout the binding medium.

A pigment is not dissolved in the binding medium and solvent. It forms a colloid of small solid particles dispersed through the whole liquid, but they are not dissolved.

Paints are applied as a thin surface which dries when the solvent evaporates.

The evaporating solvent is a pollutant. Paint manufacturers try to reduce the pollution in their paints. One way is to use emulsion paint, which contains less solvent.

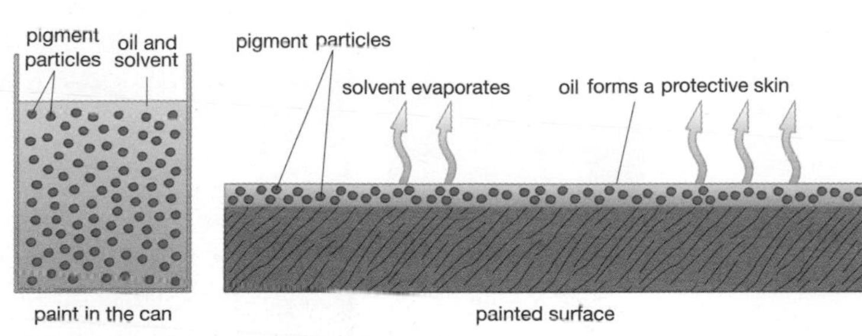

FIGURE 4: What evaporates when oil paint dries?

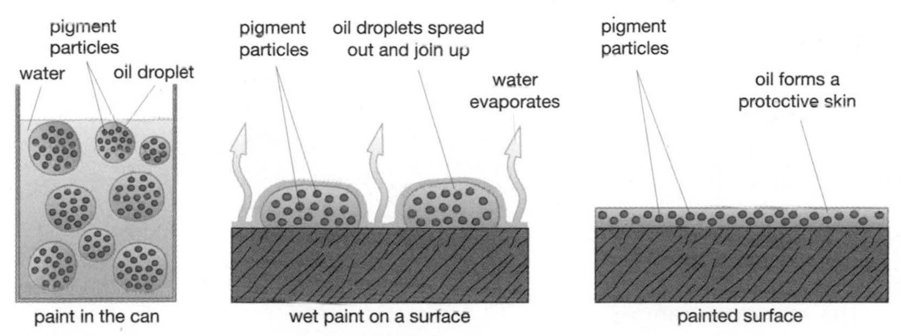

FIGURE 5: What evaporates when emulsion paint dries?

Watching paint dry

Once oil paint has been painted onto a surface:

> the solvent evaporates

> the binding medium dries, forming a skin

> the skin sticks the pigment to the surface.

Emulsion paint is a water-based paint that contains only a small amount of oil. The oil is still the binding medium with pigment particles dispersed through it. The oil is in the form of small droplets spread through water. This is because oil does not mix in water. Tiny droplets of one liquid in another liquid is called an **emulsion**.

When emulsion paint has been painted onto a surface:

> the water evaporates

> the droplets of oil join together to make a continuous film

> the skin sticks the pigment to the surface.

Questions

3 Which of the following is a colloid? • salt solution • paint • a mixture of sand and salt

4 What happens to the solvent when oil paint dries?

Colloids

Colloids are mixtures of tiny particles of one thing in another. Oil paint is a **colloid** because it is a mixture of solid particles in a liquid.

Emulsion paint is a colloid in two ways. It is a mixture of one liquid inside another – oil droplets in water. Each oil droplet also has solid particles dispersed through it.

Particles, or droplets, that are very small stay mixed in the liquid; they do not settle out. Particles in paint are small enough to stay dispersed through the liquid while it is in use, although some paints need stirring if they are left in the tin for a long time.

Drying oil paint

The oil in oil paint is very sticky and takes a long time to harden. Normal oil paints 'dry' by chemical reaction.

Once the solvent has evaporated the oil slowly reacts with oxygen in the air to form a tough, flexible film over the wood. The oil-binding medium is oxidised by the air.

Question

5 Paint kept in a closed can sometimes form a skin, but doesn't go completely hard. Suggest why:

a the paint forms a skin

b it doesn't go completely hard.

Q difference between emulsions and colloids   evaporation

You will find out:

> about thermochromic pigments

> about phosphorescent pigments

Special pigments

The pigments dissolved in substances to make thermochromic paints are very unusual.

> When they get hot they change colour.

> When they cool down they change back to their original colour.

Thermochromic pigments can be used to change colour to show if a cup is hot. They can warn if a kettle is getting hot. They can be used as bath toys to change colour in warm water.

When you were small, did you have stickers over your bed that glowed in the dark? The paint on a sticker contains a phosphorescent pigment.

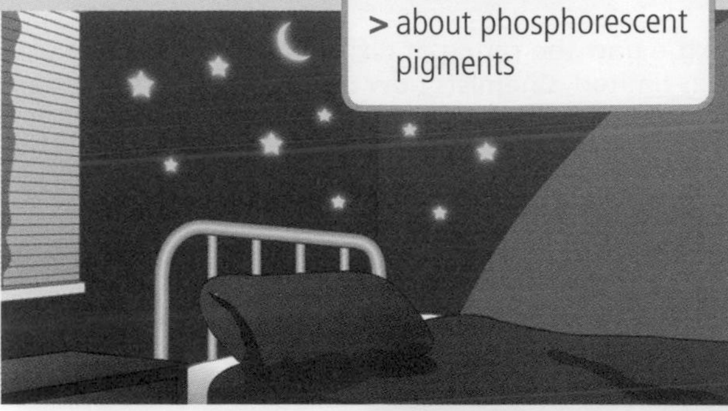

FIGURE 6: What type of pigment gives out light?

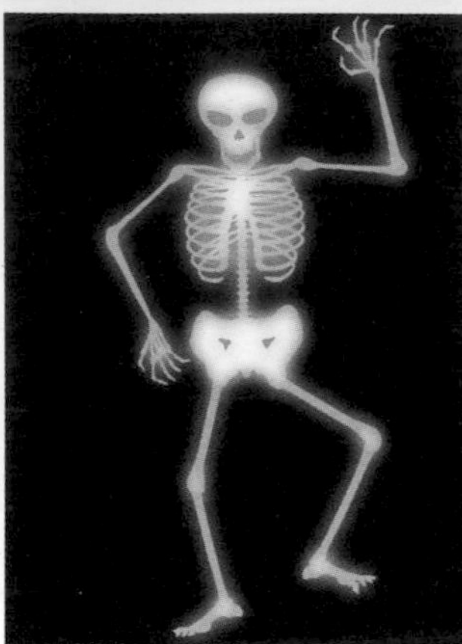

Did you know?

A new type of phosphorescent pigment has opened up possibilities for new low-level lighting systems and instrument lights. It is able to emit light for up to 10 hours.

Questions

6 What type of pigment changes colour when it is heated?

7 What do phosphorescent pigments do?

Using special pigments

How thermochromic paints work

Thermochromic pigments used in some paints are chosen for their colour and also for the temperature at which their colour changes.

Many people find that anything over 60 °C is too hot to hold.

A thermochromic pigment that changes colour at 45 °C can be used to paint cups or kettles to act as a warning.

A pigment that changes colour just above 0 °C makes a good warning paint for road signs to show if the road might freeze.

How phosphorescent pigments work

Phosphorescent pigments absorb and store energy from daylight. Over a period of time, they slowly release the energy as light.

This is why they can glow in the dark.

FIGURE 7: What type of paint has been used to make these mugs?

Questions

8 Give one use of a thermochromic pigment.

9 Suggest why phosphorescent stickers do not glow for the whole night.

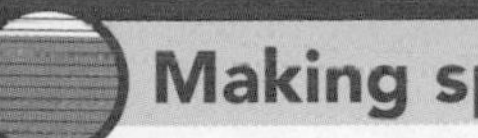

Making special pigments

How thermochromic paints are made

There are two ways to make thermochromic materials. One way is based on liquid crystals. The other way is based on leuco dyes.

Liquid crystals	Leuco dyes
These are used in precision applications	These are used in novelties where the temperature range reading is not important.
They can be engineered to give accurate temperature readings.	The temperatures that they change at are more difficult to set accurately.
The colour range is limited.	There is a wider range of colours.
They can be used for refrigerator thermometers and for medical use.	They can be used to dye T-shirts which can change colour, for bath toys in warm water and as approximate temperature indicators for microwave-heated foods.
The crystals at high temperature state will reflect blue-violet, but at low temperature state they will reflect red-orange.	The dyes are also colour-responsive to electric current as well as heat and can be used as battery testers.

Did you know?

Paint is getting 'greener'.

Modern paints are designed to give off fewer fumes

Things last longer if they have been painted, saving resources.

Most thermochromic pigments change from a colour to colourless.

Thermochromic paints come in a limited range of colours. To get a larger range of colours they are mixed with different colours of normal acrylic paints, in the same way that you mix any coloured paints.

When the mixture gets hot the blue thermochromic paint becomes colourless, so all that is seen is the yellow of the acrylic paint.

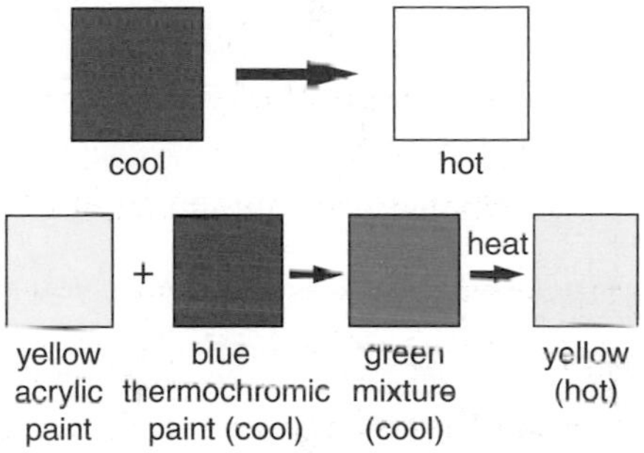

cool → hot

yellow acrylic paint + blue thermochromic paint (cool) → green mixture (cool) →heat→ yellow (hot)

FIGURE 8: Changes in colour as thermochromic paints are heated.

Phosphorescent paints

Phosphorescent pigments are sometimes used in luminous watch dials.

They are not the only type of pigment that has been used in this way. Some radioactive chemicals glow in the dark.

The people who painted the watch dials with radioactive paint used to lick their brushes to get them to a fine tip. Many of them developed cancer as a result.

Phosphorescent paints are much safer than radioactive paints.

They absorb radiation and the radiation they absorb can be re-emitted for several hours but at a low intensity. The materials can store energy for a longer time.

These materials include strontium aluminate-based pigments which are used in exit signs and pathway marking.

The first common pigment used in phosphorescent materials was zinc sulfide. This has been used for safety products for eighty years.

Questions

10 Jonathan mixes yellow thermochromic paint with red acrylic paint.

a What colour paint does he produce?

b What colour does Jonathan see when the paint is hot?

Preparing for assessment: Planning and collecting primary data

To achieve a good grade in science, you not only have to know and understand scientific ideas, but you need to be able to apply them to other situations and investigations. These tasks will support you in developing these skills.

Tasks

- Plan an investigation to see how the concentration of acid rain affects the rate of erosion.
- Once your plan has been approved, perform the investigation, record your results and write a simple conclusion.

Context

Acid rain is caused by oxides of sulfur, nitrogen and other non-metals dissolving in rain water. The oxides are produced when impurities in fossil fuels burn. The acid concentration is relatively low but erodes limestone over a long period of time.

You are asked to model this erosion more quickly, using dilute nitric acid as acid rain and marble chips instead of limestone. There are a number of ways that you can measure the time taken to erode.

Planning your investigation

You can measure the rate of reaction using a balance and a conical flask. Add marble chips to nitric acid in a conical flask and record the mass at regular intervals. The rate of reaction can be found from how much mass is lost in a given time.

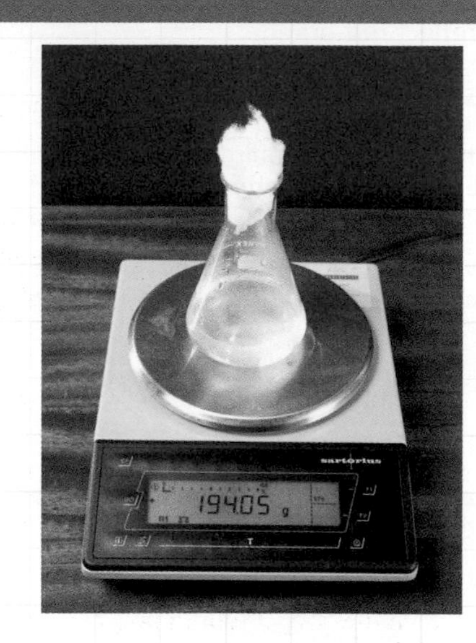

These are the things you will need to consider when planning your investigation. (You can develop your plan in groups of two or three.)

1. How much nitric acid do you need in the conical flask?

2. What mass of marble chips will be suitable?

3. How are you going to change the concentration of the acid?

4. What do you need to keep the same to make it a fair test?

5. How many different concentrations will you use before you can identify a trend?

6. Will you need to repeat your readings? If so, how many times?

7. How will you calculate the mass lost?

8. How long will you allow the reaction to proceed before starting the next test?

Remember, you will not have time for each reaction to finish.

Why does it not matter if the reaction does not finish?

9. You should carry out a risk assessment before you start the investigation.

What precautions should you take?

10. Write the plan for the investigation.

Try to write the plan in a logical order and ask yourself if someone can perform the investigation following just your plan.

Performing the investigation

Once your plan has been approved you can perform the investigation.

1. Identify all of the different acid concentrations you used.

2. If you repeated any readings, all of these will need to be recorded as well as the average result.

3. Record your results in a table like this. You may need to add extra rows or columns.

time in minutes	mass lost in g			
	acid A	acid B	acid C	acid D

4. In order to complete this as a controlled assessment you need to plot a graph and evaluate the investigation.

a. What graph would you draw?

b. What would the labels be on the axes?

c. How would you use the graph to decide on the answer to the task?

5. Is there any way in which you could have improved on how you performed the investigation?

6. What have you found out about how the concentration of acid rain affects the rate of erosion?

C1 Checklist

To achieve your forecast grade in the exam you'll need to revise

Use this checklist to see what you can do now. It gives you many of the important points you will need to know. Refer back to the relevant pages in this book if you're not sure and to see if there is anything else you need to know. Look across the three columns to see how you can progress.

Remember you'll need to be able to use these ideas in various ways, such as:

- > interpreting pictures, diagrams and graphs
- > applying ideas to new situations
- > explaining ethical implications
- > suggesting some benefits and risks to society
- > drawing conclusions from evidence you've been given.

Look at pages 278–299 for more information about exams and how you'll be assessed.

To aim for a grade E	To aim for a grade C	To aim for a grade A
know that crude oil, coal and natural gas are fossil fuels **understand** that fractional distillation works because of differences in boiling points **recognise** that LPG, petrol, diesel paraffin, heating oil, fuel oils and bitumen come from crude oil **know** that oil slicks may result from accidents with crude oil **label** the laboratory apparatus used for cracking paraffin	**explain** why fossil fuels are finite resources and non-renewable **label** a diagram of a fractionating column **explain** that crude oil can damage birds' feathers **know** that cracking converts large alkanes into smaller ones **interpret** data about the supply and demand of crude oil fractions	**discuss** the problems of finding replacements for crude oil **explain** why crude oil can be separated **understand** that intermolecular forces break during boiling **explain** the political problems with the future supply of oil **explain** how cracking matches the supply and demand of petrol
list the factors for choosing a fuel **know** why complete combustion needs plenty of oxygen **construct** word equations for complete combustion of fuels **know** that carbon monoxide is made in incomplete combustion **identify** that a yellow flame produces lots of soot	**interpret** data to choose the best fuel for a particular purpose **describe** an experiment to show the products of complete combustion are CO_2 and H_2O **construct** methane + oxygen → carbon dioxide and water **explain** the advantages of complete combustion **construct** equations for incomplete combustion	**evaluate** use of different fuels **explain** why the use increases **construct** balanced equations for complete combustion **construct** balanced equations for incomplete combustion given the product
know that air contains oxygen, nitrogen and carbon dioxide **understand** that levels of these gases are now almost constant **understand** how respiration, photosynthesis and combustion affect levels of carbon dioxide and oxygen **relate** sulfur dioxide from fossil fuels burning to acid rain which kills plants and erodes stonework **understand** catalytic converters remove carbon monoxide, which is poisonous **understand** the environmental effects of oxides of nitrogen	**know** that clean air has 21% O_2 78% N_2 and 0.035% CO_2 **know** a carbon cycle: combustion, photosynthesis and respiration **describe** how the present day atmosphere evolved **interpret** data about effects of atmospheric pollutants **know** that catalytic converters change CO to CO_2	**evaluate** the human effects on air composition by deforestation **describe** a theory of atmospheric evolution using gas levels from early volcanoes **describe** roles of oxygen-producing photosynthetic organisms **explain** why N_2 and O_2 react to make oxides in hot engines **describe** catalytic converters using equations

To aim for a grade E	To aim for a grade C	To aim for a grade A
know the two elements combined chemically to make hydrocarbons **recognise** that alkanes are hydrocarbons **recognise** that alkenes are hydrocarbons **work out** the name of a polymer given the name of a monomer **know** a polymerisation reaction is monomers making polymers	**understand** that alkenes contain at least one carbon–carbon double bond but alkanes do not **interpret** displayed formulae of alkanes **describe** that orange bromine water decolourises with alkenes **recognise** the displayed formula of a polymer **know** polymerisation needs a high pressure and a catalyst	**know** an unsaturated compound has at least one C=C bond **interpret** displayed formulae of saturated hydrocarbons **explain** that an alkene forms a colourless di-bromo with bromine **draw** the formula of a polymer from its monomer **explain** addition polymerisation
interpret information on plastics – properties and uses **explain** why a polymer is suitable for a particular use given the properties of the polymer **know** that nylon is used in clothing **understand** that many polymers are non-biodegradable **know** that waste polymers can be recycled, burned, put in landfill	**suggest** the properties a polymer should have for a particular use **compare** nylon (tough, lightweight) with breathable Gore-Tex® **explain** that some new types of polymer are being developed that dissolve or are biodegradable **explain** that landfill sites get filled quickly, burning plastics makes toxic gases	**know** atoms in plastics are held by strong covalent bonds **relate** how intermolecular forces give the properties of plastics; weak forces mean lower melting points and the polymer can be stretched **explain** why Gore-Tex® clothes are waterproof yet breathable
know a chemical change takes place if a new substance is made **explain** cooking as a chemical change because a new substance is made and the process cannot be reversed **know** emulsifiers help oil and water mix **know** that antioxidants stop food reacting with oxygen **know** carbon dioxide turns limewater milky	**recall** that when egg is cooked the protein molecules denature **explain** that the cell walls rupture in cooked potato **know** emulsifiers have two parts: hydrophilic and hydrophobic **recall** the word equation for the decomposition of baking powder **construct** equation from formulae $2NaHCO_3 \rightarrow Na_2CO_3 + CO_2 + H_2O$	**know** heated protein molecules change shape permanently **explain** that hydrophilic ends of emulsifiers bond with water **explain** that hydrophobic ends of emulsifiers bond with oil **construct** without formulae given $2NaHCO_3 \rightarrow Na_2CO_3 + CO_2 + H_2O$
know that esters are perfumes that can be made synthetically **know** that perfumes need to be non-toxic and evaporate easily **understand** solvent, solute, solution, soluble and insoluble **understand** that nail varnish remover dissolves nail varnish **know** that cosmetics need to be tested thoroughly before use	**know** that alcohols react with acids to make an ester and H_2O **explain** that perfumes evaporate so particles can reach the nose **explain** why perfumes are insoluble in water **know** that esters can be used as solvents **explain** why testing of cosmetics on animals has been banned by the EU	**explain** volatility **explain** why water will not dissolve nail varnish colours **explain** water/water attractions are stronger than varnish/water **explain** why people have differing opinions about whether the testing of cosmetics on animals is ever justified
know that solvent thins paint **know** that a binding medium sticks pigment to the surface **know** that paints are used to decorate or protect surfaces **know** thermochromic pigments change colour on heating/cooling **know** that phosphorescent pigments can glow in the dark	**describe** paint as a colloid **know** paint dries as the solvent evaporates **explain** why thermochromic pigments are suited to given uses **explain** that these pigments absorb energy, releasing it as light	**explain** why components of a colloid will not separate **explain** how oil paints dry **explain** how thermochromic pigments change acrylic paint colours **recall** phosphorescent pigments are safer than radioactive ones

C1 Exam-style questions

Foundation Tier

AO1 **1 (a)** Name three fossil fuels. [1]

AO1 **(b)** Explain why fossil fuels are said to be finite. [1]

[Total: 2]

2 The diagram shows a fractionating tower.

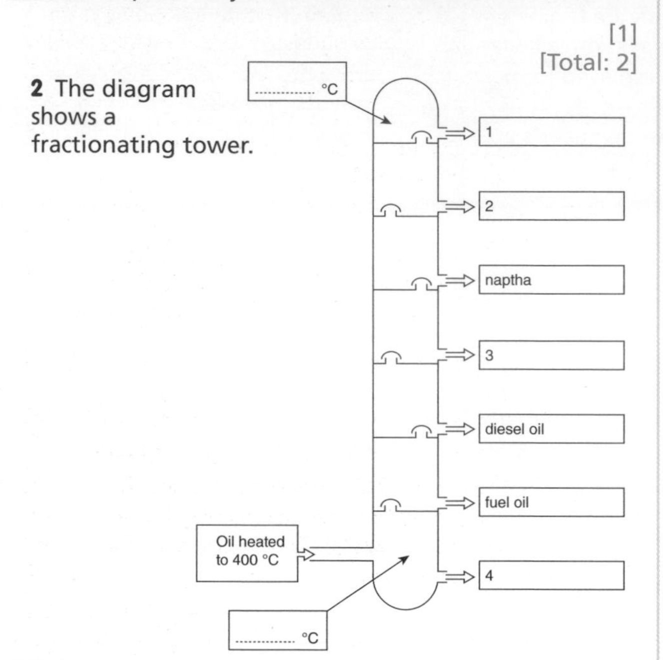

AO1 **(a)** Name the fractions collected at 1, 2, 3 and 4. [2]

AO1 **(b)** Copy the diagram and add the following labels to the correct place.
–40 to 40 300 to 350 [1]

[Total: 3]

AO1 **3 (a)** What are the proportions of the three main gases of the air (not water vapour)?

1. Nitrogen ....%
2. Oxygen ....%
3. Carbon dioxide ....% [1]

AO1 **(b)** The gas sulfur dioxide dissolves in rain water to cause acid rain.
What problems happen with acid rain? [1]

[Total: 2]

4 Look at the table below. It shows the health effects at different air qualities caused by sulfur dioxide

AO2 **(a)** What is the air quality when it is at 19 on the scale? [1]

AO2 **(b)** What is the air quality when there is noticeable damage to trees? [1]

AO3 **(c)** Sam likes gardening and has asthma. He is thinking of moving to a house where the air quality is 104. Justify why this may not be the best move for Sam. [2]

Health effects at different air qualities caused by sulfur dioxide levels		
Category of air quality	**Air quality scale**	**Health and damage effect**
Very Good	**0–15**	No health effects are shown.
Good	**16–31**	Damages some plants and trees if there is ozone too.
Moderate	**32–49**	Damages some plants and trees.
Poor	**50–99**	Smells, noticeable damage to trees.
Very Poor	**100 or over**	Problems for people with bronchitis and asthma.

[Total: 4]

AO1 **5 (a)** What is a hydrocarbon? [1]

AO2 **(b)** Look at this displayed formula of a molecule. Write down two facts about the molecule from its formula. [2]

H H
C=C
Cl H

[Total: 3]

6 Baking powder, as its name suggests, is used in cooking.

AO1 **(a) (i)** When baking powder is heated what gas does it release? [1]

AO1 **(ii)** How would you test for this gas? [2]

AO1 **(b)** When meat is cooked the shape of its protein molecules change. What is this process called? [1]

[Total: 4]

AO2 **7** Synthetic perfumes can be made from alcohols and acids. Alcohols are **not** hydrocarbons. Choose which of these may be the molecular formula of an alcohol.
C_4H_{10} C_4H_8 C_4H_9OH $CH_3CH(CH_3)CH_3$ [1]

AO1 AO2 **8** Anya is trying to decorate her room. She has a choice of paints. She wants to cover the surfaces with colour and have hanging decorations that change colour.
Write about:

- Why paints can be used to cover surfaces to decorate and protect them, how they dry and why they are called colloids.
- What she can use to make decorations change colour or glow in the dark. [6]

AO1 recall the science AO2 apply your knowledge AO3 evaluate and analyse the evidence

Worked Example – Foundation Tier

(a) The apparatus below can be used to show that when a hydrocarbon fuel is burnt in a plentiful supply of air, it produces carbon dioxide.

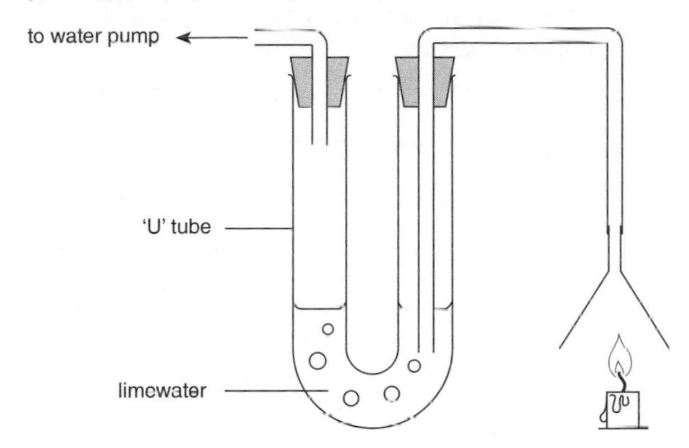

(i) Explain what will happen in such an experiment? [2]

The limewater will turn milky because carbon dioxide is being given off.

(ii) The experiment could be repeated but using dry cobalt chloride paper in the U tube. What result would you get and what would this show? [2]

The cobalt chloride paper would change colour showing that a gas was given off.

(b) (i) A candle burns with a yellow flame. What product is also made when a flame is yellow? [1]

Soot

(ii) A Bunsen burner can burn with a blue flame. Explain the advantages of this complete combustion. [3]

There is more energy released during complete combustion because no soot is made.
Carbon monoxide is not made.

(c) Write a word equation for the complete combustion of the hydrocarbon propane. [2]

propane + air ⟶ carbon dioxide + H_2O

(d) When some fuels are burned pollutants are released, including sulfur dioxide. Explain why it is important that atmospheric pollution is controlled. [3]

So that sulfur dioxide does not erode statues.

This student has scored 7 marks out of a possible 13. This is below the standard of Grade C. With more care the student could have achieved a Grade C.

How to raise your grade!

Take note of these comments – they will help you to raise your grade.

The observation and what it means are correct. 2/2

The observation was given (the colours that the paper turns do not need to be learned) but the student needs to know that the test is for water vapour. 1/2

The student is only asked to identify the product. At higher grades they would be asked to say why it is a disadvantage. This happens in the next part. 1/1

One for explaining that carbon monoxide is not made and one for soot being made. The student needs to explain that all the fuel has been used to release energy, which is why soot is made or that it is a disadvantage to make carbon monoxide as it is toxic. 2/3

Oxygen should be written not air. The equation should be in words not symbols so H_2O should be water. 1/2

The student needs to say that sulfur dioxide dissolves in rain water to make acid rain (1) one for effect of acid rain (1) and one other effect of burning fuels such as the production of toxic carbon monoxide. As there are 3 marks they must make three points. 0/3

C1 Exam-style questions

Higher Tier

AO1 **1 (a)** Explain why fossil fuels are said to be finite [1]

AO1 **(b)** Explain two problems caused by crude oil being a finite resource. [1]

[Total: 2]

AO1 **2** The diagram shows a fractionating tower. Name the fractions collected at 1, 2, 3 and 4.

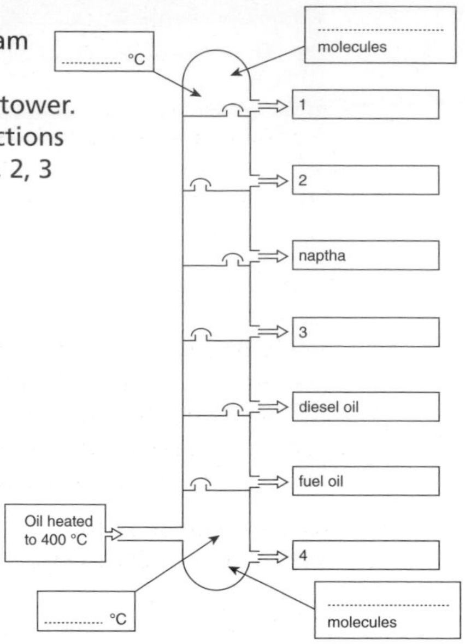

[Total: 2]

3 Look at the table. It shows the health effects at different air qualities caused by sulfur dioxide levels.

Health effects at different air qualities caused by sulfur dioxide levels		
Category of air quality	**Air quality scale**	**Health and damage effect**
Very Good	**0–15**	No health effects are shown.
Good	**16–31**	Damages some plants and trees if there is ozone too.
Moderate	**32–49**	Damages some plants and trees.
Poor	**50–99**	Smells, noticeable damage to trees.
Very Poor	**100 or over**	Problems for people with bronchitis and asthma

AO2 **(a)** What is the air quality when it is at 19 on the scale? [1]

AO2 **(b)** What is the relationship between air quality and health effects? [1]

AO3 **(c)** Sam likes gardening and has asthma. He is thinking of moving to a house where the air quality is 104. Justify why this may not be the best move for Sam. [2]

[Total: 4]

AO2 **4** Look at this displayed formula of a molecule. Write down two facts about the molecule from its formula.

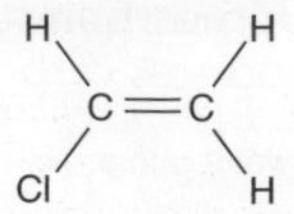

[2]

[Total: 2]

AO1 **5 (a)** Construct the balanced symbol equation of the decomposition of sodium hydrogencarbonate [2]

AO1 **(b)** When meat is cooked the shape of its protein molecules change. What is this process called? [1]

[Total: 3]

AO2 **6** There is a mis-match between supply and demand of the products in the table. Write about how this can be solved, using the data given.

Fraction	**Supply (units)**	**Demand (units)**
Petrol	**4200**	**8700**
Paraffin	**3200**	**1800**
Fuel oil	**6300**	**1200**

[2]

[Total: 2]

7 Perfumes can be made from natural products or they can be synthetic.

AO2 **(a)** Synthetic perfumes can be made from alcohols and acids. Alcohols are **not** hydrocarbons. Choose which of these may be the molecular formula of an alcohol and explain your reasons:

C_4H_{10} C_4H_8 C_4H_9OH $CH_3CH(CH_3)CH_3$ [1]

AO1 **(b)** Anya is designing a perfume. Explain why a perfume needs certain properties, and use kinetic theory to explain why a perfume is volatile. ✎ The quality of written communication will be assessed in your answer to this question. [6]

[Total: 7]

AO1 recall the science AO2 apply your knowledge AO3 evaluate and analyse the evidence

Worked Example – Higher Tier

When a hydrocarbon fuel is burnt in a plentiful supply of air, it produces carbon dioxide.

(a) If you were doing an experiment to show that combustion of a hydrocarbon produces carbon dioxide and water, explain how would you test these? [2]

You would use limewater because it turns milky when carbon dioxide passes through it. You would test the water vapour.

(b) Construct the balanced symbol equation for the incomplete combustion of propane, C_3H_8, when carbon is made. [2]

$C_3H_8 + O_2 \longrightarrow 3C + 4H_2O$

(c) Incomplete combustion can also produce carbon monoxide. A catalytic converter changes carbon monoxide into carbon dioxide.

Explain how this happens. [3]

When an engine is hot, nitrogen from the air reacts with oxygen to make oxides of nitrogen. These oxides react with carbon monoxide in the catalytic converter to make carbon dioxide. $CO + NO \longrightarrow N_2 + CO_2$

(d) (i) Propane is an alkane. Propene is an unsaturated compound.

What is an unsaturated compound? [1]

It has at least one double bond.

(ii) If the displayed formula of propene is

```
H     H
 \    |
  C = C     H
 /     \   /
H       C
       / \
      H   H
```

Draw the displayed formula of poly(propene). [3]

```
 [  H    H    H    H    H    H  ]
 [  |    |    |    |    |    |  ]
——C————C————C————C————C————C——
 [  |    |    |    |    |    |  ]
 [  H   CH3   H   CH3   H   CH3 ]n
```

How to raise your grade!

Take note of these comments – they will help you to raise your grade.

One mark awarded as the reason for using limewater was clearly known. The student needs to explain how to test for water vapour using cobalt chloride. 1/2

The formulae of the reactants and products are correct. The student has then tried to balance the number of molecules by writing 3 in front of C and 4 in front of H_2O. However, they have forgotten to balance the number of oxygen molecules by using 2 molecules, so no score.

This is a common mistake.
$C_3H_8 + 2O_2 \longrightarrow 3C + 4H_2O$ 1/2

This is a good answer that describes the first stage of the reaction chain well.

An equation is then given to explain what goes on in the catalytic converter. However, for a top level answer the equation should be balanced.
$2CO + 2NO \longrightarrow N_2 + 2CO_2$ 2/3

This is nearly correct but does not say that the double covalent bond is between carbon atoms. 0/1

The correct formula, 3 carbon atoms and alternate CH_3 groups is given.

The start and end brackets for polymers and *n* after the bracket indicating many monomers, not just the three unit samples are also given. 3/3

This student has scored 7 marks out of a possible 11. This is below the standard of Grade A. With more care the student could have achieved a Grade A.

C2 Chemical resources

Ideas you've met before

The Earth and raw materials

The rock cycle – how the three main rock types are formed.

How weathering and erosion destroys rocks.

Describing materials as atoms, compounds or mixtures.

 **What are the three main rock types?**

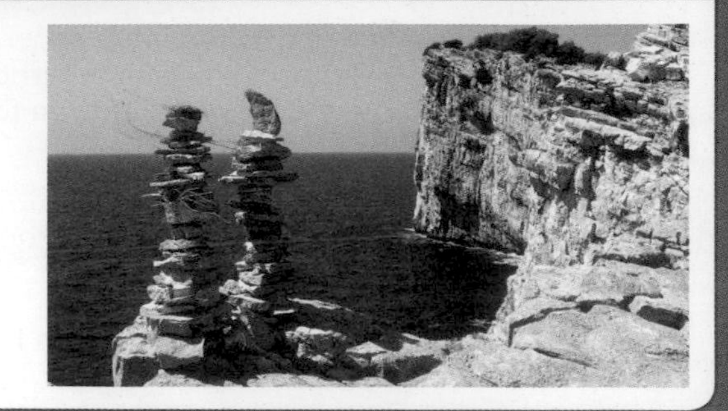

The properties of metals

The difference between metals and non metals.

How metal ores can be extracted, purified and used.

Separation techniques, such as dissolving, filtering and evaporating

 **What are the main properties of metals?**

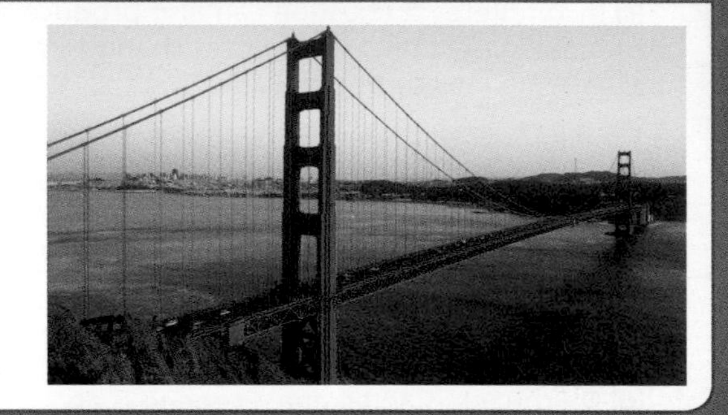

Chemicals from the air and sea

The composition of the atmosphere.

How the atmosphere has changed over time.

The structures of solids, liquids and gases.

Changing state.

 **How can you change a solid to a liquid and back again?**

Acids and alkalis

Common acids and alkalis in the home.

How to measure pH level using Universal Indicator.

Neutralisation.

 **Write the word equation to show an acid reacting with an alkali.**

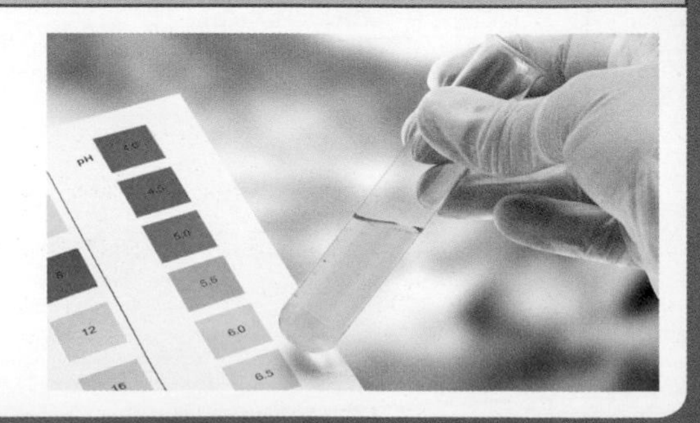

In C2 you will find out about...

- the structure of the Earth
- plate tectonics theories
- scientific evidence
- the main construction materials and their properties
- constructing word and symbol equations

- how to extract a metal from its ore by reduction
- using electrolysis to purify and recycle metals
- alloys and their uses
- rusting and corrosion
- deciding the best materials to make cars

- how ammonia is made
- the world's need for fertilisers
- factors that affect the cost of making new materials
- problems and benefits with fertilisers
- how to make a fertiliser

- neutralisation in terms of ions
- the reactions of metal oxides, metal hydroxides and carbonates with acids
- how electrolysis works
- uses of sodium chloride
- the electrolysis of sodium chloride and the uses of the products

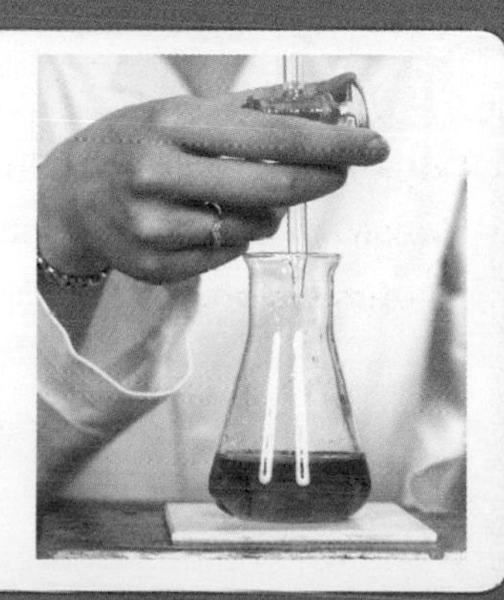

The structure of the Earth

You will find out:
- > about the structure of the Earth
- > about plate tectonics
- > what happens when two plates collide

Does the Earth move?

The Earth does move, which can make life very dangerous.

Scientists weren't really convinced until they had an explanation of just how it does move.

FIGURE 1: This seismograph shows tremors within the Earth.

What is the Earth made from?

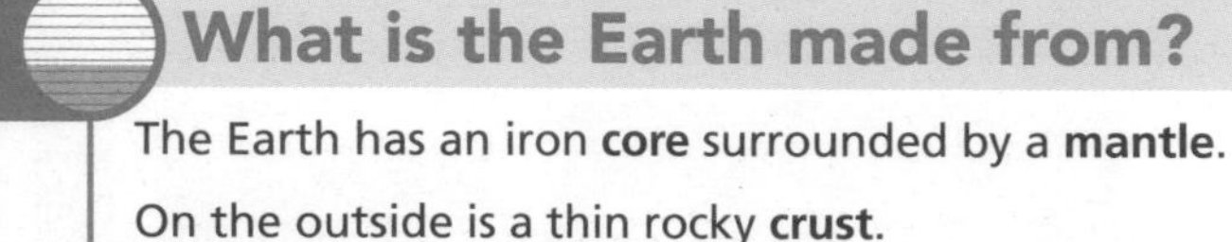

The Earth has an iron **core** surrounded by a **mantle**.

On the outside is a thin rocky **crust**.

The outer layer of the Earth is made from **tectonic plates**.

Tectonic plates move very slowly.

Their movement causes earthquakes and volcanoes at the plate boundaries.

There have been many theories to explain why the surface of the Earth is like it is. For example, evidence suggests South America and Africa were joined. Today, most scientists agree that movement of tectonic plates is the cause. It has taken millions of years for the continents to move apart.

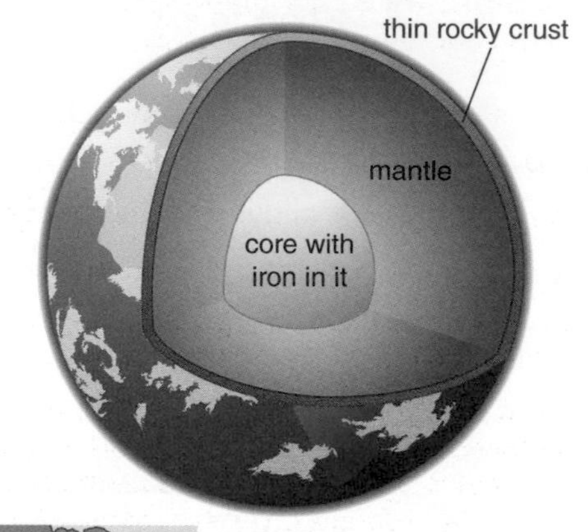

FIGURE 2: What part of the Earth has iron in it?

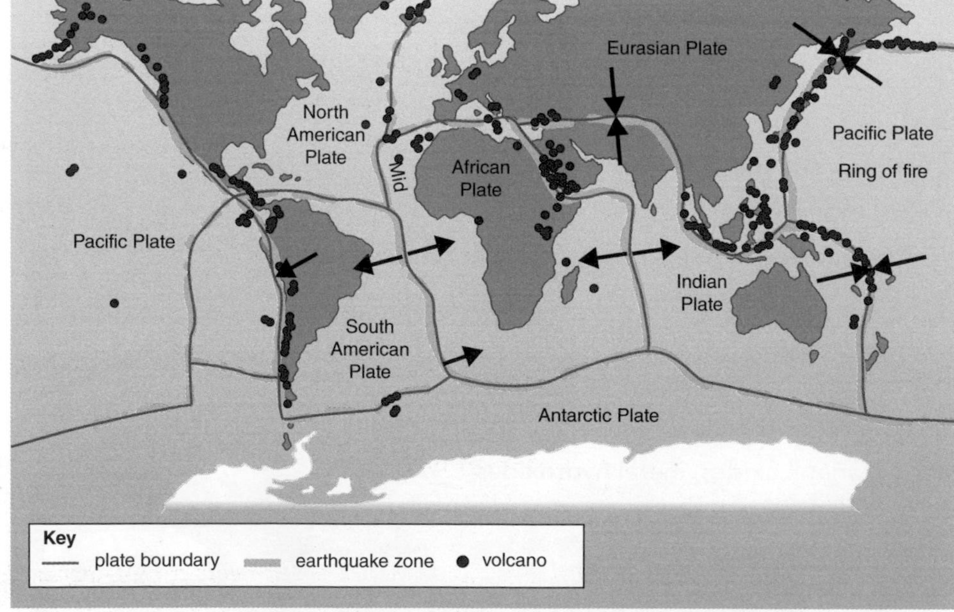

FIGURE 3: What do you notice about the positions of the plate boundaries and the volcanoes and earthquake zones?

Did you know?

The trip from Europe to Disneyworld in Florida, America is getting longer. America and Europe are moving apart by about 2.5 cm a year.

Question

1 Where on the Earth's crust do most volcanoes and earthquakes happen?

Structure of the Earth

The outer layer of the Earth is called the **lithosphere**. It is relatively cold and rigid. It is made of the crust and the part of the mantle that lies just underneath.

The tectonic plates that make up the Earth's crust are less dense than the rest of the mantle below. The theory of plate tectonics is now widely accepted because:

> it explains the evidence

> it has been tested and the results discussed by many scientists.

The crust is far too thick for anyone to be able to drill through it – yet. Most of our knowledge comes from measuring **seismic waves** produced by earthquakes and by artificial explosions. (For more on measuring seismic waves see Stable Earth, pages 216–219.)

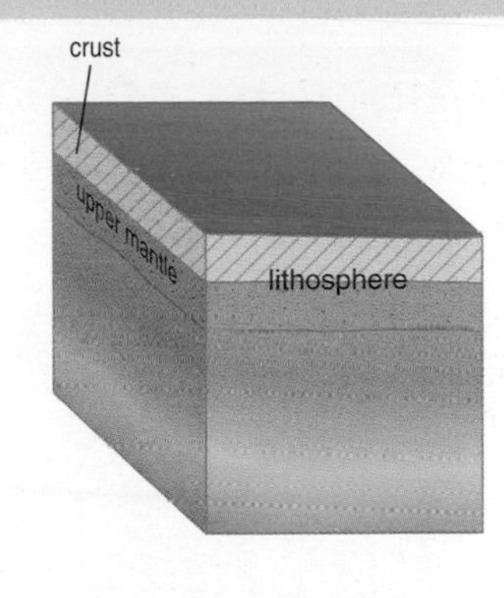

FIGURE 4: What parts of the Earth make up the lithosphere?

The measurement of seismic waves improved in the 1960s when scientists were developing ways of detecting nuclear explosions.

Questions

2 Give two ways that seismic waves are produced.

3 Explain why tectonic plates do not sink into the Earth's mantle.

More on the Earth's structure

The Earth's crust and the upper part of the mantle together form the lithosphere, which is rigid and brittle. The lithosphere is divided into tectonic plates.

Even though these plates are rigid, they can move. Just below the crust the mantle is cold and rigid, but further down it is hotter and there is a semi-rigid layer (called the asthenosphere). This hot, semi-rigid layer is more like plasticine, which can 'flow' even though it is solid. Energy from the Earth's core is transferred towards the surface by **convection currents** in the mantle. These convection currents slowly move the plates.

Tectonic plates can:

> move apart

> collide

> scrape sideways past each other.

There are two types of plate, with different densities.

> Oceanic plates are denser than continental ones.

> Oceans accumulate on top. Continental plates float higher in the mantle.

The margins of an oceanic plate are cooler than the centre, so it is even denser at its edges. When an oceanic plate collides with a continental plate the dense oceanic plate sinks below the continental plate. As it sinks it pulls more of the oceanic plate down after it. This is known as **subduction**.

As the sinking plate descends, it may get stuck and then suddenly move, which causes an earthquake.

Further down, it partially melts to form **magma**. Some of the magma works its way up to the surface and creates a chain of volcanoes along the edge of the plate.

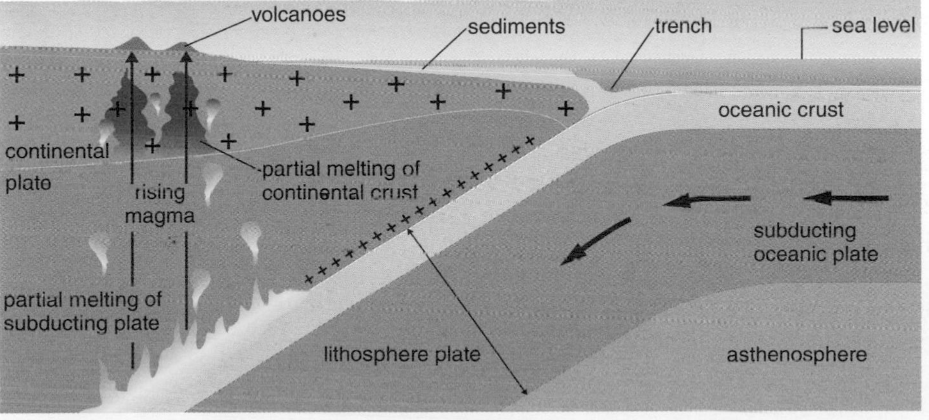

FIGURE 5: What is subduction?

Question

4 Suggest what happens when two continental plates collide.

What happens to molten rock?

You will find out:

- > what happens to molten rock
- > how to tell how fast a rock cooled
- > about evidence for moving plates

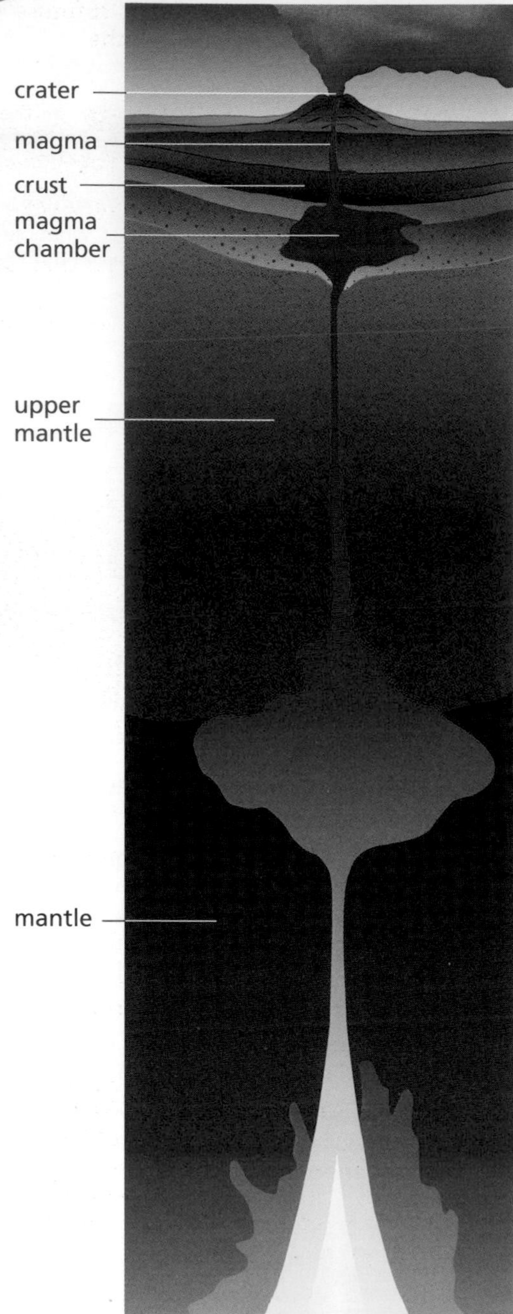

FIGURE 6: What happens to magma at the Earth's surface?

Underneath the surface of the Earth most of the rock is solid.

But some of the rock does melt and is called magma. It slowly moves up to the surface of the Earth through weaknesses in the crust. Eventually it cools down and solidifies to make **igneous rock**.

We can tell how an igneous rock cooled by looking at its crystals.

> Igneous rock that cools rapidly (close to the surface) has small crystals.

> Igneous rock that cools slowly (further from the surface and better insulated) has large crystals.

a 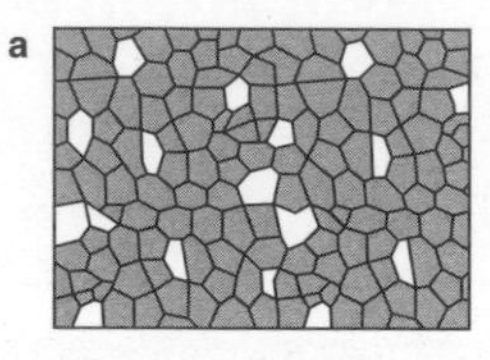 b 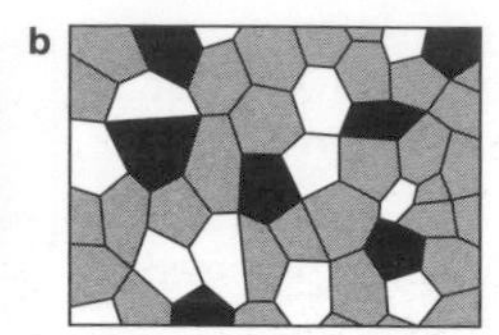

FIGURE 7: **a** small crystals, rapid cooling; **b** large crystals, slow cooling.

Molten rock that reaches the surface of the Earth is called lava and it comes out of a volcano.

Volcanoes that produce runny lava are often fairly safe.

If the lava is thick and sticky then an eruption can be explosive and the volcano is much more dangerous.

Why do people live near dangerous volcanoes?

In AD 79 the Roman town of Pompeii was destroyed when Vesuvius erupted.

People moved back and built a modern town even nearer to the volcano. Vesuvius erupts fairly often, but usually not so violently.

The ash from volcanoes makes a rich soil that is good for growing things. Most volcanic eruptions are not that dangerous, so people think it is worth moving back.

FIGURE 8: Why do some people choose to live near to active volcanoes?

Questions

5 What is the difference between magma and lava?

6 Some volcanoes produce runny lava and some produce thick sticky lava. Which is the safest type of lava?

magma and lava volcano pictures USGS dynamic Earth

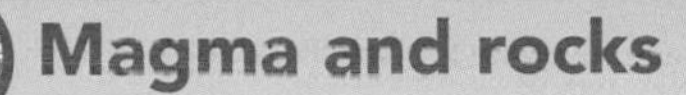

Magma and rocks

Magma rises through the Earth's crust because it is less dense than the crust.

Magma cools and solidifies into igneous rock either after it comes out of a volcano as lava, or before it even gets to the surface.

Volcanoes erupt in different ways. Different compositions of magma produce different types of lava, which causes different types of eruption.

Scientists study volcanoes to help them predict future eruptions, and also to find out more about the structure of the Earth.

Question

7 Why do geologists study volcanoes? Give two reasons.

Evidence for plate tectonics

A bit of history

People noticed how well the coastline of Africa matches that of South America.

In 1914, Alfred Wegener suggested that the continents were formed from one 'supercontinent' that was splitting apart at the time of the dinosaurs. This was called the continental drift theory.

However, nobody could explain how the continents were able to move through the rocks of the ocean floor so the theory was not accepted.

Scientists began to accept the theory in the 1960s, when geologists found a huge ridge in the middle of the Atlantic Ocean. The rocks around the ridge are very young.

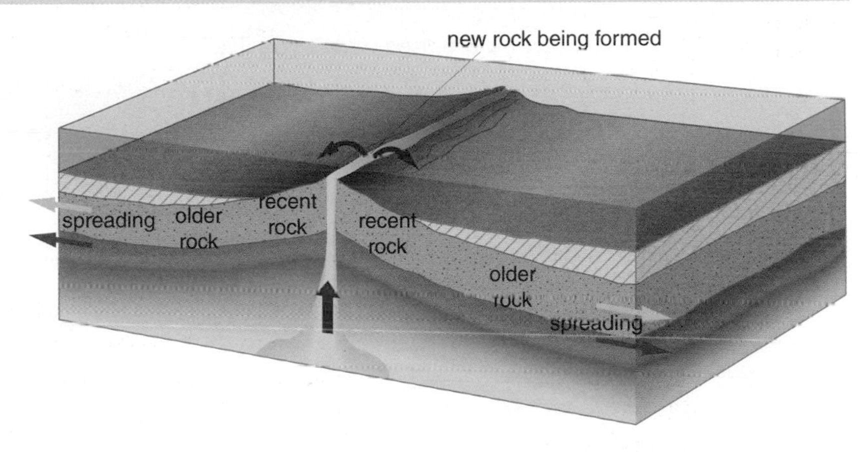

FIGURE 9: Does rock get older or younger as the distance from the central ocean ridge increases?

The further from the ridge the older and cooler the rocks are. We now know that the sea floor is spreading outwards at this point. This is where plates are being formed.

Subduction zones are where plates are being destroyed.

Subsequent research all fits with the idea of plate tectonics, and slowly the theory was accepted.

Magma and volcanoes

Differences in the composition of magma do not just lead to different types of rock. If magma reaches the surface of the Earth, then different types of magma can cause different types of eruption.

> Iron-rich magma (called **basaltic** magma) tends to be runny. Volcanoes with lava made from this sort of magma are often fairly 'safe'. The lava tends to spill over the edges of a volcano and people who live nearby can get out of the way.

> Silica-rich magma is less runny. It produces volcanoes that may erupt explosively. Silica-rich magma forms **rhyolite** if it cools rapidly.

This happens when dissolved gases in the magma have no time to escape from the stiff liquid. The magma erupts like spray from a shaken can of fizzy drink. It shoots out as clouds of searingly hot ash and pumice. The falling ash buries houses and people before they have time to escape. Rain turns the ash into mudslides which can trap and kill.

Geologists are continually developing their measurement techniques and theories of how volcanoes erupt. This means that they can forecast volcanic eruptions more accurately than before. Unfortunately, these forecasts are still not 100% accurate, and disasters still occur.

Questions

8 Explain why silica-rich magma is more dangerous than iron-rich magma.

igneous rock   subduction zone   mid-Atlantic ridge

Construction materials

You will find out:
- > the names of some construction materials
- > where the materials come from
- > some differences between limestone, marble and granite

What is your house built of?

The buildings in your local town started as raw materials beneath or on the surface of the Earth.

FIGURE 1: Brick is a traditional building material.

Building materials

What construction materials are used where you live?

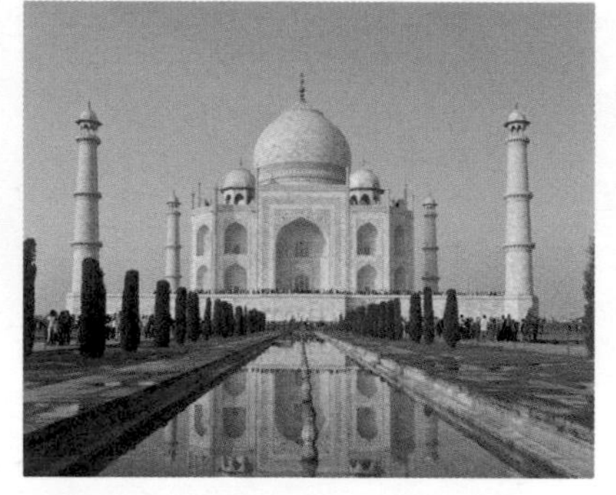

FIGURE 2: The Taj Mahal Tomb in Agra, India is inlaid with marble.

FIGURE 3: These steel frames will be hidden inside the walls.

FIGURE 4: The National Theatre in London is made from concrete.

FIGURE 5: This church is built from granite.

FIGURE 6: Huge amounts of aggregates go into roads.

FIGURE 7: St Paul's Cathedral in London is built from limestone.

Remember!

Make sure you know the names of these construction materials:
granite, marble, limestone, aluminium, iron (steel), concrete, aggregate.

Questions

1 Name five construction materials.

2 Make a list of the materials used to build your school.

3 Find out if your school has a steel frame.

The raw materials

Most modern buildings are made from materials dug out of the Earth.

Stone, such as **limestone, marble** and **granite**, is cut out of the ground and used in buildings. Blocks of stone are expensive to quarry and are only used for special buildings.

Stone buildings, such as cathedrals, are normally made from limestone. Limestone is easier to cut into blocks than marble or granite.

Marble is much harder than limestone.

Granite is harder still and is very difficult to shape.

Some buildings look as if they are built completely from stone, but they aren't. They are only lined on the outside with stone and have a different material on the inside. The stone is used as an attractive 'front'.

Smaller buildings, such as houses, are normally built from brick.

Larger structures are made from a steel frame or **reinforced concrete**. The walls are then built inside the frame. These walls can be made of brick, concrete, aluminium or glass.

Brick, concrete, steel, aluminium and glass also come from the ground, but they need to be manufactured from raw materials.

FIGURE 8: What building material is made from the clay in this pit?

Raw material	clay	limestone and clay	sand	iron ore	aluminium ore
Building material	brick	cement	glass	iron	aluminium

Questions

4 Look at the headstones in your local cemetery. They all have dates on them. How well have the different types of materials used for the headstones weathered?

5 List granite, limestone and marble in order of hardness, starting with the softest.

Rock hardness

Igneous and metamorphic rocks are normally harder than sedimentary rocks.

> Granite is an **igneous** rock and is very hard.

> Marble is a **metamorphic** rock. It is not as hard as granite, but it is harder than limestone.

> Limestone is a **sedimentary** rock and is the softest.

Igneous rock is formed out of liquid rock that cools and forms interlocking crystals as it solidifies. It is this interlocking structure that gives the rock its hardness.

Metamorphic rock is rock that has been changed. Marble is metamorphic; it is a form of limestone that has been subjected to heat and pressure in the Earth's crust, which makes it harder than the original limestone.

Sedimentary rock is made of fragments that have settled into layers. Limestone is a sedimentary rock made from the shells of dead sea-creatures that have slowly stuck together.

Question

6 Explain why granite is harder than marble and marble is harder than limestone. Include how each rock is formed in your answer.

Cement and concrete

You will find out:

- how cement is made
- about thermal decomposition
- about the difference between cement and concrete
- how reinforced concrete works
- about some of the environmental problems of quarries

Cement is made from limestone.

Limestone and marble are both forms of calcium carbonate. When calcium carbonate is heated it thermally decomposes to form calcium oxide and carbon dioxide.

Concrete is an artificial rock. Just mix cement, sand, aggregate (gravel) with water and allow to set.

Reinforced concrete has steel rods inside it and is much stronger than normal concrete. The concrete is poured around steel rods and left to set.

FIGURE 9: What will happen to this limestone quarry at the end of its useful life?

Getting building materials out of the ground

Every year millions of tonnes of limestone and other rocks are mined from quarries.

Without these quarries there would not be any buildings or roads. But quarries are often in areas of outstanding natural beauty and so can create environmental problems.

In the photograph look for:

- dust pollution
- damaged landscape (compare the quarry with the fields beyond).

Think about other problems:

- noise from explosives and machinery
- more lorries on country lanes
- what happens to the quarry after the stone has been taken out?

When a quarry closes the owners have to landscape the area.

The land can be:

- covered with soil and planted with grass
- used as a rubbish tip and then covered with soil
- left to fill with water and used for fishing and sailing.

Questions

7 What is the chemical name for limestone?

8 Suggest two ways of dealing with a quarry once it is disused.

Cement and concrete

Calcium carbonate thermally decomposes at very high temperatures.

The word and symbol equations for the reaction are:

calcium carbonate ⟶ calcium oxide + carbon dioxide

$$CaCO_3 \longrightarrow CaO + CO_2$$

Thermal decomposition is the chemical breakdown of a compound into at least two other compounds under the effect of heat.

To make **cement** limestone is heated with clay.

Cement mixed with sand and water is called mortar. It is very good for sticking bricks together, but it is not strong enough to use on its own. If gravel (aggregate) is added to the mixture it makes concrete, which is much stronger.

Reinforced concrete has steel rods or steel meshes running through it and is even stronger. It is a composite material.

FIGURE 10: Steel meshes are used to reinforce concrete. Steel is flexible – why is this an advantage?

FIGURE 11: Why do you think this road bridge is made from reinforced concrete and not ordinary concrete?

Did you know?

The Romans invented concrete.

They could even make a concrete that set under water.

Question

9 Explain why the first process below is an example of thermal decomposition and the other two are not:

- When limestone is heated it forms calcium oxide and carbon dioxide.
- When magnesium is heated in air it forms magnesium oxide.
- When salt is heated it melts.

More on reinforced concrete

Concrete is hard. It is very strong under compression (squashing force). It is much weaker under tension (pulling force).

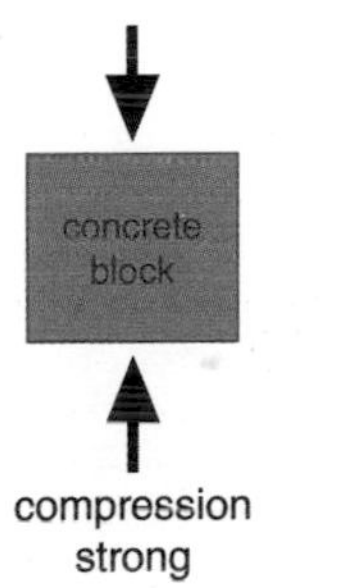

FIGURE 12: Is concrete stronger under compression or tension?

If a heavy load is put on a concrete beam it will bend very slightly. When a beam bends its underside starts to stretch. This puts it under tension and cracks start to form.

Steel is flexible, and it is strong under tension. Steel rods in reinforced concrete stop the concrete stretching.

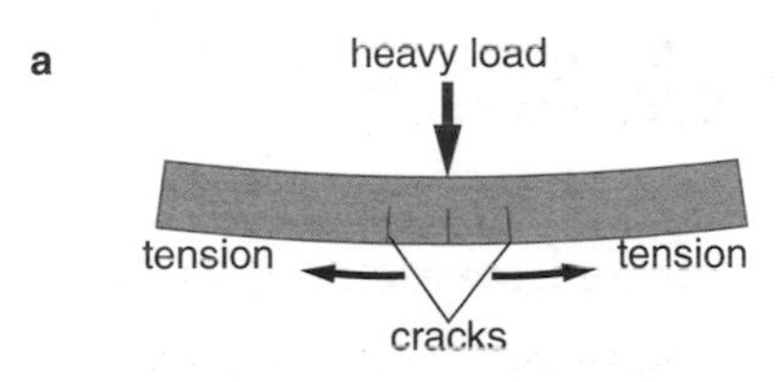

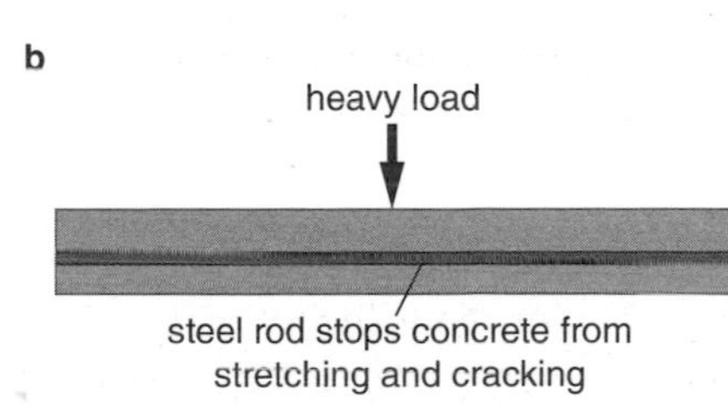

FIGURE 13: Why does putting steel rods or mesh in concrete stop cracks appearing in it?

Question

10 Concrete is hard, but this is not enough to make it a really useful building material. Explain why.

Metals and alloys

The Bronze Age

About 6000 years ago humans learnt how to extract copper from its ore.

This discovery marked the start of the Bronze Age.

You will find out:

> that copper can be extracted using carbon

> about purifying copper by electrolysis

> about recycling copper

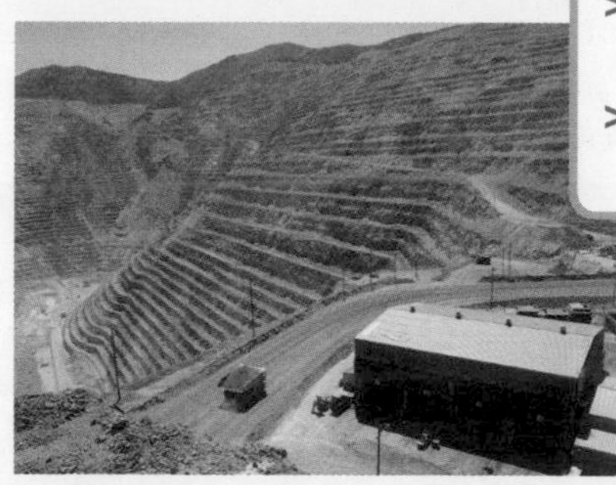

FIGURE 1: Huge amounts of rock are dug from copper mines to get enough ore for modern needs.

Making copper

Extracting copper from its ore

Copper is a metallic element. Ores are compounds.

One way to extract copper is to convert the ore to copper oxide.

The copper oxide can be heated with carbon to remove the oxygen. Removal of oxygen is called **reduction**.

Making copper pure

Smelted copper is not very pure. We use **electrolysis** to purify it. Millions of tonnes of copper are produced each year, so the electrolysis is done on a large scale.

Recycling

Recycling copper is cheaper than extracting it from its ore. This is because:

> it saves resources – we don't need so many huge copper mines

> it saves the energy needed to crush rocks and to operate smelters and electrolysis cells.

FIGURE 2: Copper smelters work at high temperatures. Why does smelting copper need so much energy?

FIGURE 3: Electrolysis of smelted copper. The size of the person gives you an idea of scale.

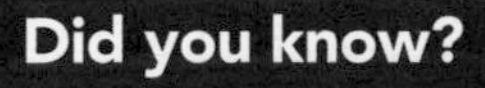

Did you know?

We need tiny amounts of copper in our diet to stay healthy.

Bananas give us copper in our diet – but not enough to smelt!

Questions

1 What can you heat with copper ore to get the copper out?

2 Why can smelted copper not be used to make things?

Extracting copper from its ore

Impure copper can be purified in the laboratory using an electrolysis cell.

In an electrolysis cell the **anode** is impure copper. The anode dissolves into the **electrolyte** and pure copper coats the **cathode**. The cathode is 'plated' with new copper.

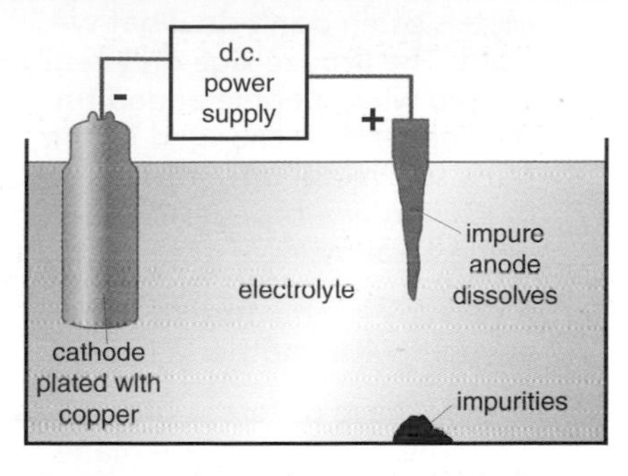

FIGURE 4: What is the anode made from in this electrolysis cell?

Recycling copper

Copper has a fairly low melting point. It is easy to melt down and recycle, which makes it cheaper than extracting copper from the ore.

However, there are problems.

> getting people to recycle it in the first place

> sorting the copper from other metals

> even the copper samples must be sorted. Valuable 'pure' copper scrap must not be mixed with less-pure scrap. Copper with solder on it will contaminate the metal when the copper is melted.

Contaminated copper can still be used to make alloys, such as brass or bronze, or it may have to be electrolysed again to purify it.

FIGURE 5: Copper used in electric wires must be very pure. How can copper become contaminated?

Question

3 Draw a labelled diagram of the apparatus needed to purify copper by electrolysis.

4 Suggest one problem linked to recycling copper.

More on purifying copper

Electrolysis is the break-up of a chemical compound (the electrolyte) when you use an electric current. In this case, something else happens as well.

To purify copper we use an electrolyte of copper(II) sulfate solution. The impure copper is the anode, and a sheet of pure copper is used for the cathode.

Ions from the copper(II) sulfate electrolyte gain electrons and deposit onto the copper cathode.

$Cu^{2+} + 2e^- \longrightarrow Cu$

The cathode gains mass as pure copper is plated onto it. Gain in electrons is called **reduction**.

The impure anode dissolves. Copper atoms in the anode turn into ions, putting copper ions back into the electrolyte.

$Cu - 2e^- \longrightarrow Cu^{2+}$

The anode dissolves, so it loses mass. Loss of electrons is called **oxidation**.

The impurities from the copper anode fall to the bottom of the cell.

(For more on oxidation and reduction in electrolysis, see Chemicals from the sea, pages 172–175.

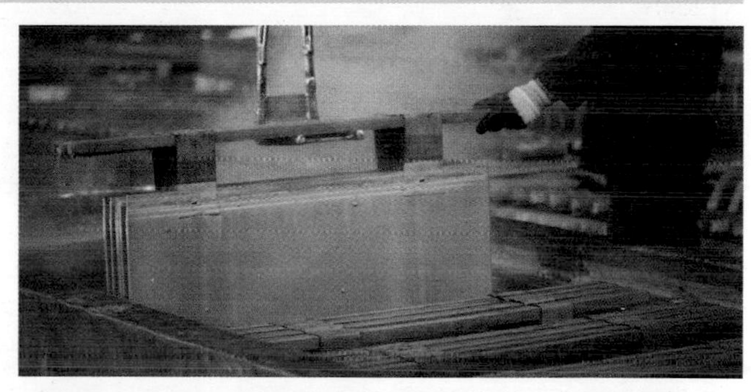

FIGURE 6: Copper cathodes being lifted from an electrolysis cell. Are they pure or impure?

Questions

5 During the electrolytic purification of copper, what happens at:

a the cathode?

b the anode?

electrolysis cell electrolyte element oxidation reduction pure recycled

You will find out:
- what alloys are
- why alloys are useful
- about 'smart' alloys

Tweaking metals

Metals often don't do what we want. The Bronze Age only really started when people added tin to the copper. This changed it from a fairly soft metal into something much harder – bronze. They had made an **alloy**.

> An alloy is a mixture of a metal element with another element.

> Alloys have different properties from the metals they are made from, which is why we use them.

Different alloys have different uses, such as amalgam used by the dentist to fill your teeth. (For more on alloys, see Making cars, pages 154–157.)

FIGURE 7: Brass is an alloy used to make taps, door handles and musical instruments. Why is brass used for door handles?

FIGURE 8: What is solder used for?

Did you know?

FIGURE 9: Copper alloys are used in coins.

In 2002, Europe introduced the new Euro coin. 184 000 tonnes of copper were needed to make them.

Questions

6 Look at the descriptions of alloys below.

- alloy A – strong, cheap
- alloy B – low melting point
- alloy C – soft when made, hardens quickly

Which alloy would be good for:

a sticking two metals together.

b making car bodies.

c filling teeth?

copper alloys properties and uses

Alloys and their uses

Most metals form alloys. Here are just a few:

> **brass** contains copper and zinc
> **solder** contains lead and tin
> **steel** contains iron
> **amalgam** contains mercury.

Different alloys have different properties. We match the properties of the alloy to the job we use it for.

Bronze

Bronze is harder than copper, hard enough to make ploughshares (for ploughing fields) and swords. Bronze is one of the few metals that shrinks very slightly when it solidifies. This makes it easy to cast, which is one reason why statues are often made of bronze.

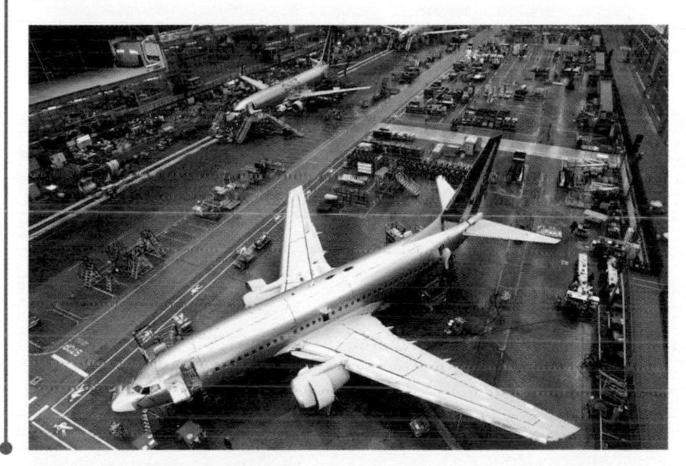

Steel

Steel is much stronger than iron, and stainless steels do not rust.

Solder

Alloys can have lower melting points than the pure metal. This property is very useful. An alloy of lead and tin melts so readily that it can be used to join metals together. This alloy is called solder.

Alloys are often more useful than the original metals, though nowadays pure copper is more important than bronze or brass. Pure copper conducts electricity so well that vast amounts are turned into electric wire.

Questions

7 Heavy lorries take cargo to a port to be loaded onto ships. The lorries are driven out onto a jetty. The legs of the jetty are made of an alloy and stretch down into the seawater. Suggest two properties of the alloy.

8 What one property of solder makes it a useful alloy for sticking metals together?

FIGURE 10: Suggest what properties make an alloy of aluminium and copper suitable for making aeroplanes.

Smart alloys

Do you wear glasses? Can the frames be bent without breaking them? (Do not try this!) If so, the frames are probably made from a smart alloy.

Nitinol is a **smart alloy** made from nickel and titanium.

Smart alloys are more bendy than steel so are harder to damage.

However, this is not why metal alloys are called 'smart'. Smart alloys can change shape at different temperatures. This is called 'shape memory'. Surgeons can put a small piece of metal into a person's blocked artery and then warm it slightly. As it warms up it changes shape into a much larger tube that holds the artery open and reduces the risk of the person suffering a heart attack. Smart metal alloys are also used in shower heads to reduce the water supply if the temperature gets so hot that it might scald.

Smart alloys are becoming more important as new ways of using them are discovered.

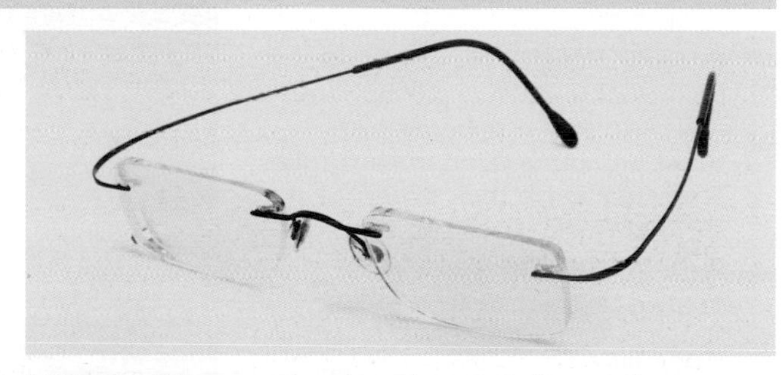

FIGURE 11: What property does the smart alloy used to make the arms of these glasses have?

Question

9 Why is the usage of smart alloys increasing?

Making cars

You will find out:
- > about materials used in manufacturing motor vehicles
- > about rusting and corrosion of metals

What happens to cars?

A car is mostly made from iron. Iron is a very valuable resource. Today a lot of aluminium is also used to make cars. Aluminium is an even more valuable resource. Plastics, glass and other metals are also used to make cars.

There are millions of cars in this country. Some people replace their cars after a few years. What happens to those cars? Many cars are sold and millions are scrapped each year. The scrap metal merchants keep the valuable metals and dispose of all the other materials.

Almost all of these materials could be recycled.

FIGURE 1: A scrap metal yard.

Rusting and corrosion

Most cars have a metal body. Steel or aluminium is normally used to make a car body. Steel is an alloy that contains mostly iron.

All cars made with steel **rust**.

> Iron needs oxygen and water to rust.

> Rust flakes off the surface of the iron. This allows more rusting to take place.

Addition of oxygen is oxidation. Reaction with oxygen is oxidation.

Cars made with aluminium do not **corrode** when oxygen and water are present.

Georgia and Ellie investigate the rusting of iron. They put two iron nails into four numbered test tubes and put a bung in each tube. They then leave their experiment for two weeks. Each test tube has different conditions:

1 distilled water

2 dry air

3 water with no dissolved air

4 salt water.

After two weeks only the nails in test tubes 1 and 4 have rusted.

Test tubes 2 and 3 did not have both oxygen and water in them. This experiment shows that oxygen and water are needed for rusting.

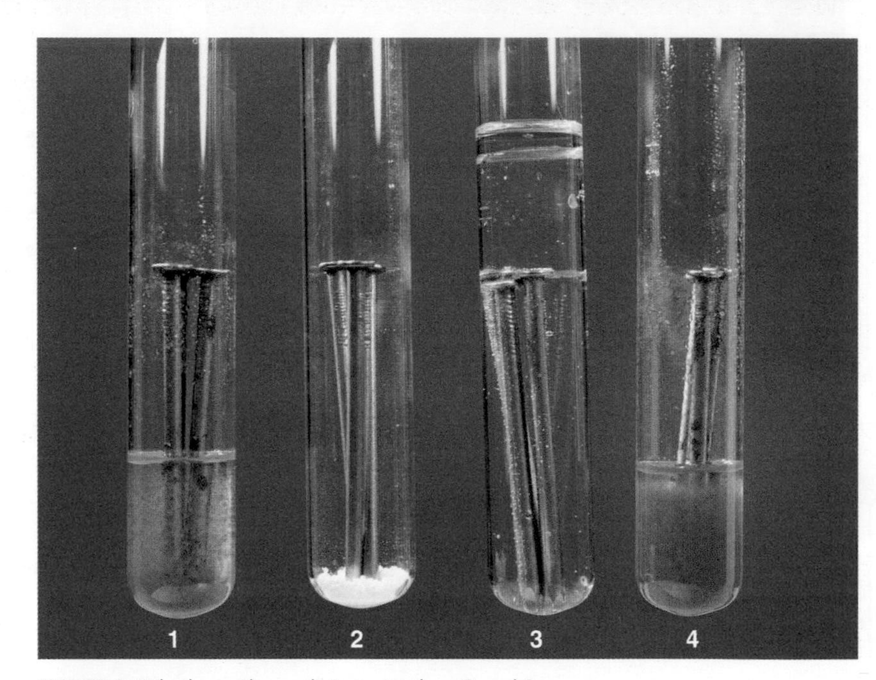

FIGURE 2: Why have the nails in test tubes 2 and 3 not rusted?

Questions

1 Look at the photograph of Georgia and Ellie's results. What conclusion can you make about the effect of salt on the rate of rusting?

2 Car bodies made of iron are painted. How does this help to prevent a car body from rusting?

More on rusting and corrosion

Rusting is a chemical reaction between iron, oxygen and water. The chemical name for rust is **hydrated iron(III) oxide**.

iron + oxygen + water ⟶ hydrated iron(III) oxide

Any reaction in which oxygen is added is called an oxidation reaction.

Rate of rusting

Salt water makes the rusting reaction much faster. In winter icy roads are treated with salt. This means that car bodies rust quicker. Acid rain also increases the rate of rusting.

Aluminium is useful because it does not corrode in moist air, so car bodies made from aluminium do not need much protection against corrosion. In air, a protective oxide layer forms on the surface of aluminium. Rust is also an oxide layer, but it does not protect the rest of the iron. It flakes off, leaving more exposed iron to rust.

Corrosion of metals other than iron

Today there is far more atmospheric pollution. Acid rain can have a pH value as low as 3. This means that many metals corrode quickly in moist air. Georgia and Ellie investigate the effect of different atmospheric conditions on metals. They leave strips of metal for two weeks and then look at the appearance of the metal. They show their results in a table.

Condition / Metal	At start	Dry air	Moist clean air	Moist acidic air	Moist nitrogen	Moist alkaline
Aluminium	shiny silver	shiny silver	shiny silver	dull silver	shiny silver	dull silver
Copper	shiny salmon-pink	shiny salmon-pink	small patches of green on surface	green layer on surface	shiny salmon-pink	small patches of green on surface
Iron	shiny silver	shiny silver	small patches of brown on surface	lots of brown flakes on surface	shiny silver	small patches of brown on surface
Magnesium	silver	silver	dull silver	greyish layer on surface	silver	dull silver
Lead	shiny silver	shiny silver	dull silver	dark, almost black layer on surface	shiny silver	dull silver
Silver	shiny silver	shiny silver	shiny silver	dull silver	shiny silver	shiny silver
Zinc	shiny silver	shiny silver	dull silver	greyish layer on surface	shiny silver	dull silver

Georgia and Ellie conclude that more corrosion happens in moist acidic air than in any other conditions. They also find that oxygen is needed for corrosion.

Questions

3 Which metal tested by Georgia and Ellie corroded the least?

4 Explain how Georgia and Ellie were able to conclude that water was needed for corrosion.

Materials in a car

You will find out:

> about the properties of iron and aluminium
> how to relate the properties of materials to their uses
> about the advantages of recycling materials used in manufacturing cars

A car is made from lots of different materials. A material is chosen because of its properties. Glass is transparent. Imagine a car with copper windows.

Some of the materials used are:

> metals and alloys – copper, steel and aluminium

> plastics

> glass

> fibres.

Iron and aluminium

We use different materials because they have different properties.

The table shows some properties of aluminium and iron. If something is 'malleable' it can be beaten into a thin sheet.

Property	Aluminium	Iron
Corrosion in moist air	no obvious corrosion	rusts rapidly
Density in g/cm^3	2.7	7.9
Malleability	malleable	malleable
Electrical conductivity	good	good
Magnetism	not attracted to a magnet	attracted to a magnet

Most metals used in cars are alloys. Alloys have different properties to those of the metal they are made from, which can make them more useful.

Steel is an alloy of iron. Not only is steel harder and stronger than iron, but also it does not corrode as easily.

Old cars are often dumped and taken to a scrap-metal yard. Here their parts are recycled.

It is easy to separate iron from aluminium. Iron is magnetic and aluminium is not.

The car body is cut into smaller pieces and an electromagnet is used to attract iron or steel parts.

Recycling makes sense because:

> extracting iron and aluminium from their ores takes enormous amounts of energy

> it avoids environmental damage due to mining and quarrying

> it reduces the amount of rubbish that goes into landfill sites.

FIGURE 3: Iron filings are attracted to a magnet.

FIGURE 4: What scrap metal does this electromagnet collect?

Question

5 Describe how you would separate a mixture of iron filings and aluminium powder.

how electromagnets work

Materials used in cars

Different materials are used in a car for different reasons.

Material and its use	Reasons material is used
aluminium in car bodies and wheel hubs	does not corrode, low density, malleable, quite strong
iron or steel in car bodies	malleable, strong
copper in electrical wires	ductile, good electrical conductor
lead in lead–acid batteries	chemical reaction with lead oxide produces electricity
plastic in dashboards, dials, bumpers	rigid, does not corrode, cheap
pvc in metal wire coverings	flexible, does not react with water, electrical insulator
glass and plastic/glass composite in windscreens	transparent, shatterproof (may crack)
fibre in seats	can be woven into textiles, can be dyed, hard-wearing

Steel or aluminium car bodies?

An alloy has different properties to those of the pure metals that are used to make it. Alloys are used when they have better or more appropriate properties than those of the pure metal. Steel is stronger and harder than iron. It does not corrode as easily as pure iron.

Steel and aluminium can both be used to make car bodies. Each material has its particular advantages.

> Steel is stronger and harder than aluminium.

> Aluminium is more **malleable** than steel.

> Aluminium does not corrode as easily as steel.

> Aluminium is less dense than steel. This means that the mass of a car body is much less than that of the same car body made from steel.

> Steel needed for a car body is cheaper than aluminium.

Recycling

European Union law requires 85% of a car to be recyclable. This percentage will increase to 95% in the future. These figures include all parts of a car, including the batteries. Technology has to be developed to separate all the different materials used in making a car. This is a challenge to the motor industry, but recycling makes sense for everybody.

> More recycling of metals means that less metal ore needs to be mined. This saves finite resources and reduces environmental damage.

> Recycling of iron and aluminium saves money and energy compared with making them from their ores.

> The benefits of recycling plastics include less crude oil being used to make plastics and less non-biodegradable waste being dumped.

> Recycling of glass has been happening for many years so the technology is well established.

> Recycling batteries reduces the dumping of toxic materials into the environment.

Question

6 Compare and contrast the properties of aluminium and steel.

More on steel versus aluminium in car bodies

Increasing numbers of car bodies are made from aluminium than from steel. The advantages of aluminium are beginning to outweigh those of steel:

> Steel is cheaper than aluminium, so cars made from steel cost less to buy.

> A car with an aluminium body is lighter, so it has a better fuel economy than a similar car made from steel.

> Aluminium does not corrode as easily as steel so the car body has a much longer lifetime.

FIGURE 5: An aluminium car body
What are the advantages of aluminium over steel when making car bodies?

Question

7 The cost of making 1 kg of aluminium is more than that for making 1 kg of steel. Explain why this is not a tremendous disadvantage for using aluminium in making car bodies.

Q advantage aluminium car body

Preparing for assessment: Applying your knowledge

To achieve a good grade in science, you not only have to know and understand scientific ideas, but you need to be able to apply them to other situations and investigations. These tasks will support you in developing these skills.

Vorsprung durch Technik – progress through technology

Since 1994 Audi has been building cars with aluminium alloy bodies. It is used not only in the panels but also in the frame below, called the ASF, or Audi Space Frame. First Audi built the luxury A8 saloon. More recently, in 2009 the Audi Aluminium A5 Coupe started production reducing the car body mass by 40%, and reducing the overall body mass by 100 kg.

By 2001, over 150 000 aluminium Audis had been constructed, rising to over 278 000 in 2009.

Two major factors that influence car design are fuel economy and passenger safety. With fuel becoming more expensive and pollution more of a problem, car designers have to find ways of making cars travel further for each litre of fuel used.

Cars also have to be safer and each design is thoroughly crash-tested before it is put on sale.

Cars have to be light and strong and Audi has found that a good way of achieving this is to make both the frame structure and the body panels from aluminium. Aluminium does not rust and so does not need the protection that steel does.

Task 1

A more common material used to make car bodies is steel (mainly made up of iron). Discuss and decide in your group whether steel or aluminium, or both, would satisfy the following design features.

> Car bodies should conduct electricity as this can then form part of the electrical circuits such as to light bulbs.

> Car bodies should not corrode as it weakens them and spoils their appearance.

> Car bodies have to be shaped from flat sheets so they have to be workable (this is called malleability).

Write down a summary of your group discussions and the decisions.

Task 2

A car designer considers both materials and tests them. He comes to the conclusion that the aluminium-bodied car would be lighter, corrosion free and more expensive. Explain why each of these factors is significant.

Task 3

Another motor company decides to follow the example of Audi and build its car bodies from aluminium. It does this and after several years it notices that its cars:

> last longer than other cars

> use less fuel than similar steel cars

> are more expensive to make.

Explain why the three observations are true. Then try to decide whether the motor company is likely to be pleased with each of its findings.

Maximise your grade

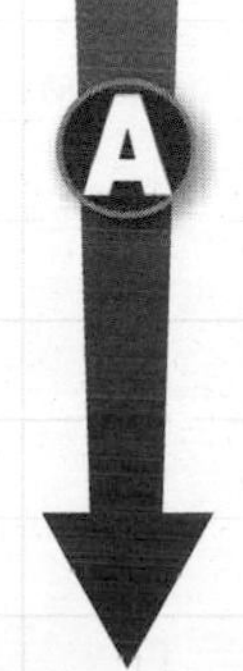

Answer includes showing that you can...
Describe what an alloy is.
Describe one similarity or difference between steel and aluminium.
Describe similarities and differences between steel and aluminium.
Name the conditions needed for iron to rust.
Suggest which material would be better to build a car body from.
Describe advantages and disadvantages of building car bodies from aluminium or from steel.
Explain advantages and disadvantages of building car bodies from aluminium or from steel.
Evaluate information from the text. Try this: > The Audi Aluminium A5 Coupe body mass was reduced by 40%. > The overall body mass is 1310 kg. Explain why the overall mass only reduced by 100 kg. If current production continues at the same rate, estimate how many aluminium cars would have been produced now.

Manufacturing chemicals – making ammonia

You will find out:

> how ammonia is made
> about reversible reactions
> why special conditions are used in the manufacture of ammonia

Ammonia is the key to reducing starvation

About 100 years ago people found out how to make ammonia from the air.

Since then, the use of ammonia-based fertilisers on crops has increased yields and prevented millions of people from starving.

FIGURE 1: The number of starving people in the world has fallen because of the use of fertilisers.

Ammonia

The formula for ammonia is NH_3. Ammonia is made by joining nitrogen and hydrogen.

> Nitrogen comes from air.

> Hydrogen is made from natural gas or by cracking oil fractions.

Nitrogen is unreactive, but in 1908, Fritz Haber found that he could make nitrogen react with hydrogen if he used an iron catalyst. Haber's discovery is now used to make millions of tonnes of ammonia every year.

Haber's process has two reactions.

A forward reaction

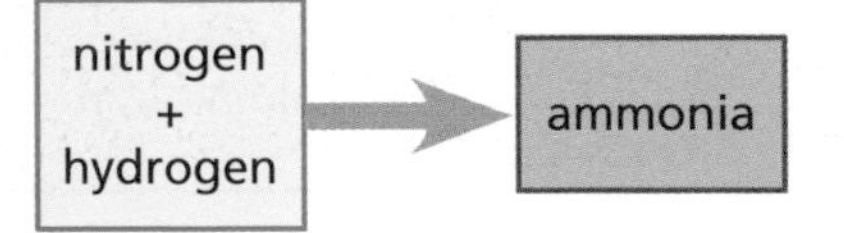

A backward reaction

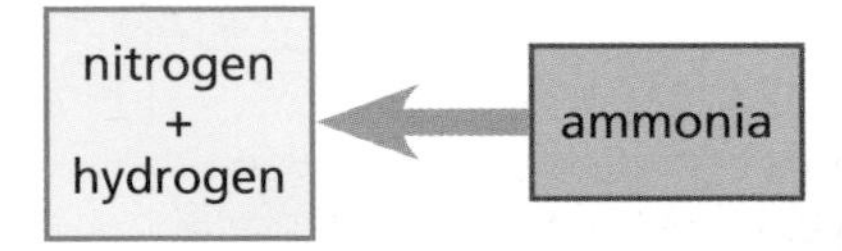

The reaction can go in both directions – at the same time.

Reactions that can go in both directions are called **reversible** reactions.

To write an equation for a reversible reaction we use a special arrow, $\rightleftharpoons$, to show that the reaction goes both ways.

nitrogen + hydrogen $\rightleftharpoons$ ammonia

Did you know?

A factory that makes chemicals is called a 'plant'. An ammonia plant makes ammonia.

Did you know?

A catalyst is something which speeds up a reaction, and can be recovered at the end of the reaction.

Questions

1 Give two features of a catalyst.

2 Write a word equation for the reaction between nitrogen and hydrogen.

3 What does the symbol $\rightleftharpoons$ mean?

4 Fritz Haber solved the problem of making nitrogen and hydrogen react. What was the problem?

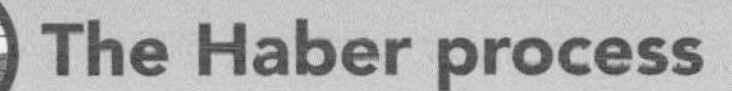

The Haber process

nitrogen + hydrogen ⇌ ammonia

$$N_2 + 3H_2 \rightleftharpoons 2NH_3$$

Every bit of grain crop grown in the world needs nitrogen. Over half of that nitrogen comes from artificial fertilisers. The main way these fertilisers are made starts with the **Haber process**.

Ammonia manufacture has to be efficient to produce large quantities. To work most efficiently the Haber process uses:

> an iron **catalyst**

> a high pressure of around 200 atmospheres

> a temperature of around 450 °C

> a **recycling** system. Not all the nitrogen and hydrogen react, so the nitrogen and hydrogen that do not react are sent back into the reaction vessel.

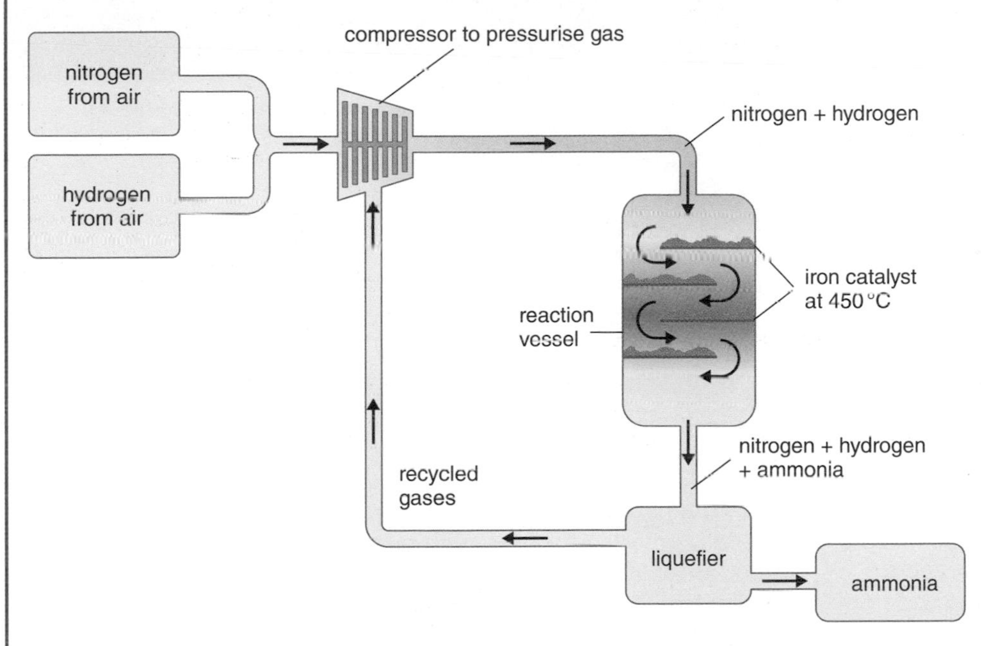

FIGURE 2: The Haber process. Suggest how the compressor helps to make the reaction more efficient.

Questions

5 What is ammonia made from?

6 Write down the four important conditions used in the industrial manufacture of ammonia.

7 What happens when hydrogen and nitrogen are passed over an iron catalyst?

More on the Haber process

As the reaction is reversible, the percentage yield for the reaction cannot be 100 per cent. Surprisingly, the conditions inside the Haber plant do not even give as high a **percentage yield** as they could.

> The high pressure increases the percentage yield – but even higher pressures could be used.

> The high temperature decreases the percentage yield. However, high temperatures do make the reaction go faster.

> 450 °C is an **optimum** temperature – the yield is not as good, but that yield is made faster, so it produces more ammonia in a day.

> Catalysts do not affect the yield – they just make the reaction go faster.

Question

8 Ammonia can be removed by cooling the gases coming out of the reaction vessel. Explain why cooling the gases that are to be recycled increases energy consumption.

You will find out:

> what affects the costs of making a chemical

What affects the costs of a big chemical plant?

Modern ammonia plants make 1500 tonnes of ammonia every day. These plants are carefully designed to keep the cost of making ammonia as low as possible.

What affects the cost of making ammonia?

- The cost of building the plant in the first place.
- The labour costs – paying people's wages.
- The cost of the chemicals – the hydrogen and nitrogen.
- The energy costs.
- How fast the reaction will go – the catalyst makes it go faster, so more ammonia can be made, but the catalyst itself costs money.

What is ammonia used for?

When Fritz Haber first did his experiments, he wanted the ammonia to make explosives.

These days:

- over 80 per cent of all ammonia goes into fertilisers
- ammonia is also used to make nitric acid – most of which goes into more fertilisers
- very small amounts of ammonia are used to make household cleaners.

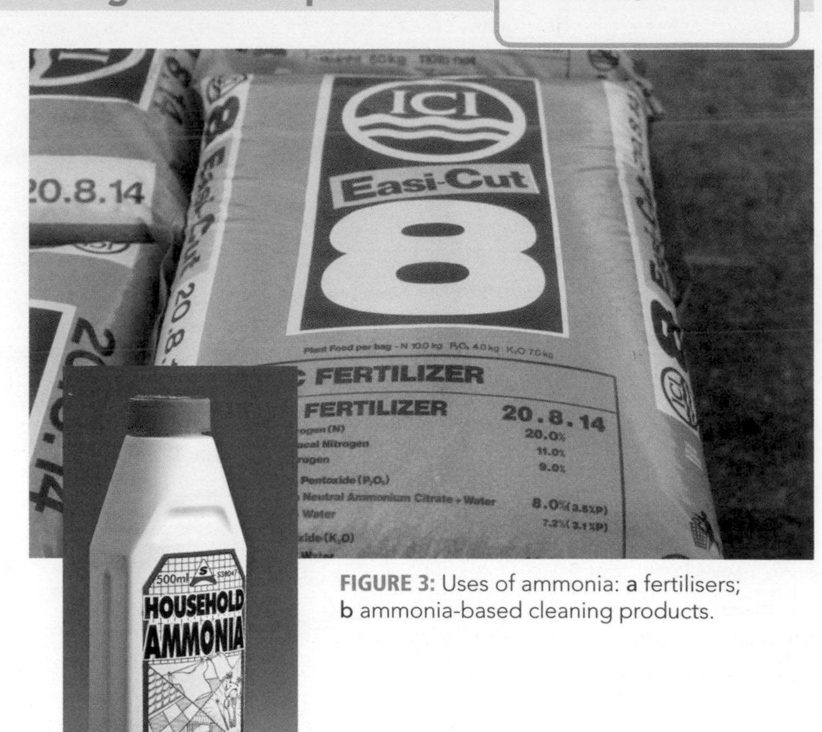

FIGURE 3: Uses of ammonia: **a** fertilisers; **b** ammonia-based cleaning products.

Questions

9 What is the main use of ammonia?

10 What would happen to the costs if ammonia could be made at a lower temperature? Explain your answer.

What affects the costs of ammonia production?

- Labour costs. Many chemical plants are now heavily automated and they need very few people to operate them.
- The cost of the reactants. Hydrogen is made from natural gas or by cracking oil fractions, which costs money. Nitrogen comes from the air, but it is not free. It has to be cleaned, dried and compressed.
- Recycling unreacted materials means that money is not wasted on unused reactants.
- High pressure makes the reaction work better, but costs more. The pipes and fittings have to be made of stronger steel. Designers use a pressure that gives a good yield, but that does not make it impossibly expensive to make the equipment.

FIGURE 4: A chemical plant. Can you list the factors that affect the cost of producing ammonia?

uses of ammonia

> There are energy costs – the higher the temperature, the more fuel is needed.

> How fast the reaction goes. The faster the reaction, the more product is made from the same equipment, so the cheaper it is. Chemists are always trying to improve the catalyst so that the reaction will go even faster.

> Pollution-control costs. Reducing pollution is expensive and manufacturers have always tried to make their chemicals as cheaply as possible. Antipollution laws now mean that the chemicals cost a little bit more, but the plants do much less damage to the environment.

The table shows percentage yield under different conditions.

Pressure in atmospheres	Temperature			
	100 °C	200 °C	300 °C	400 °C
50	95%	74%	40%	15%
100	97%	82%	53%	25%
200	99%	89%	67%	39%

Questions

11 What conditions in the table give the lowest yield of ammonia?

12 What conditions in the table give the highest yield of ammonia? Explain why these conditions are not used.

Did you know?

Heating anything is expensive. Chemical plants work at as low a temperature as they can get away with.

Optimum conditions

Chemical plants do not work at the conditions that produce the highest percentage yield for a reaction. A low yield is acceptable if the daily output of the product is high enough. The plants are designed to work at the conditions that will make the chemical most cheaply. These are known as the optimum conditions.

Temperature is especially important. High temperature should be bad for this reaction, yet it runs at 450 °C. This sort of temperature means higher energy costs and also lower yields. However, the increase in rate more than compensates; the plant produces more ammonia in a day at this temperature than it would at lower, more 'efficient' temperatures.

The reaction has a low percentage yield. However, as long as the unreacted chemicals can be recycled, they can go back into the reaction vessel. It is the combination of yield and rate that must give enough product.

Total energy costs are not solely due to heating costs. The plant needs compressors and pumps to achieve a high pressure and to move chemicals through pipes.

Questions

13 Haber plants are often built next to oil refineries. Explain why this reduces costs.

14 Many chemists have tried to find new catalysts that would make the Haber process operate more quickly. Suggest two ways in which the costs would be less if a Haber plant could be built with better catalysts.

ammonia optimum temperature GCSE

Acids and bases

You will find out:
- about pH numbers
- the differences between an alkali and a base
- about neutralisation
- what makes an acid acidic

Why does it taste so good?

Why does orange juice taste tangy? Why do we like vinegar with chips? What is the secret ingredient in fizzy drinks?

The answer is acids. We have been swallowing acids since time began.

Have you ever felt ill after eating too much fruit? You've probably eaten too much acid. We think that it was people who ate too much acid who first discovered the effects of neutralisation.

FIGURE 1: Why might you feel unwell if you eat too much blackberry and apple pie?

Opposites

The opposite of an **acid** is a **base**. **Alkalis** are bases that will dissolve in water. Acids turn litmus red, alkalis turn it blue.

Adding an acid to a base

When an acid and a base or alkali meet they **neutralise** each other.

pH scale

The **pH** scale shows how acidic or alkaline a substance is. A reading of pH = 0 shows a strong acid. A reading of pH = 14 shows a strong alkali.

Remember!
The pH increases when an alkali is added to a solution.
The pH decreases when an acid is added to a solution.

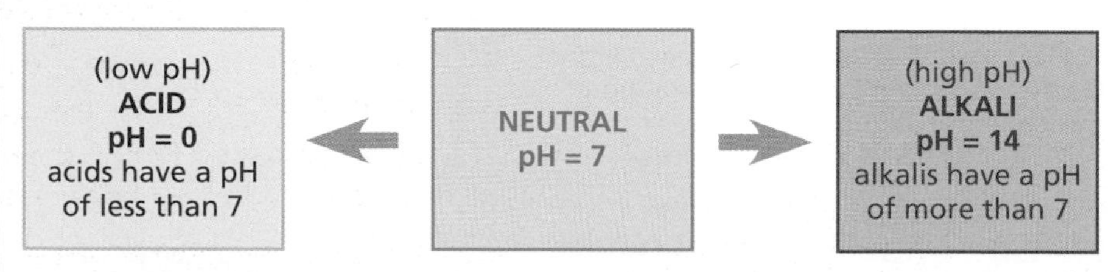

FIGURE 2: The pH scale.

Questions

1 What neutralises a base?

2 The table below shows information about pH changes. Fill in the empty boxes.

pH changes from	7 to 3	10 to 8	5 to 8	1 to 4	10 to 7
Is this an increase or a decrease in pH?					
Is the final solution acid, alkaline or neutral?					

pH scale acids and bases

Bases and alkalis

When is a base an alkali? When it dissolves in water. An alkali is just a soluble base. Copper oxide is an example of a base that is not an alkali because it does not dissolve in water. However, it still reacts with acids.

Neutralising acids

A base neutralises an acid. The most common bases are metal oxides and metal hydroxides.

sulfuric acid + sodium hydroxide ⟶ sodium sulfate + water

acid + base ⟶ salt + water

What makes a solution acidic?

All acidic solutions contain hydrogen ions, H^+.

Indicators and pH

Indicators are substances which change colour when pH changes.

Indicators such as litmus change colour suddenly at one pH value only.

Indicators such as universal indicator change gradually over a range of different pH values.

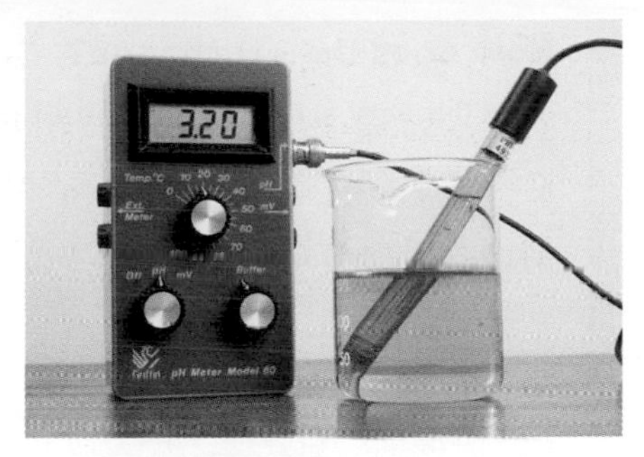

FIGURE 3: A pH meter. What does the reading show?

Questions

3 What does pH measure?

4 What is formed when an acid reacts with a base?

Hydrogen ions

An acid solution contains hydrogen ions, H^+. The hydrogen ions are responsible for the reactions of an acid.

An alkali solution contains hydroxide ions, OH^-.

When an acid neutralises an alkali, the hydrogen ions react with the hydroxide ions to make water.

$H^+ + OH^- \longrightarrow H_2O$

Remember!

Make sure you use the correct colour chart when measuring acidity using Universal indicator solution. Different makes of Universal indicator solution produce slightly different colours.

Questions

5 What happens to the hydrogen ions when an acid is neutralised?

6 Write an ionic equation to show the neutralisation of hydrogen ions.

7 The formula for sulfuric acid is H_2SO_4. Suggest how many hydrogen ions each molecule will form when sulfuric acid dissolves in water.

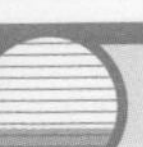

Changing pH

You will find out:
- what a salt is
- how to predict which salts form when different acids react

How does the pH change?

pH = 0 is very acidic. A reading of pH = 14 is very alkaline.

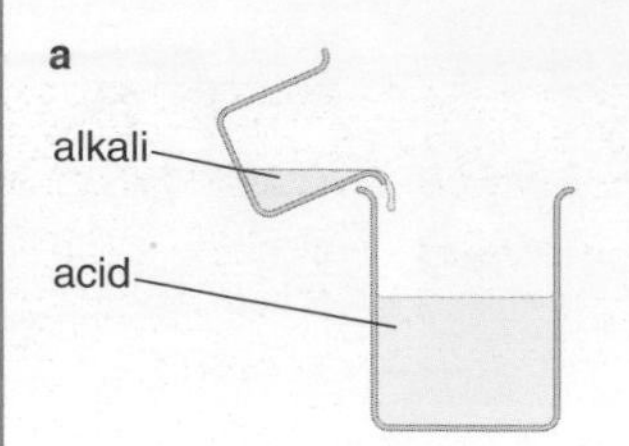

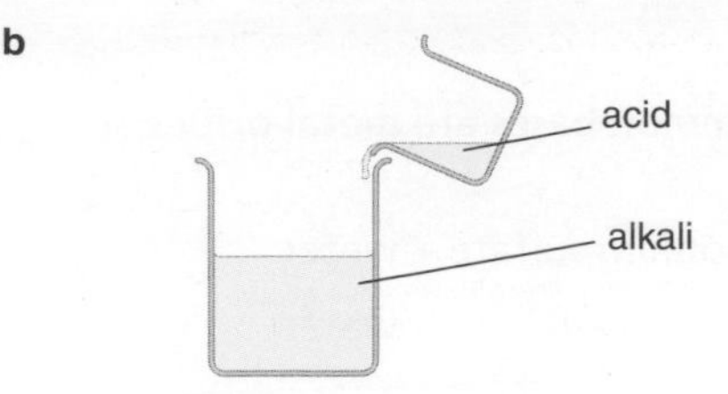

FIGURE 4: How pH changes: **a** adding an alkali to an acid; and **b** adding an acid to an alkali.

a
- the pH at the start is low
- the pH rises as the alkali neutralises the acid
- when neutral, the pH = 7
- when more alkali is added the pH rises above 7

b
- the pH at the start is high
- the pH falls as the acid neutralises the alkali
- when neutral, the pH = 7
- when more acid is added the pH falls below 7

What does universal indicator do?

Many indicators will tell you if something is acid or alkali.

Universal indicator tells you more; it shows just how acidic something is.

Universal indicator is a mixture of indicators, each one changing at a different pH. Different makes of universal indicator give slightly different colours, so universal indicator always comes with a colour chart to compare with.

A few drops are added to the test solution and then the colour of the solution is compared to a standard colour chart.

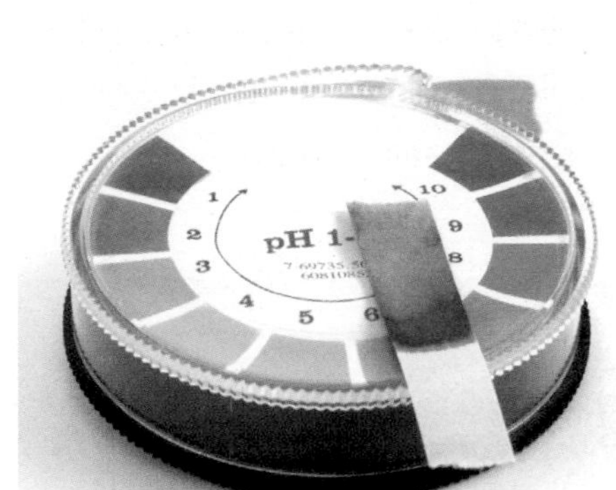

FIGURE 5: What pH does this strip of indicator show?

Salts

Reaction of acid with metal oxides or metal hydroxides

Metal oxides and hydroxides are bases, so they neutralise acids.

acid + oxide ⟶ salt + water

acid + hydroxide ⟶ salt + water

Reaction of acid with metal carbonates

The reaction of a metal carbonate with an acid is similar to that above, but a gas is also given off:

acid + carbonate ⟶ salt + water + carbon dioxide

What is a salt?

A **salt** is made from part of a base and part of an acid. When sodium hydroxide reacts with hydrochloric acid, sodium chloride is formed.

Did you know?

The word alkali comes from the Arabic al kali.

The Arabs were the first proper chemists.

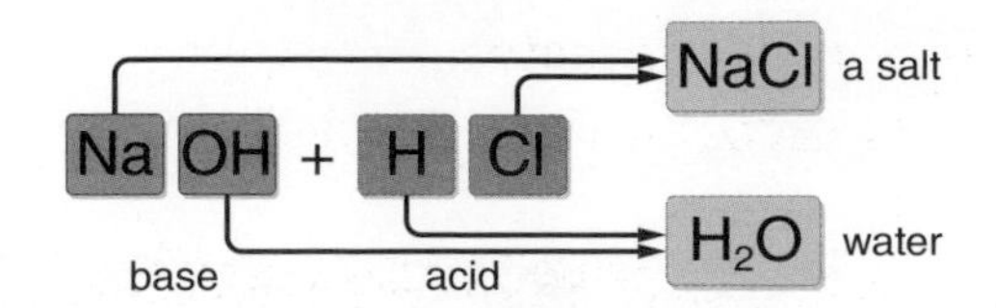

FIGURE 6: The reaction of sodium hydroxide with hydrochloric acid produces sodium chloride.

balanced symbol equation

To work out the name of a salt, look at the base and acid it was made from. Salt names have two parts, the first part from the base and the second part from the acid.

For example, when sodium hydroxide reacts with nitric acid, the salt formed is sodium nitrate (Figure 7).

sodium	nitrate
from the base	from the acid

FIGURE 7: The name of the salt produced when sodium hydroxide reacts with nitric acid.

Examples of salt names are:

> nitrates come from nitric acid

> chlorides come from hydrochloric acid

> sulfates come from sulfuric acid

> phosphates come from phosphoric acid.

Did you know?

Acids are easy to find in nature, they are in fruit. Alkalis are hard to find. They are in the ashes from wood fires.

Questions

8 What salt is formed when sulfuric acid reacts with copper oxide?

9 What salt is formed when nitric acid reacts with potassium hydroxide?

10 What salt is formed when hydrochloric acid reacts with magnesium carbonate?

11 What salt is formed when nitric acid reacts with calcium carbonate?

More on acids and bases

Some common acids and bases are shown in the table.

Acid	Formula
hydrochloric acid	HCl
nitric acid	HNO_3
sulfuric acid	H_2SO_4
Base	
sodium hydroxide	$NaOH$
potassium hydroxide	KOH
copper II oxide	CuO
ammonia	NH_3
Carbonate	
sodium carbonate	Na_2CO_3
calcium carbonate	$CaCO_3$

Equations

The following are the **balanced symbol equations** for the reactions between some of the acids and bases listed above.

Acid plus base (metal hydroxide)

$HCl + NaOH \rightarrow NaCl + H_2O$

$H_2SO_4 + 2KOH \rightarrow K_2SO_4 + 2H_2O$

Acid plus base (metal oxide)

$2HCl + CuO \rightarrow CuCl_2 + H_2O$

$2HNO_3 + CuO \rightarrow Cu(NO_3)_2 + H_2O$

Acid plus carbonate

$2HCl + CaCO_3 \rightarrow CaCl_2 + H_2O + CO_2$

Did you know?

It isn't just metal compounds that react with acids. Often, the metal element reacts as well. This is a different type of reaction; the metal does not act as a base.

Questions

Write a balanced symbol equation for each of the following reactions:

12 HNO_3 and KOH.

13 H_2SO_4 and NH_3.

Fertilisers and crop yield

You will find out:

- > about fertilisers and what they do
- > about eutrophication
- > how to calculate the percentages of elements in a fertiliser

Fertilisers make plants grow

Plants make most of their food from water and carbon dioxide in the air. They make it by photosynthesising.

They only need small amounts of other elements, which their roots take from the soil.

Fertilisers give plants elements that may be in short supply in the soil.

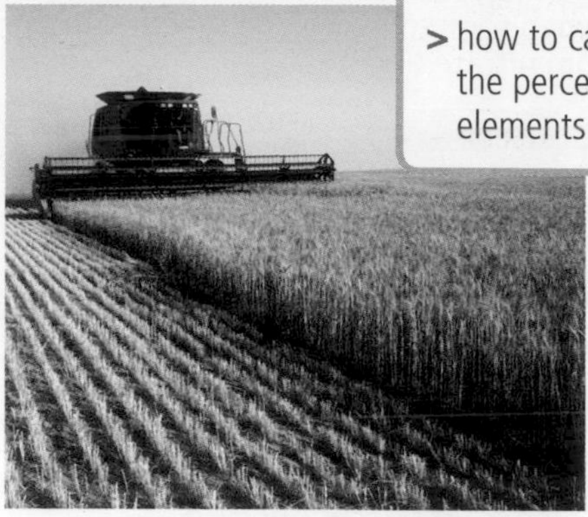

FIGURE 1: Fertilisers increase crop yields.

Fertilisers

Carbon dioxide and water do not give a plant all the elements that it needs. The plant does still need tiny amounts of other elements. It gets these essential elements by taking in minerals through its roots.

Fertilisers increase crop yields. They help plants grow bigger and faster because they contain the **essential elements** plants need.

What are these essential elements? Three of the most important ones are:

> nitrogen, N

> phosphorus, P

> potassium, K.

Fertilisers that contain these elements are often called NPK fertilisers. The formula of a fertiliser gives the essential elements it contains. Just look for the symbols N, P and K. Fertilisers benefit us by increasing the food supply. Farmers do need to be careful because if fertilisers run off the land into streams or rivers they can kill the aquatic organisms – the plants and animals that live in the water. This is called **eutrophication**.

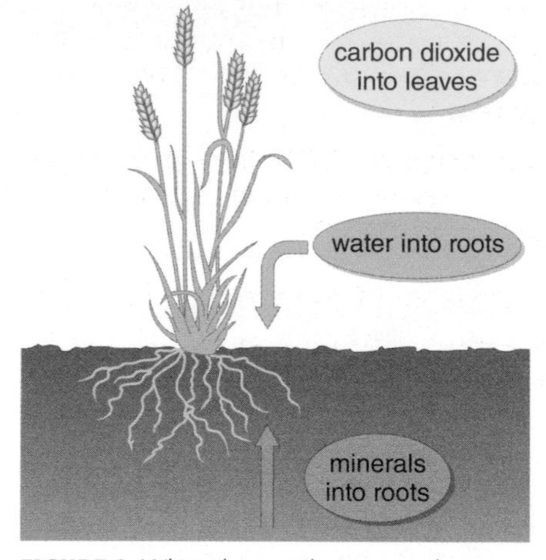

FIGURE 2: What three substances does a plant need to grow?

FIGURE 3: How much N, P and K is there in these fertilisers?

Questions

1 Name the essential elements in potassium nitrate, KNO_3.

2 Name the essential elements in ammonium phosphate, $(NH_4)_3PO_4$.

More on fertilisers

Farmers use fertilisers to increase their crop yields. This gives them more grains of wheat, larger grains of wheat, or both.

Even though plants need certain essential elements, it is no use putting them into the soil as pure elements. They have to be turned into chemical compounds first. The plants absorb these compounds through their roots. They can only do this if the compounds dissolve in water. Most fertilisers dissolve easily, but some dissolve slowly – so that the chemicals are let out into the soil over a long time period.

The world population is increasing and there is a need to produce more food. However, if too much fertiliser is used or if it is applied incorrectly it runs off into rivers, and so pollutes water supplies and causes eutrophication.

Question

3 What must happen to fertilisers in the soil before they can be absorbed by plants?

Why farmers use fertilisers

Plants normally grow well in ordinary soil. The chemicals that the plants need are already dissolved in the soil water. However, a farmer's crop may well need more dissolved minerals than are present in the soil. When crops are grown in a field year after year the minerals are gradually removed by the plants. It takes time for more of the essential elements to dissolve. The farmer uses fertiliser to put back the missing essential elements into the soil quickly.

Nitrogen is especially important to plants. Proteins are essential for growth, and protein molecules all contain nitrogen atoms. If a plant gets more nitrogen it makes more protein, so it can grow more.

Fertilisers in the wrong place – eutrophication

Fertilisers must be applied carefully. Rain water dissolves fertilisers which run off into rivers and lakes. When the level of fertilisers in the water rises too high, eutrophication occurs.

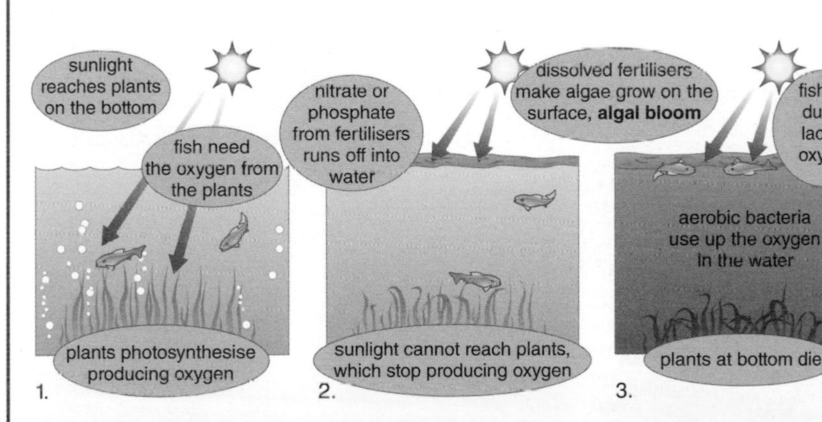

FIGURE 4: What causes eutrophication?

Questions

4 Suggest why ammonium nitrate might make a more useful fertiliser than sodium nitrate.

5 Why might farmers use potassium nitrate instead of ammonium nitrate as a fertiliser for their crops?

Making fertilisers

You will find out:

- how a fertiliser is made from an acid and an alkali
- the names of some nitrogenous fertilisers

Fertilisers can be made by neutralising alkalis with acids.
A fertiliser that contains nitrogen is called a **nitrogenous fertiliser**.

1. Use a measuring cylinder to pour alkali into a conical flask.

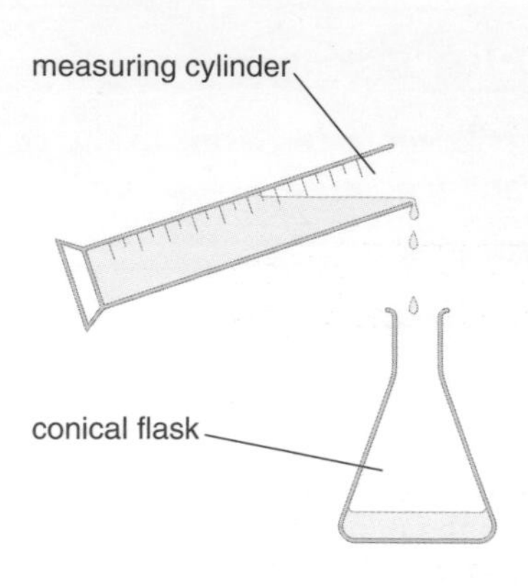

2. Add acid to the alkali until it is neutral.

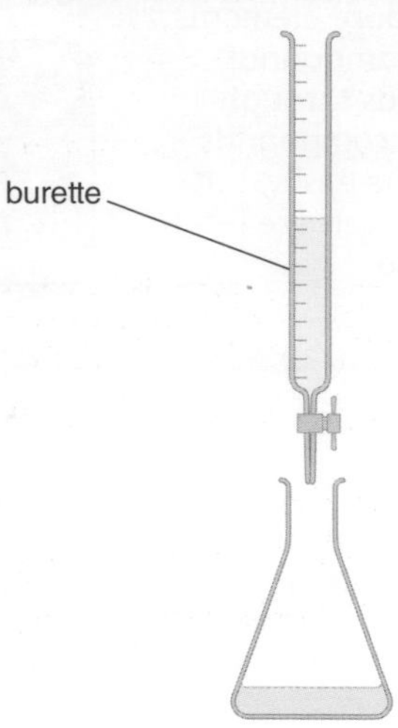

3. Evaporate.

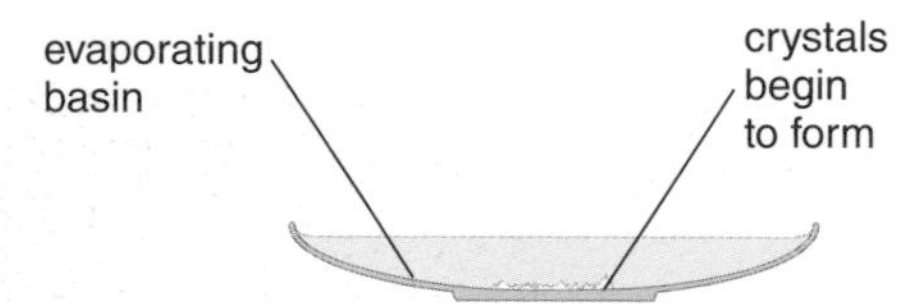

4. Filter off the crystals.

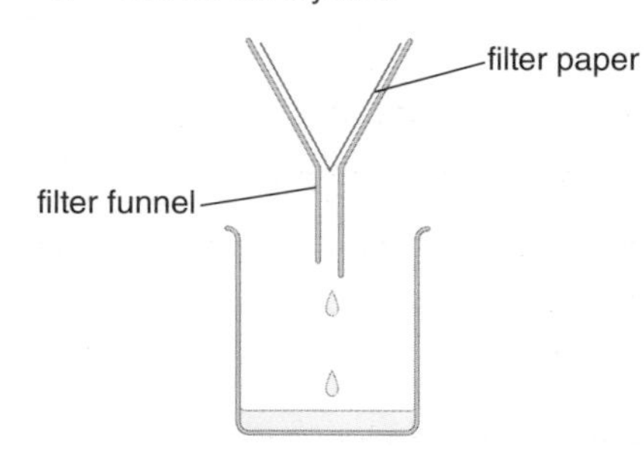

FIGURE 5: Making a fertiliser using a neutralisation reaction. Why is the solution filtered?

Here are the names of some nitrogenous fertilisers:

- ammonium nitrate
- ammonium phosphate
- ammonium sulfate
- urea.

Questions

6 What does a nitrogenous fertiliser contain?

7 Which of the following are nitrogenous fertilisers?

- ammonium sulfate
- potassium chlorate
- potassium nitrate
- sodium sulfate

8 Ammonium sulfate is a nitrogenous fertiliser. Which part of the name tells you that it contains nitrogen?

nitrogenous fertilisers   neutralisation fertilisers

More on making fertilisers

Many fertilisers are salts, so they can be made by reacting acids with alkalis.

acid	alkali	fertiliser
nitric acid	potassium hydroxide	potassium nitrate
nitric acid	ammonia	ammonium nitrate
sulfuric acid	ammonia	ammonium sulfate
phosphoric acid	ammonia	ammonium phosphate

Questions

9 Which acid and base react to make ammonium sulfate?

10 Which acid and base react to make potassium nitrate?

11 What substance is produced when nitric acid reacts with ammonia?

12 What substance is produced when phosphoric acid reacts with ammonia?

Making ammonium sulfate in the laboratory

React sulfuric acid (acid) with ammonia (base).

sulfuric acid + ammonia → ammonium sulfate

To find out exactly how much to use, a titration is carried out before mixing the main batch of chemicals.

Method

> Titrate the alkali with the acid, using an indicator.

> Repeat the titration until consistent results are obtained.

The acid and alkali have now reacted completely. This makes a neutral solution of ammonium sulfate fertiliser, but the fertiliser is contaminated with indicator.

> Use the titration result to add the correct amounts of acid and alkali together without the indicator.

> The fertiliser you have made is dissolved in water, so evaporate most of the water using a hot water bath.

> Leave the remaining solution to crystallise.

> Filter off the crystals.

FIGURE 6: Carrying out a titration in the laboratory. How can the end point of a titration be detected?

Question

13 Suggest how a solid sample of ammonium phosphate could be made. Outline all the main stages.

Chemicals from the sea: the chemistry of sodium chloride

You will find out:
> why salt is important
> how salt is mined

Salt of the Earth

Sodium chloride is the chemical name for common salt. Salt mines in the UK produce 6 million tonnes of it every year. Only a fraction of this is used for food.

FIGURE 1: Very little of the salt produced in this country is used for food.

Where does it come from?

Salt is a raw material

> Salt is used in food as a **preservative** and to give flavour.

> Large amounts of rock salt are scattered on roads every winter.

> Even more salt is purified and used where you do not see it – it is used as a raw material in the chemical industry to make other chemicals. The UK's salt-producing region is also a centre for the chemical industry.

FIGURE 2: Salt can be mined from underground salt deposits.

Questions

1 Give the chemical name for common salt.

2 State three ways in which salt is used.

FIGURE 3: In hot countries it is easy to get salt from the sea.

salt mining in Cheshire   salt evaporation ponds

Mining and subsidence

Rock formations containing salt deposits exist in several places in the UK; Cheshire is the most famous, and has the main salt mines.

> Salt can be extracted from rock by using underground cutting machines. The salt mined in this way is impure. It is called rock salt and is used for gritting roads.

> Another way of extracting the salt is to drill a borehole into the salt layer and pump water down. The salt dissolves and can be pumped back up to the surface. The water can then be evaporated to leave solid table salt.

The nearby chemical industries do not need solid salt; they use salt solution, so the liquid coming out of the mine can be sent in tankers direct to where it is needed.

Once the salt is removed from a mine, the ground may subside. **Subsidence** used to be a major problem. Some houses in these areas had steel frames to protect them if the ground started to move. Mining is now much more carefully controlled.

FIGURE 4: A cutting machine in a salt mine. What is impure salt used for?

FIGURE 5: Evaporating water from salt solution. What is the solid salt produced used for?

FIGURE 6: Subsidence used to be a major problem around mines. What is subsidence?

Did you know?

Roman soldiers were given money so that they could buy salt. This is where the word 'salary' comes from.

Questions

3 Suggest why chemical industry often grows up near salt mines.

4 It is cheaper for salt-solution mines not to have to evaporate the salt. Why?

5 What is the main disadvantage of rock salt?

6 Explain what causes ground to subside in mining areas.

rock salt uses   solution mining salt

Electrolysis of sodium chloride solution

You will find out:
- about the electrolysis of salt solution
- a chemical test for chlorine
- why the electrolysis of salt solution is important

Salt is an important source of chlorine and sodium hydroxide. Household bleach, PVC and solvents are all made from substances derived from salt.

Test for chlorine

Chlorine gas is produced when concentrated salt solution or melted salt is electrolysed. A simple laboratory test for chlorine is to hold moist litmus paper in a test tube of the gas. The litmus paper turns white. Chlorine is a bleach – it bleaches the colour of the indicator.

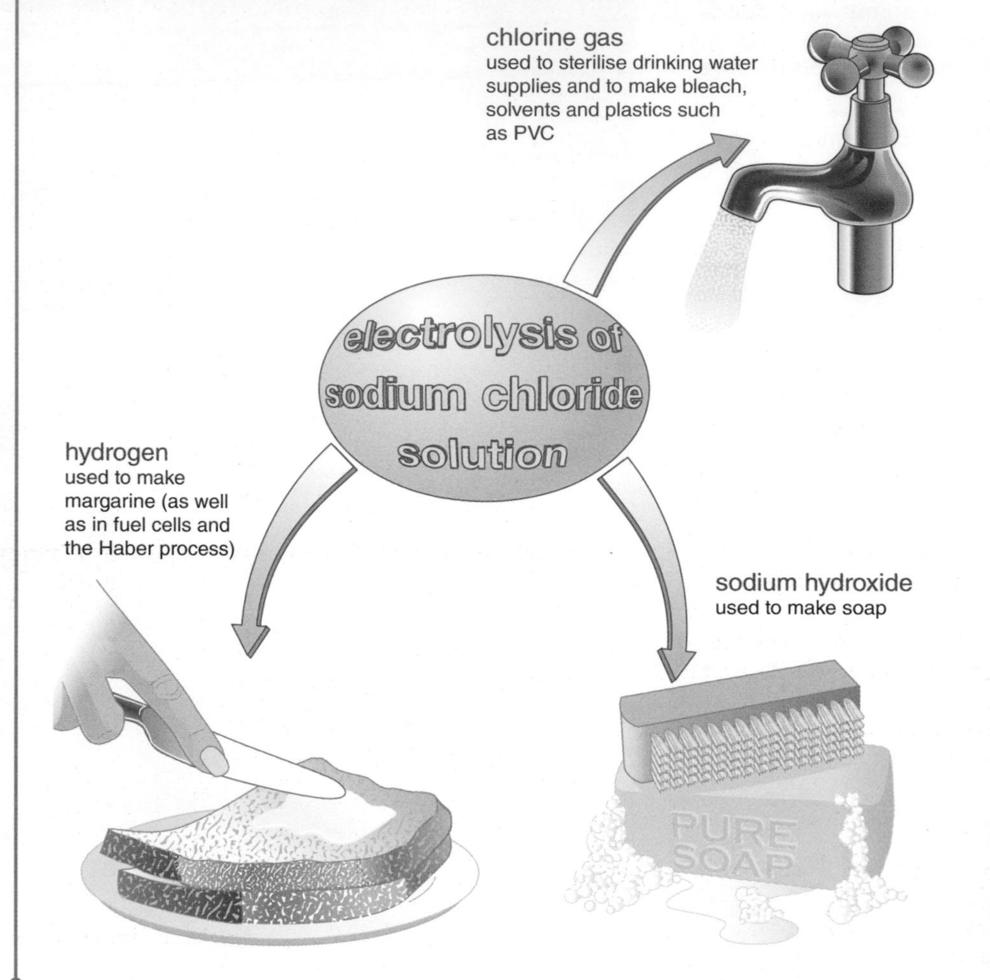

FIGURE 7: Electrolysis of a concentrated sodium chloride solution gives three products.

Questions

7 State one use of hydrogen.

8 State one use of sodium hydroxide.

9 Give the chemical test for chlorine.

Electrolysis of sodium chloride solution

Concentrated sodium chloride solution is also known as brine. Much of the brine produced by salt mines goes straight to different chemical works in the area to be electrolysed. The electrolysis discharges hydrogen from the water and chlorine from the sodium chloride, leaving sodium hydroxide in solution.

water + sodium chloride → hydrogen + chlorine + sodium hydroxide

Hydrogen and chlorine are reactive, so it is important to use inert electrodes so that the products don't react before they are even collected.

electrolysis of brine   uses of chlorine

In summary:

- the sodium chloride must be a concentrated solution in water
- the electrodes must be inert
- hydrogen is produced at the cathode
- chlorine is produced at the anode
- sodium hydroxide solution is formed in the cell.

Questions

10 During electrolysis of sodium chloride solution, what is formed at the anode?

11 Suggest what the sodium hydroxide produced might be contaminated with.

12 Suggest why the electrodes must be inert.

13 What two chemicals can be reacted together to make domestic bleach?

Bleach

Chlorine is a powerful bleach. It is also a highly poisonous gas, so it cannot be used by itself in the home. However, chlorine can be reacted with sodium hydroxide to make a much safer substance that releases chlorine easily when it is needed. Household bleaches containing chlorine are made in this way.

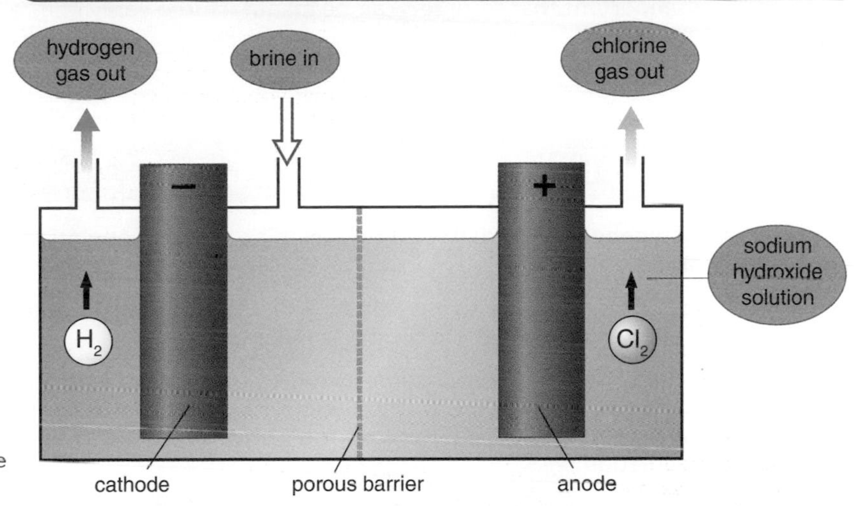

FIGURE 8: Why must the electrodes be inert?

More on solution electrolysis

Concentrated sodium chloride has two possible ions that can be liberated at the cathode: Na^+ ions from salt and H^+ ions from water. It is much easier to push electrons onto hydrogen ions than onto sodium ions, so hydrogen is discharged. The half reaction at the cathode is:

$2H^+ + 2e^- \longrightarrow H_2$

Gain of electrons is reduction.

The ions that could be liberated at the anode are Cl^- ions from salt and OH^- ions from water. It is easier to discharge the Cl^- ions, so chlorine gas is formed. The half reaction at the anode is:

$2Cl^- - 2e^- \longrightarrow Cl_2$

Loss of electrons is oxidation.

The Na^+ and OH^- ions are not discharged, they are left behind, making sodium hydroxide solution.

Chlorine is only formed at the anode if the solution is concentrated enough. If the solution is dilute, oxygen forms instead.

The chlor-alkali industry

Chlorine and sodium hydroxide are raw materials for making over half the chemicals that we need, from solvents, plastics and paints, and soaps to medicines and food additives.

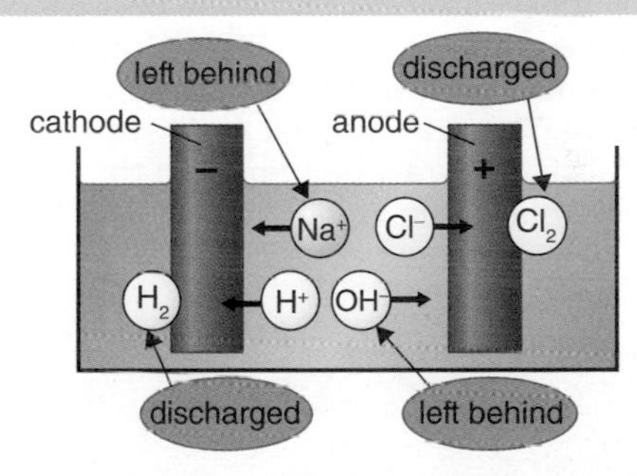

FIGURE 9: How ions move in electrolysis of sodium chloride.

Questions

14 Suggest why it is more expensive to produce chlorine by electrolysing molten sodium chloride than sodium chloride solution.

15 Which electrode gives the same reaction as it did with molten sodium chloride?

16 Write an equation to show how water splits into ions.

17 Explain why a barrier in the middle of the cell must be porous.

Preparing for assessment: Research and collecting secondary data

To achieve a good grade in science, you not only have to know and understand scientific ideas, but you need to be able to apply them to other situations and investigations. These tasks will support you in developing these skills.

Tasks

Why is it important that fertilisers can be manufactured easily and relatively cheaply?

Feeding the world

800 million people – one sixth of the developing world's population – suffer from hunger and the fear of starvation.

Food production will have to increase by 70% over the next 40 years to feed the world's growing population, the United Nations food agency predicts.

Many factors affect food production, including conflict and wars, climate, the availability and cost of seed, the type of land and pests. The climate in Australia is similar to Africa, but Australia can overcome problems by spending money on irrigation, fertilisers and pesticides, providing enough food for its population.

Wheat grows in a variety of climates, but it is best when not too hot and when it gets regularly watered. Wheat also needs three main minerals to grow well, and these are called NPK. The letters stand for the elements. Nitrogen is needed for green growth, potassium for root growth, and phosphorus for seed production – the part we use to make bread. Fertilisers contain NPK but chemical fertilisers are expensive. One of the main chemicals in fertilisers is ammonia. This is made by the Haber process, by reacting nitrogen from the air with hydrogen obtained from natural gas. An alternative is to use manure. This is an organic fertiliser, from animals or composted plants. Animal manure contains urine and faeces on straw bedding and this can be collected and spread onto fields. If intensive animal farming takes place, the liquid waste or slurry can be collected then spread.

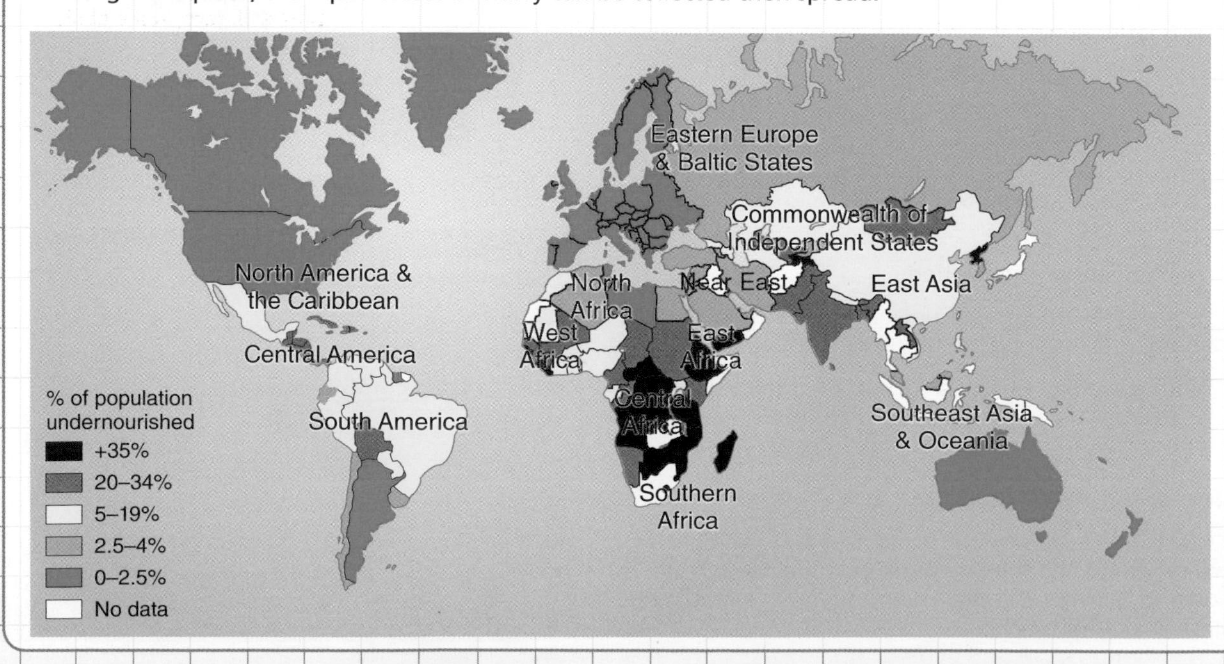

Amare is an Ethiopian farmer in East Africa. He grows wheat to make bread. To grow well, the wheat needs sunlight and water. Sunlight is not usually a problem in Africa, but if the temperature is too high the crop can suffer from heat stress. Rainfall is a bigger problem. There are two rainy seasons, a short one in March and a longer season between June and September. These seasons are not always reliable, and in some years they fail causing drought. The longer season is when the wheat is sown. It is harvested in October or November. Amare does not have any cattle, so he cannot put manure on his land, and he cannot afford to buy chemical fertilisers. His average yield is 1.4 tonnes/hectare. In the UK the average is 7.5 tonnes/hectare. Amare knows that growing his crop takes nutrients out of his land. Adding chemical fertilisers would provide these minerals, increase the yield, but Amare cannot afford to buy chemical fertilisers.

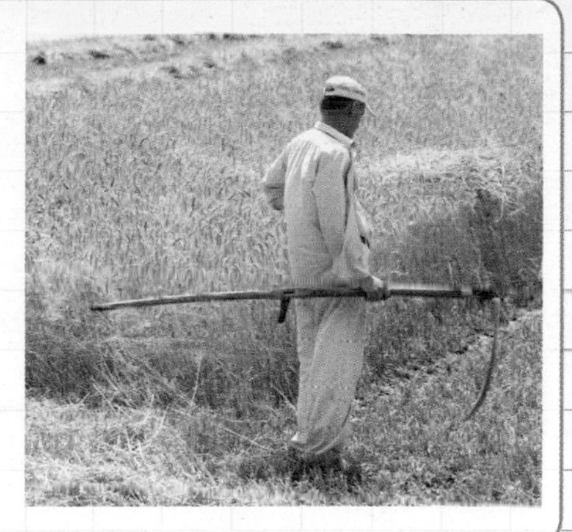

Research and collecting secondary data

Use the information provided and secondary data to answer these questions.

1. What are chemical fertilisers and how do they work?
2. What are the alternatives to chemical fertilisers?
3. Why are chemical fertilisers expensive?
4. What alternatives are there to chemical fertilisers?
5. How can you make a fertiliser in the laboratory?

Before you start, plan how you are going to collect this information. You need to:

> Write down how you found the information.

> Write down a list of all the sources of your information.

> Clearly present the information so it could be used to plan an actual investigation.

You can answer the questions using the information provided, but it is much better if you do your own research. Remember, there are secondary sources other than the internet.

Think carefully about the questions asked and make sure you answer the questions. What exactly do you have to research?

Remember to provide book title and author; newspaper title and date; full website address.

How are you going to record all this information?

General rules

1. You may work with other students but your written work should be done on your own.
2. You cannot get detailed help from your teacher.
3. You are not allowed to redraft your work.
4. Your work can be hand written or word processed.
5. It is expected that you complete this task in two hours.
6. You are allowed to do this research outside the laboratory.
7. You must be aware of and mention any health and safety issues.

C2 Checklist

To achieve your forecast grade in the exam you'll need to revise

Use this checklist to see what you can do now. It gives you many of the important points you will need to know. Refer back to the relevant pages in this book if you're not sure and to see if there is anything else you need to know. Look across the three columns to see how you can progress.

Remember you'll need to be able to use these ideas in various ways, such as:

> interpreting pictures, diagrams and graphs
> applying ideas to new situations
> explaining ethical implications
> suggesting some benefits and risks to society
> drawing conclusions from evidence you've been given.

Look at pages 278–299 for more information about exams and how you'll be assessed.

To aim for a grade E	To aim for a grade C	To aim for a grade A
describe the structure of the Earth – rocky crust, mantle and iron core **understand** most scientists accept that very slow plate movements over millions of years result in volcanoes and earthquakes **describe** how crystal size in magma is linked to cooling time **understand** volcanoes can release runny lava slowly, or thick lava violently	**describe** the lithosphere as the outer mantle and crust, composed of tectonic plates less dense than the mantle **explain** how seismic waves are used to study the Earth's structure, and why the evidence is widely accepted **understand** the type of volcanic eruption is linked to the composition of the lava **explain** geologists study volcanoes to forecast eruptions better and to reveal information about the Earth's structure	**link** the properties of the mantle to how it moves **describe** the development of plate tectonic theories, in terms of energy transfers, convection currents and subduction zones **explain** how igneous rocks are formed – 'safe' runny iron rich basalt and thick silica rich rhyolite from explosive eruptions **explain** why scientists are better able to predict eruptions
recall the main rocks used for construction and problems of mining or quarrying **recall** limestone and marble are calcium carbonate, limestone thermally decomposes into calcium oxide and carbon dioxide **describe** how concrete is made and reinforced	**relate** the main construction materials to substances found in the Earth's crust **know** how to compare hardness, i.e. limestone is softer than marble which is softer than granite **construct** word and balanced symbol equations to describe the thermal decomposition of limestone **recall** how cement is made	**explain** why granite, marble and limestone have different hardnesses **construct** balanced symbol equations (formula not given) for the decomposition of limestone **explain** why reinforced concrete is a better construction material
understand that copper can be extracted from its ore by heating with carbon, and that removal of oxygen is reduction **recall** alloys are mixtures containing one or more metals, and give one use for amalgams, brass and solder **recognise** alloys have different properties from their metals, and be able to interpret data on the main properties	**label** the apparatus needed for electrolysis **explain** the advantages and disadvantages of recycling copper **recall** the main metals in amalgams, brass and solder. **explain** why metal or alloy are suited to given use	**describe** how electrolysis is used to purify copper, and be able to give ionic half equations to explain oxidation and reduction **evaluate** the suitability of metals for different uses when given data **explain** how the use of 'smart alloys' like nitinol for spectacle frames have increased alloy applications

To aim for a grade E	To aim for a grade C	To aim for a grade A
recall rusting needs iron, water and oxygen and that adding oxygen is oxidation **compare** the properties of iron and aluminium **recall** the main materials needed to make a car	**understand** how salt water and acid rain affect rusting **understand** rusting is oxidation and construct the word equation **explain** why aluminium does not corrode **describe** advantages and disadvantages of building cars from aluminium and steel	**explain** the advantages of aluminium in cars: better fuel economy and longer lifetime.
recall the Haber process makes ammonia from the air and from hydrogen that comes from cracking oil or gas **describe** the factors that add to the cost of making a new substance **understand** that reversible reactions proceed in both directions	**describe** the conditions needed to make ammonia in the Haber process and construct the balanced symbol equation. **interpret** data about percentage yield in reversible reactions, and recognise the importance of ammonia in world food production	**explain** why the conditions used in the Haber process are needed **explain** the economic considerations in manufacturing ammonia **interpret** data about rate, percentage yield and costs for alternative industrial processes
describe how universal indicator can be used to estimate pH levels **recall** an alkali is a soluble base **understand** that an acid can be neutralised by an alkali	**recall** that in neutralisation, acid + base → salt + water **recall** that all acid solutions contain H^+ ions, and that pH is determined by concentration of H^+ ions **explain** how metal oxides and hydroxides, and carbonates react with acid, and construct word equations. Predict the names of salts from laboratory acids	**explain** neutralisation in terms of $H^+ + OH^- \rightarrow H_2O$ **construct** balanced equations for the neutralisation of common acids by bases and carbonates
recall nitrogen, phosphorus and potassium are the three essential minerals plants need **understand** the benefits and problems of using fertilisers **identify** the apparatus needed to prepare a fertiliser by neutralising an acid with an alkali	**explain** why fertilisers need to be soluble to be absorbed by plants **identify** arguments for and against using fertilisers **predict** the names of the acids and alkalis needed to make different fertilisers	**explain** how fertilisers increase crop yield in terms of providing and replacing essential elements **explain** the process of eutrophication **describe** in detail the preparation of a named synthetic fertiliser by the reactions of an acid and an alkali
recall that sodium chloride is an important raw material obtained from the sea or from buried salt deposits **recall** electrolysis of salt solution gives chlorine and hydrogen, and recall chlorine bleaches moist litmus paper **recall** sodium chloride is used as a preservative and as a flavouring **recall** uses for chlorine, hydrogen and sodium hydroxide	**describe** how salt can be mined as rock salt if extracted by solution mining **explain** how mining can cause subsidence **recall** the products of brine electrolysis **explain** the need for inert electrodes **describe** how household bleach is made by reacting sodium hydroxide and chlorine	**explain** the products of brine electrolysis using a balanced equation, and give the ionic equations at the anode and cathode **explain** why the electrolysis of sodium chloride involves both oxidation and reduction **explain** the economic importance of the chlor-alkali industry

C2 Exam-style questions

Foundation Tier

1 The diagram shows the structure of the Earth.

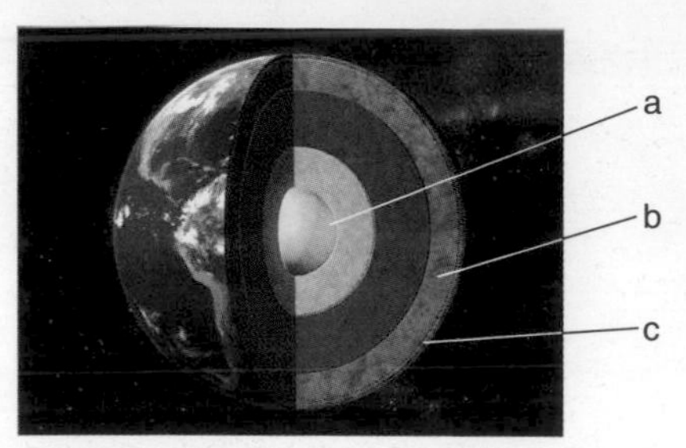

AO1 **(a)** Which letter represents:
(i) the Earth's core? [1]
(ii) magma? [1]

AO1 **(b)** We live on the Earth's thin crust. What is it made from? [1]

AO1 **(c)** Suggest the risks and benefits of living near to a volcano. [2]

AO1 **(d)** What is the difference between magma and lava? [2]

AO1 **(e)** Granite is a rock with large crystals. Describe how these large crystals form. [4]

[Total: 11]

AO1 **2 (a)** Copper oxide can be changed into copper by heating it with carbon. What is this method called? [1]

AO1 **(b)** Give one use of the following alloys:
(i) solder [1]
(ii) amalgam [1]

AO1 **(c)** Suggest **two** reasons why recycling copper is cheaper than extracting it from its ore. [2]

[Total: 5]

AO1 **3 (a)** Name the **two** main raw materials needed to make ammonia. [2]

AO2 **(b)** State **three** factors that affect the cost of making a new chemical. [3]

AO1 **(c)** Name **two** essential elements needed for plant growth. [2]

AO1 **(d)** Which part of a plant absorbs these essential elements? [1]

[Total: 8]

4 The table shows the colours that four different indicators turn.

Indicator	In pH 2	In pH 5	In pH 7	In pH 10
Litmus	red	red	purple	blue
Methyl orange	red	orange	orange	orange
Phenol-phthalein	milky	milky	milky	pink
Universal	red	orange	green	blue

AO2 **(a)** Which indicators turn orange in weak acids? [1]

AO2 **(b)** Which indicators are useful for finding a neutral solution? [1]

AO3 **(c)** Many people believe universal is the most useful indicator. Suggest why. [2]

[Total: 4]

5 Car bodies rust if the paint gets scratched.

AO1 **(a)** Name **two** substances needed for iron to rust. [2]

AO1 **(b)** What type of chemical reaction is rusting? [1]

AO1 **(c)** Aluminium cars do not rust. Explain why. [2]

AO1 **(d)** Suggest why rust-preventing paint only works if all the rusty flakes are removed first.
✎ The quality of written communication will be assessed in your answer to this question. [6]

[Total: 11]

AO1 recall the science AO2 apply your knowledge AO3 evaluate and analyse the evidence

Worked Example – Foundation Tier

Cars are made from many different materials. The table shows some properties of aluminium and iron.

property	aluminium	iron
density in g/cm^3	2.7	7.8
corrodes	no	yes
melting point in °C	660	1538
cost / tonne in £	1466	321

a Use the table to suggest the advantages and disadvantages of making a car body from aluminium. [6]

It's light weight and not corrosive.

It's got a lower melting point.

Aluminium is almost 5 times more expensive than iron.

b Alloys are often used in cars. Describe what the word alloy means, and give an example. [2]

Alloys are metals.

Steel is a strong alloy.

How to raise your grade!

Take note of these comments – they will help you to raise your grade.

These longer 6-mark answers usually have marks awarded for the Quality of Written Communication shown by the symbol ✏ so answers need planning, and care is needed with spelling, punctuation and grammar. The answer uses the information, but the first two lines only repeat it. It would be better if the answer was set out as advantages and disadvantages. To improve the answer, add reasons, such as linking lighter weight to fuel economy. Corrosion could be linked to a longer lifetime. Lower melting points may be a disadvantage if close to the engine. 4/6

There is not enough here. It is important to learn the key definitions for exams. 1/2

This student has scored 5 marks out of a possible 8. This is below the standard of Grade C. With more care the student could have achieved a Grade C.

C2 Exam-style questions

Higher Tier

1 The surface of the Earth's lithosphere is composed of slowly moving tectonic plates.

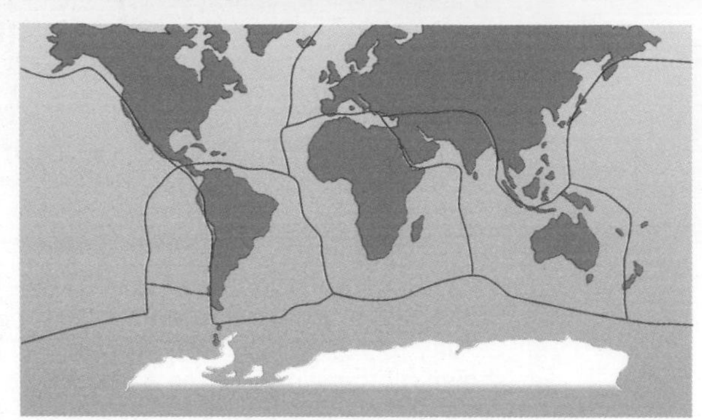

AO1 **(a)** Describe the evidence for Alfred Wegner's continental drift theory and how it explains mountain formation. ✐ The quality of written communication will be assessed in your answer to this question. [6]

AO2 **(b)** In 1914, few scientists supported Wegner. Why do most scientists now support his theory? [2]

[Total: 8]

2 Copper can be purified by electrolysis.

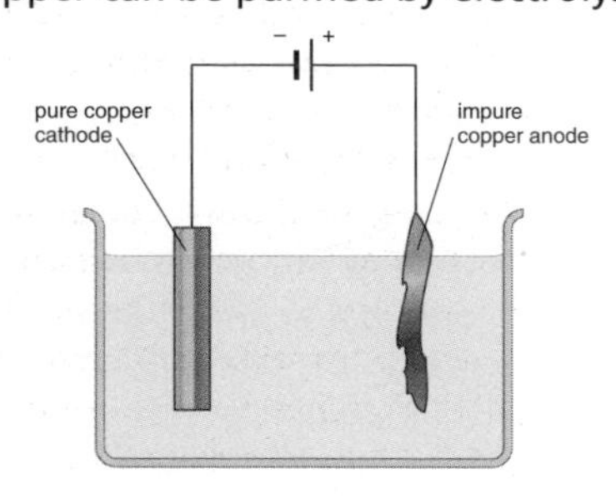

AO1 **(a)** Give an advantage and a disadvantage of recycling copper. [2]

AO1 **(b)** Using electron half equations, explain why the electrical purification of copper involves both an oxidation and a reduction process. ✐ The quality of written communication will be assessed in your answer to this question. [6]

[Total: 8]

3 Limestone is calcium carbonate. It has the chemical formula $CaCO_3$. Limestone can also be decomposed by heating into calcium oxide and carbon dioxide.

AO1 **(a)** Construct a balanced equation to show the decomposition of calcium carbonate. [1]

AO1 **(b)** Concrete is made by mixing cement, salt, small stones and water. This table shows some information on different concrete mixtures. The same amount of small stones and water was added each time.

Sample	Sand	Cement	Strength (N)
1	40	34	42
2	41	33	29
3	43	22	21
4	45	28	26
5	60	50	42

(i) Explain why sample 1 and 5 give the same strength. [1]

(ii) Plot a graph to show the amount of cement affects the strength. [3]

(iii) Describe the relationship shown by the graph. [1]

(iv) Concrete can be reinforced by pouring it into a steel cage. Explain why this makes a better building material. [2]

[Total: 8]

AO1 **4 (a)** Describe **two** ways salt can be extracted. [2]

AO1 **(b)** In the electrolysis of concentrated salt solution, what is made at the anode, at the cathode and left in solution? [3]

AO1 **(c)** Describe why inert electrodes are needed for this electrolysis. [2]

AO2 **(d)** One of the products can be used to make hydrochloric acid.

(i) What is made when hydrochloric acid reacts with sodium carbonate? [2]

(ii) Construct a balanced symbol equation for this reaction. [2]

[Total: 11]

AO1 recall the science AO2 apply your knowledge AO3 evaluate and analyse the evidence

Worked Example – Higher Tier

The Haber process makes ammonia, which is used to manufacture fertilisers.

(a) Give the balanced symbol equation for the manufacture of ammonia. [2]

$N_2 + H_2 \rightarrow NH_3$

(b) Describe three conditions needed to produce a high yield. [3]

High pressure

Increase the temperature

Use a catalyst

(c) Fertilisers can create problems like eutrophication. Explain the process of eutrophication. [6]

Fertilisers can get washed in ponds.

They make bacteria grow, using up all the oxygen, so things rot.

The light is blocked by algae growth.

Fish die.

How to raise your grade!

Take note of these comments – they will help you to raise your grade.

Symbols and equation correct, but not balanced. 1/2

High pressure is correct but the figure would be a useful addition – 200 atm. Increased temperature needs a value for the mark –450 °C is the optimum. Catalyst is mentioned, but may need to name it as iron for the mark. 2/3

These longer 6-mark answers usually have marks awarded for the Quality of Written Communication shown by this symbol ✐ so answers need planning, and care is needed with spelling, punctuation and grammar. For the most part the information is relevant and presented in a structured and coherent format. Specialist terms are used for the most part appropriately. There are occasional errors in grammar, punctuation and spelling. 4/6

This student has scored 7 marks out of a possible 11. This is below the standard of Grade A. With more care the student could have achieved a Grade A.

P1 Energy for the home

Ideas you've met before

Temperature is a measure of how hot things are

Temperature is measured on the Celsius scale using a thermometer.

Water freezes at 0 °C and boils at 100 °C.

There are many different types of thermometer.

Energy is measured in joules.

Suggest a typical value for room temperature.

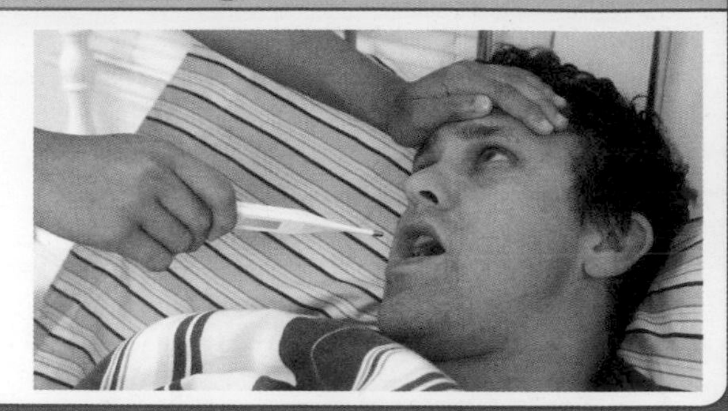

Why fuels are useful

Fuels are substances which burn to release energy.

Heat is a form of energy.

Fossil fuels have been formed over many millions of years.

Supplies of fossil fuels are running out and they must be conserved.

Name two examples of fossil fuels.

Energy transfer

Heat flows as a result of a difference in temperature.

Metals are good conductors and non-metals are good insulators.

Energy can be transferred by conduction, convection and radiation.

Solids, liquids and gases can change state when energy is added or removed.

Energy transfers can be represented by flow diagrams (Sankey diagrams).

State one example of a good conductor and one example of a good insulator.

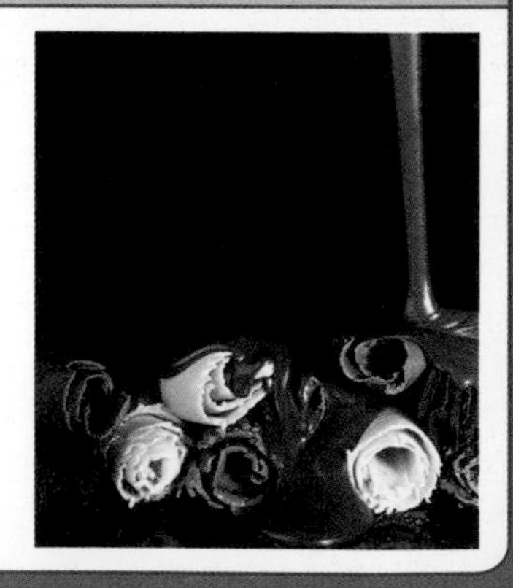

Behaviour of light

Light travels in straight lines.

Light may be absorbed, transmitted or reflected when it hits an object.

Light can change direction at a boundary between two different media.

White light can be dispersed to give a range of seven different colours.

List the colours of the spectrum in order.

In P1 you will find out about...

- the difference between heat and temperature
- the energy needed to change the temperature of a body
- the energy needed to change the state of a body

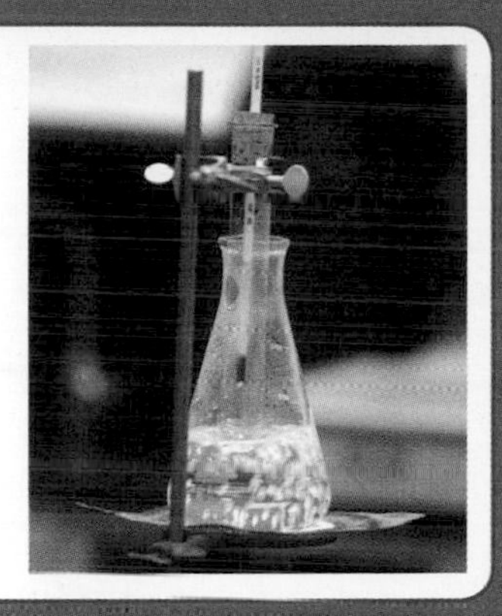

- the use of infrared radiation for cooking, data transmission and heat sensors
- the use of microwave radiation for cooking and communication
- how radio signals can be either analogue or digital
- the possible dangers to health from overuse of mobile phones

- the causes of suntan, sunburn and skin cancer
- how the effects of ultraviolet radiation can be reduced
- how ozone protects the Earth from the effects of ultraviolet radiation
- the ozone layer getting thinner due to CFCs

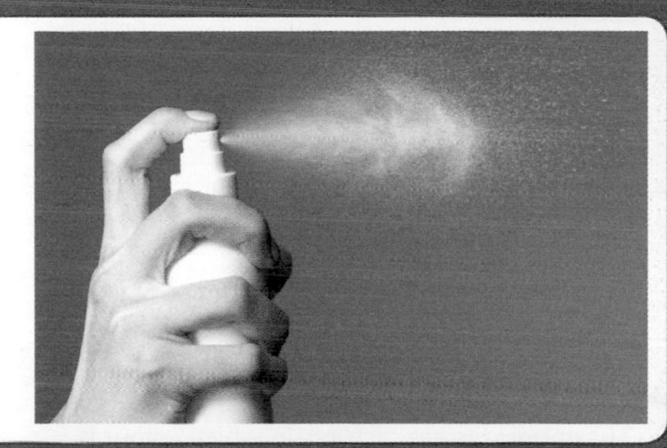

- wave frequency, wavelength and wave speed
- the electromagnetic spectrum as a family of waves
- waves showing reflection, refraction, diffraction and interference
- total internal reflection of waves
- laser light as a narrow beam with a single frequency

Heating houses

You will find out:

> about the difference between heat and temperature
> how energy flows
> how to read a thermogram

Heat transfer

If you can manage to keep a piece of chocolate on your tongue long enough for it to melt, your tongue gets colder.

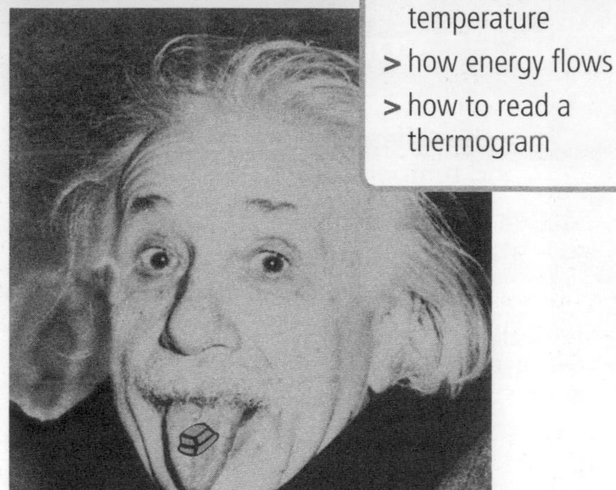

FIGURE 1: Do you think Einstein would have learned anything from this experiment?

Hot and cold

Hot objects have a high **temperature** and usually cool down. Temperature is a measure of how hot something is.

An object that is 50 °C above room temperature cools down at a faster rate than an object that is only 5 °C above room temperature. Figure 3 shows a typical cooling curve. Eventually, the objects reach the temperature of their surroundings.

Cold objects have a low temperature and usually warm up.

Cold solid objects may melt. Eventually, they reach the temperature of their surroundings.

In the laboratory, temperature is normally measured with a thermometer.

About heat

> The unit of temperature is the degree Celsius (°C).

> Heat is a measurement of **energy**.

> The unit of energy is the **joule** (J).

Temperature pictures

Thermograms are pictures in which colour is used to represent temperature.

Questions

1 What is the unit of temperature?

2 What is the unit of heat?

3 Mrs Collins takes a loaf of bread out of the freezer an hour before tea. Why does she take the bread out so early?

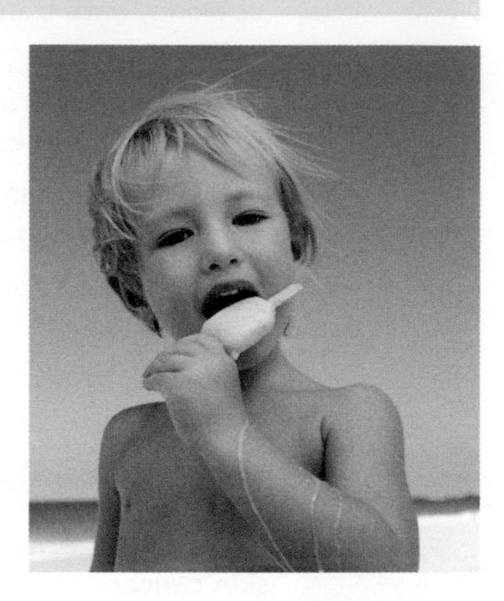

FIGURE 2: Why does the ice cream melt?

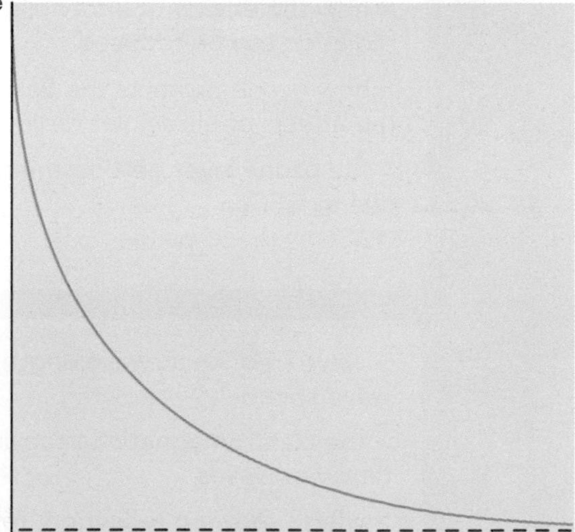

FIGURE 3: What happens to the slope of the graph as the temperature falls?

thermogram thermography

Energy flow

> If you leave the front door open on a cold winter's day, the house gets colder.

> If you put an ice cube on your hand, your hand gets colder.

Energy, in the form of heat, flows from a warmer to a colder body.

Inside the house is warmer than outside, and your hand is warmer than the ice cube. Energy flows from the inside of the house to the outside and from your hand to the ice cube.

When energy flows away from a warm object, the temperature of that object decreases. The house and your hand become colder. Figure 3 shows how the temperature of something hot changes as it cools to the temperature of its surroundings.

Temperature in a thermogram is measured using colour as the scale.

FIGURE 5: In a thermogram, white, yellow and red represent the hottest areas. Black, dark blue and purple represent the coldest areas.

FIGURE 4: Why does the temperature of a house fall when the front door is left open on a cold day?

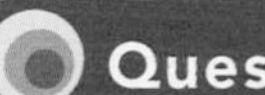

Questions

4 When you blow warm air onto a wet patch of skin, your skin cools down. Suggest why.

5 Nitrogen freezes at –210 °C and boils at –196 °C. When a lump of ice at –5 °C is dropped into liquid nitrogen, the nitrogen boils. Explain why.

6 The bonnet of one car is red in the thermogram. Suggest why.

Measuring temperature

Temperature is a measure of 'hotness'. It allows one object to be compared with another.

Temperature does not have to be measured on the degree Celsius scale. On a cold winter's day you do not use a thermometer to help you decide what to wear.

Temperature can be measured on an arbitrary or chosen scale.

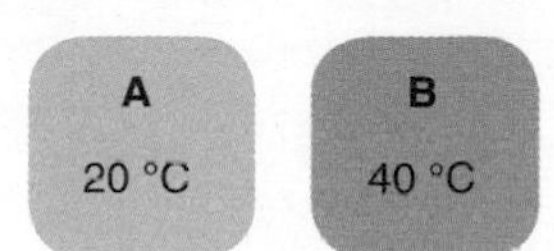

FIGURE 6: B is not twice as hot as A.

The particles in a substance are constantly vibrating. When the substance is heated and its temperature increases, the particles vibrate faster. The average kinetic energy increases. Temperature is a measure of the average kinetic energy of the particles.

Heat is a measure of the total internal energy of an object. It is measured on an absolute scale.

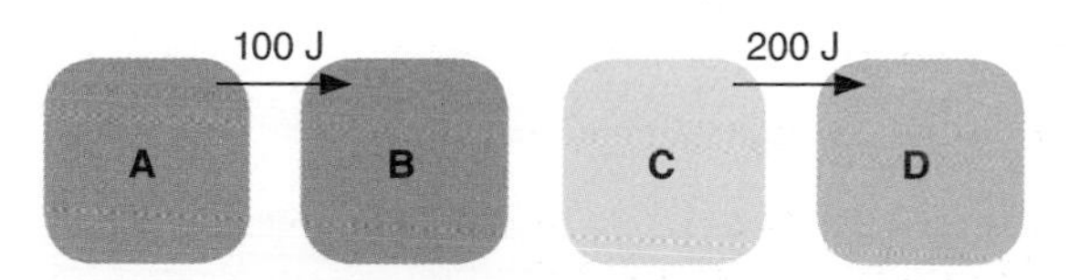

FIGURE 7: Twice as much energy is transferred from C to D as from A to B.

Question

7 How does the thermogram in Figure 5 show that the roof of one of the houses is not very well insulated?

Does heating mean hotter?

You will find out:

> about factors that affect the amount of energy needed to change the temperature of a substance

> about factors that affect the amount of energy needed to change the state of a substance

A liquid in a beaker is heated to a certain temperature. The amount of energy needed depends on:

- > how much liquid there is in the beaker (its mass)
- > what the liquid is
- > the rise in temperature.

The energy needed to change the temperature of the liquid could be measured by using an electrical immersion heater connected to a joulemeter.

FIGURE 8: In a second experiment more liquid was added to the beaker and the liquid was heated to the same temperature. Would the amount of energy needed to heat the liquid be different from that needed in the first experiment?

Changes of state

If you heat a solid, such as a lump of ice, its temperature rises until it changes to a liquid. The temperature then stays the same until all of the solid has changed to a liquid.

The temperature of the liquid then rises until it changes into a gas. The temperature then stays the same until all of the liquid has changed to a gas.

The temperature of the gas then rises.

- > Heat is needed to **melt** a solid at its **melting point**.
- > Heat is needed to **boil** a liquid at its **boiling point**.

Latent heat is the heat needed to change a state without a change in temperature.

Remember!

Heat and temperature are different.
Energy transfer does not always involve a rise or fall in temperature.

the solid is melting
the liquid is boiling
temperature
time

FIGURE 9: Why does the line on the graph flatten out during changes of state?

Did you know?

The syrup in a steamed syrup pudding appears to be hotter than the sponge even though they have been cooked at the same temperature. This is because the specific heat capacity of the syrup is greater than the specific heat capacity of the sponge.

Questions

8 What is the melting point of ice?

9 What is the boiling point of water?

latent heat everyday applications

Specific heat capacity

When an object is heated and its temperature rises, energy is transferred.

The amount of energy needed to change the temperature of an object depends on the material the object is made from. All substances have a property called **specific heat capacity**. The specific heat capacity is:

> the amount of energy needed to change the temperature of one kilogram of a substance by one degree Celsius
>
> the specific heat capacity unit is joule per kilogram degree Celsius (J/kg °C)

Specific latent heat

The amount of energy needed to change the state of an object without a change in temperature depends on the material the object is made from. This means that energy is transferred even though there is no temperature change when things boil, condense, melt or freeze.

All substances have a property called **specific latent heat**. The specific latent heat is:

> the amount of energy needed to change the state of one kilogram of a substance without a change in temperature
>
> the specific latent heat unit is joule per kilogram (J/kg)

Calculations

Specific heat capacity:

energy transferred = mass × specific heat capacity × temperature change

Example

> Calculate the energy transferred when the 80 kg of water in a tropical freshwater fish tank is heated from 10 °C to 25 °C.
>
> energy transferred = mass × specific heat capacity × temperature change
> energy transferred = 80 × 4200 × (25 – 10)
> energy transferred = 80 × 4200 × 15
> energy transferred = 5 040 000 J = 5040 kJ

The tables show some specific heat capacities and specific latent heats.

Substance	Specific heat capacity in J/kg °C
copper	390
mercury	140
rubber	1600
seawater	3900
water	4200

Change of state	Specific latent heat in J/kg
ice to water	340 000
water to steam	2 260 000

Specific latent heat:

energy transferred = mass × specific latent heat

Example

> Calculate the energy transferred when the 1.5 kg of water in a kettle changes from liquid to gas at 100 °C.
>
> energy transferred = mass × specific latent heat
> energy transferred = 1.5 × 2 260 000
> energy transferred = 3 390 000 J = 3390 kJ

Questions

10 Mohammed says that when you heat an object it gets hotter. Anne says it does not. Who is correct? Explain your answer.

11 The specific heat capacity of oxygen is 913 J/kg °C. How much energy is needed to raise the temperature of 1 kg of oxygen by 1 °C?

12 Calculate the energy transferred from a glass of coke to melt 100 g of ice cubes at 0 °C.

13 A copper saucepan has a mass of 1.5 kg. It is used to cook potatoes. Calculate how much extra energy is needed to raise the temperature of the saucepan from 15 °C to 100 °C.

Hidden heat

When a substance changes state from solid to liquid or liquid to gas, energy is needed to break the bonds that hold the molecules together. This is why there is no change in temperature.

Questions

14 190 MJ of energy are needed to melt 500 kg of aluminium at 660 °C. Calculate the specific latent heat of aluminium.

15 97.5 MJ of energy are supplied to 1000 kg of seawater at 5 °C. Calculate the final temperature of the seawater. (Refer to table above)

Keeping homes warm

You will find out:

- about the insulating properties of air
- how energy is transferred by conduction, convection and radiation
- how insulation reduces energy loss by conduction, convection and radiation

Insulation and protection

When astronauts go outside for a spacewalk they need to wear a special spacesuit. The surface of the suit facing the Sun might reach 120 °C. The other side of the suit may be as cold as –160 °C. The spacesuit keeps the astronaut's body at normal Earth temperature.

FIGURE 1: Why do astronauts need a special suit to walk in space?

Energy transfer

Insulation

It is very cold in the Arctic. Explorers keep warm because they have fur coats.

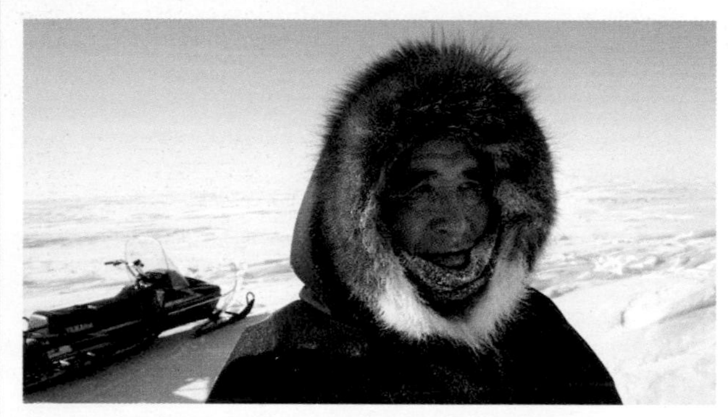

FIGURE 2: How does his fur coat keep him warm?

Air is a good **insulator**. Because there is a lot of air trapped inside the coat, it does not allow energy to transfer from a warm body to cold surroundings.

Conduction

Whenever anything is heated, the particles move more quickly.

The particles in a solid are close together so energy is easily transferred as the particles vibrate.

The particles in a gas are far apart, so energy cannot be transferred easily from particle to particle.

There are no particles in a vacuum so no energy can be transferred by conduction.

Radiation

In some countries, **infrared radiation** from the Sun is reflected from a shiny reflector. The heat is used for cooking or generating electricity. A device at the focus of the reflector converts heat energy into electricity.

Infrared radiation is absorbed by dull or rough surfaces. This is why solar water heaters on the roof of a house are painted black.

FIGURE 3: This solar reflector is at a power station in Australia. What happens to the radiated heat energy reflected from the surface?

FIGURE 4: Why are the pipes of the solar water heater painted black?

Questions

1 In winter it is better to wear several layers of thin loose-fitting clothes rather than one thick tight-fitting layer. Suggest why.

2 Explain why a vacuum does not transfer energy by conduction.

3 Suggest why there is little demand for solar power stations in the UK.

Remember!

Dark, black, rough surfaces absorb heat. They **do not** attract heat.

Practical insulation in the home

Double glazing

A double-glazed window has a space between two panes of glass. The space is filled with air or argon or has a vacuum.

> Air and argon are good insulators. A vacuum is the best insulator.

> A solid is a good **conductor** of heat.

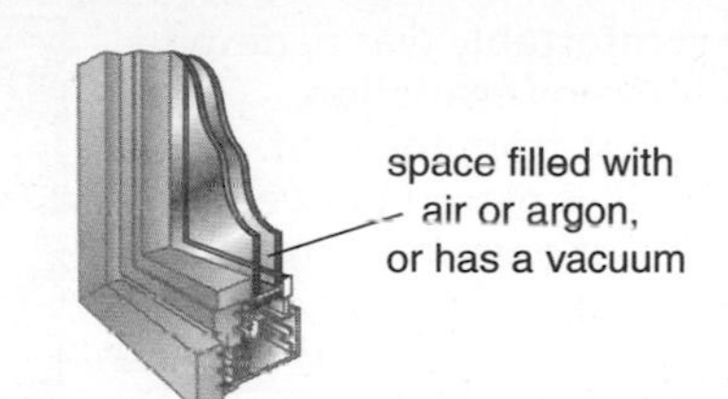

FIGURE 5: What is best to put in the space between the panes of glass?

Foam in a cavity contains a lot of trapped air. The foam reduces energy loss by conduction.

The blocks that are used when building new homes have shiny foil on both sides. This means that energy transfer by radiation is reduced. In winter, the energy in the home is reflected, keeping the home warmer. In summer, the energy from the Sun is reflected, keeping the home cooler.

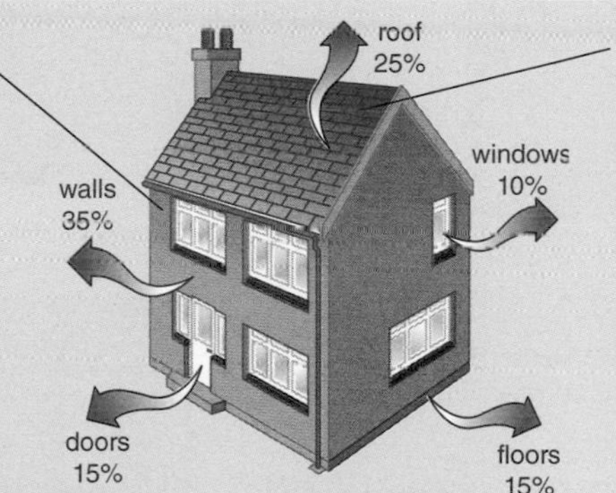

FIGURE 6: Heat losses from a house

Warm air in the room is in contact with the ceiling. The ceiling is a conductor and energy is transferred into the loft.

If there is no insulation, the air at the base of the loft warms up and then a **convection** current of warm air moves through the roof space and escapes through the tiles.

If the loft is insulated, the air in the insulation cannot move around the loft by convection.

Questions

4 Why is glass a better conductor of heat than argon?

5 Why do the foam blocks used for cavity wall insulation have shiny foil on both sides?

Did you know?

Heating bills can be reduced by 20% if cavity wall insulation is installed.

Conduction, convection and radiation

Conduction

Metals are the best conductors. Electrons in metals can leave their atoms and move about as free electrons. The atoms become charged ions. These ions are very close together and they are continually vibrating. When a solid metal is heated, the kinetic energy of the vibrations increases.

The energy is transferred from hot areas of the metal to cooler areas by the free electrons. The electrons move through the material, colliding with the ions.

Why convection?

When a gas is heated it expands. The same mass of gas occupies a greater volume.

As the volume increases, the density decreases. This is why hot air rises.

$$\text{density} = \frac{\text{mass}}{\text{volume}}$$

The unit of density is g/cm^3 or kg/m^3.

If air is such a good insulator, why do we need cavity wall insulation? Why is the air in the cavity not good enough?

Air is a very good **convector**.

> As air in the cavity is warmed it rises through the cavity.

> Colder air falls and a convection current is formed. The convection current transfers energy.

> The air trapped in the foam stops convection from taking place.

Radiation and space

Infrared radiation is an electromagnetic wave. Energy transfer by infrared radiation does not need a material. Energy can be transferred through a vacuum as in space.

Questions

6 Explain what is meant by a convection current.

7 Explain how cavity wall insulation reduces energy transfer by conduction, convection and radiation.

You will find out:
- > how energy is lost from uninsulated homes
- > about payback time
- > about energy efficiency
- > about Sankey diagrams

Saving energy

The people in this log cabin are comfortably warm, despite the snow. The roof of the cabin contains **insulation**. Insulation in a loft reduces energy loss from the roof. Fibreglass wool, or a similar material, is placed between the joists. Fibreglass wool contains a lot of gaps in its structure. Air is trapped in these gaps. Air is a good insulator.

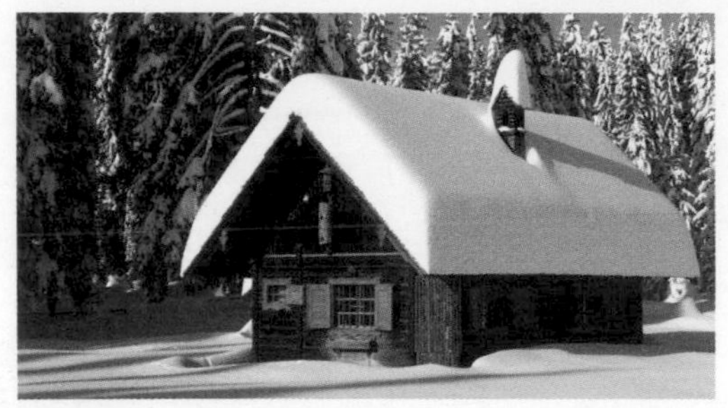

FIGURE 7: How can you tell this cabin is well insulated?

FIGURE 8: Why is fibreglass wool a good insulator?

> As new homes are built, solid foam boards are put inside the cavity walls. Older homes, with cavity walls, can have the foam injected.

> Loft insulation and cavity wall insulation contain air. Air is a poor conductor. Poor conductors are good insulators.

> Closing the curtains and double glazing cuts down energy loss through windows.

> Some people put aluminium foil behind the radiators.

> Draught strips around doors and windows are cheap and effective.

A traditional coal fire set in a wall of a room is not the most efficient way to heat a room.

> For every 100 J of energy stored in coal and released as heat, about 25 J are transferred to the room. The remaining 75 J are lost to the surroundings.

> The fire has an **efficiency** of 0.25 (or 25%).

Not all fires are so inefficient.

> A fire in the centre of a room can be more than 50% efficient.

FIGURE 9: Why is a coal fire in the centre of a room more efficient than one against an outside wall?

Questions

8 Derek's wood burning stove transfers 1000 J of energy stored in the wood into 700 J of useful heat. Calculate the efficiency of the stove.

9 What property of aluminium foil makes it suitable to use behind radiators to reduce energy loss? Choose one word from this list.

- cuts easily
- light
- metal
- shiny

Energy loss

Energy efficiency

$$\text{efficiency} = \frac{\text{useful energy output} \ (\times 100\%)}{\text{total energy input}}$$

Sankey diagrams

We sometimes show energy transformations as Sankey diagrams. These show energy transformation with the help of an arrow dividing into two or more similar arrows. The width of each arrow depends on the amount of energy transferred to each element.

This Sankey diagram shows how to represent the energy transformation from the coal fire set in the wall of a room.

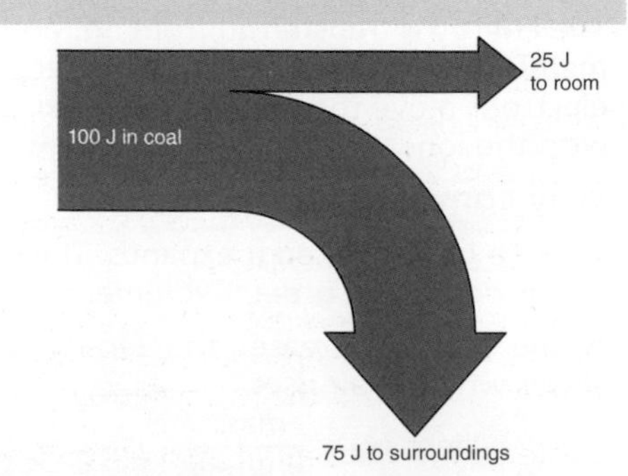

FIGURE 10: How many times wider is the 75 J arrow compared to the 25 J arrow?

energy efficiency reducing energy loss Sankey diagram

Homes lose energy from the roof, walls, doors, floors and windows. There is a saying, 'Don't heat the street'. In general, energy is lost from a 'source' to a 'sink'. The home is the energy source and the atmosphere is the sink. A sink is a very large object that does not change its temperature by a noticeable amount when it absorbs energy from the source.

To reduce energy loss, uninsulated homes need to be insulated. This can be quite expensive.

To work out which is the most cost-effective type of insulation, the **payback time** is calculated.

$$\text{payback time (years)} = \frac{\text{cost of insulation}}{\text{annual saving}}$$

Look at the row for cavity wall insulation in the table. This means that after two years, the homeowners have recovered in fuel saving the cost of the cavity wall insulation. After that they are £125 a year better off. As fuel costs rise, they could save even more.

The savings, in terms of energy resources and money, depend on what type of insulation is fitted.

Insulation	Typical cost in £	Typical annual fuel saving in £
cavity wall	250	125
double glazing	5000	100
draught–proofing	100	25
loft	250	150

Questions

Use the table above to help you answer the following questions.

10 Which insulation saves the most money per year?

11 Calculate the payback time for loft insulation.

12 Calculate the payback time for draught-proofing.

13 Mr Collins sees an advertisement for a gas fire.

Energy Efficient Gas Fire	Output 5600 J/s Only 1400 J/s wasted

a Calculate the efficiency of the gas fire.

b Draw a Sankey diagram to represent the energy transformation.

Did you know?

A staggering 25% of all carbon dioxide emissions in the UK are as a result of heating and lighting our homes.

Trapped air reduces energy loss by conduction and convection.

Energy transfer processes and design

Energy transfer processes

> Energy from a coal fire warms the room by radiation.

> The hot air rises up the chimney by convection.

> Energy is transferred through the surrounding bricks by conduction.

Buildings that are energy efficient are well insulated and allow as little energy as possible to be lost to the surroundings.

Designers and architects have to consider the best way to heat a home and the best way to make sure that energy is not wasted.

Energy efficiency is not just about heating homes. Everything that transfers energy wastes some of the energy as heat to the surroundings.

Designers of household appliances have to consider where energy may be lost and how they can reduce the loss.

Remember!

A question may require you to change the subject of an equation. Practise rearranging equations until you are really confident.

Question

14 Mrs Collins' coal fire is 20% efficient. It burns 10 kg of coal each day. The energy output to the room is 57.6 MJ. Calculate the energy in 1 kg of coal.

payback time energy transfer process

A spectrum of waves

You will find out:
- about the features of a transverse wave

The electromagnetic spectrum

There are transverse waves that form a family of waves called the electromagnetic spectrum.

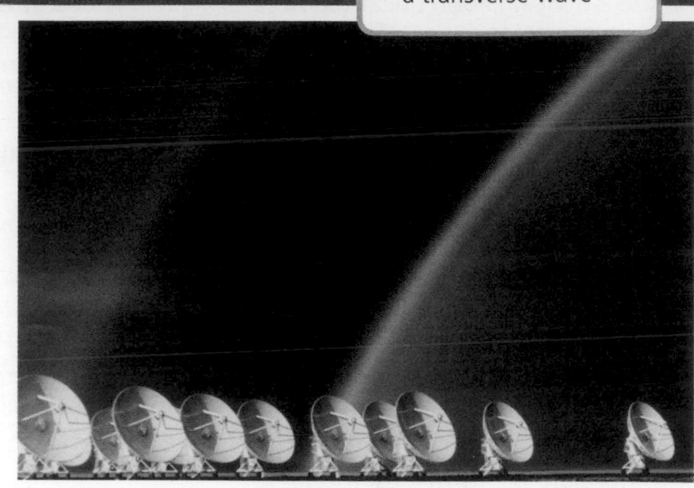

FIGURE 1: Where else could you see the colours of visible light of the electromagnetic spectrum?

Transverse waves

> A Mexican wave ripples round a stadium as people stand up and sit down.

> Water particles move up and down as a wave spreads out from where a pebble is dropped into water.

A Mexican wave and a water wave are **transverse waves**. A transverse wave travels in a direction at right angles to the wave vibration. Light is another example of a transverse wave.

The speed of light, and all waves in the **electromagnetic spectrum**, is 300 000 km/s in space or in a vacuum.

FIGURE 2: A transverse wave spreads out from where a pebble is dropped into water. Does the water move out in circles or go up and down?

FIGURE 3: The parts of a transverse wave.

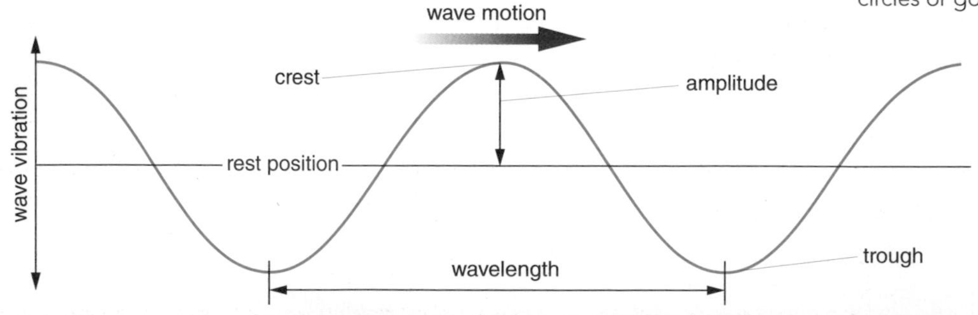

Questions

1. What is the speed of radio waves?
2. What is the speed of microwaves?
3. The speed of sound is 300 m/s. In a thunderstorm, why do we see lightning before we hear thunder?
4. wave speed = frequency × wavelength. A guitar string is plucked. The frequency of the note produced is 200 Hz and the wavelength is 1.5 m. Calculate the speed of sound.

electromagnetic spectrum   transverse waves   wave motion

Wave properties

> The **amplitude** of a wave is the maximum displacement of a particle from its rest position.

> The **wavelength** of a wave is the distance between two successive points on the wave having the same displacement and moving in the same direction. The unit of wavelength is the metre (m).

> The **frequency** of a wave is the number of complete waves passing a point in 1 second. The unit of frequency is the **hertz** (Hz).

> Surfers ride the crest of a wave. This is the point of maximum displacement above the rest position.

> The trough of a wave is the maximum displacement below the rest position.

> The speed of a wave is calculated using the equation:

wave speed = frequency × wavelength

The unit is metres per second (m/s).

Example

A girl throws a pebble into a pond. The distance between the ripples is 0.25 m and five waves reach the edge of the pond each second. Calculate the speed of the water wave.

wave speed = frequency × wavelength

wave speed = 5 × 0.25

wave speed = 1.25 m/s

Remember!

The abbreviation for a unit begins with a capital letter only if it is named after a person.

FIGURE 4: What part of a wave does a surfer ride?

Questions

5 The distance between the crest of a wave and the next trough is 4 m. What is the wavelength of the wave?

6 The height of a wave from crest to trough is 30 cm. What is the amplitude of the wave?

7 A dolphin produces a sound with a frequency of 120 000 Hz. The wavelength is 1.25 cm (0.0125 m). Calculate the speed of sound in water.

Radio waves and earthquake waves

Remember!

At higher tier you may be expected to change the subject of an equation and use scientific notation.

Questions

8 Earthquake waves of wavelength 500 m travel through the Earth at 7000 m/s. Calculate their frequency.

9 Radio waves travel at 300 000 000 (300×10^6) m/s. Radio 4 broadcasts at a frequency of 200 kHz. Calculate the wavelength of Radio 4.

amplitude frequency wavelength wave speed

The electromagnetic spectrum

You will find out:
- about the waves in the electromagnetic spectrum
- about how the behaviour of the waves helps to determine their use

There are seven types of transverse wave in the electromagnetic spectrum.

> Radio waves have the lowest frequency and gamma rays the highest.

> All electromagnetic waves travel in straight lines through a particular medium.

> They can be reflected. Light is reflected from a mirror. The microwave signals from satellites are reflected by satellite dishes.

When a ray of light strikes a mirror, the angle of incidence is equal to the angle of reflection.

> They can be **refracted**.

When light passes from one medium into another at an angle to the boundary, it changes direction. This sometimes produces odd effects at the boundary between the two mediums.

> Some electromagnetic waves are used for communication.

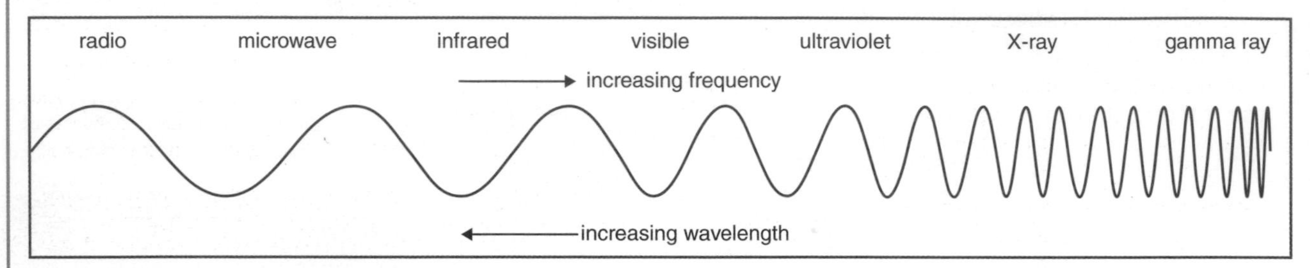

FIGURE 5: The electromagnetic spectrum

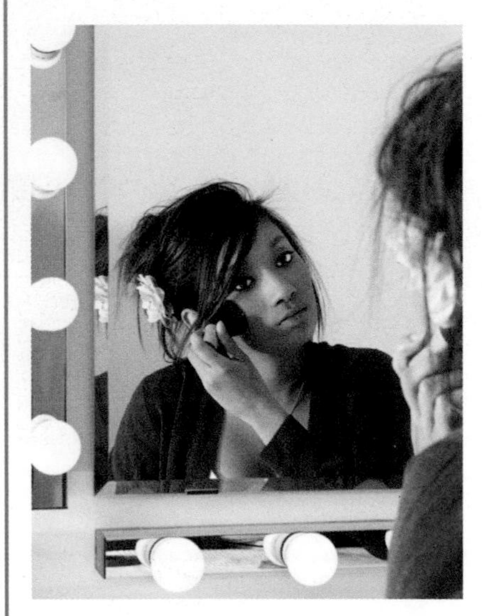

FIGURE 6: What happens to light when it strikes a mirror?

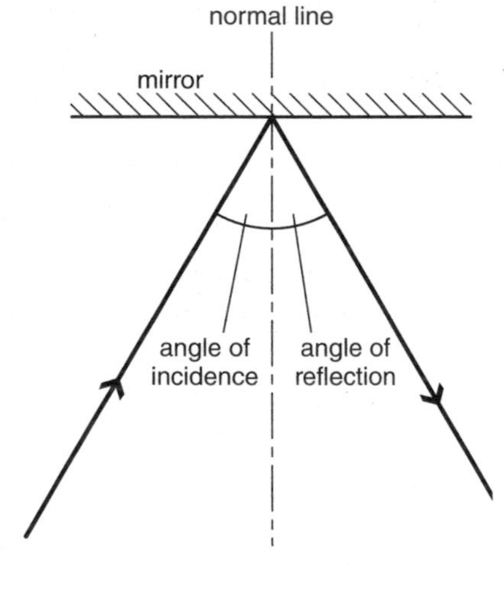

FIGURE 7: How does the angle of incidence compare to the angle of reflection?

FIGURE 8: This effect is caused by refraction of light at the surface of the water.

radio

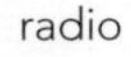

infrared

microwave

light

FIGURE 9: Are there other ways in which waves are used in everyday life?

Questions

10 A ray of light strikes a flat mirror with an angle of incidence of 50°. What is the angle of reflection?

11 The angle between a flat mirror and a ray of light is 30°. What is the angle of reflection?

12 What type of electromagnetic radiation is used by a remote controller for an air conditioning system?

angle of incidence angle of reflection

Getting the message across

Some optical instruments use more than one plane mirror. The simple periscope is one example.

When a wave travels from one medium into another, the frequency of the incident wave does not change. The wavelength does change. This is because the speed of the wave changes. Electromagnetic waves slow down in a denser medium.

When they are incident at an angle to the boundary, this change in speed causes the wave direction to change. This is refraction.

In communications, such as radio, the size of the receiver depends on the wavelength of the wave. The longer the wavelength, the longer the receiver needs to be.

When waves pass through an opening, they spread out. This is known as **diffraction**.

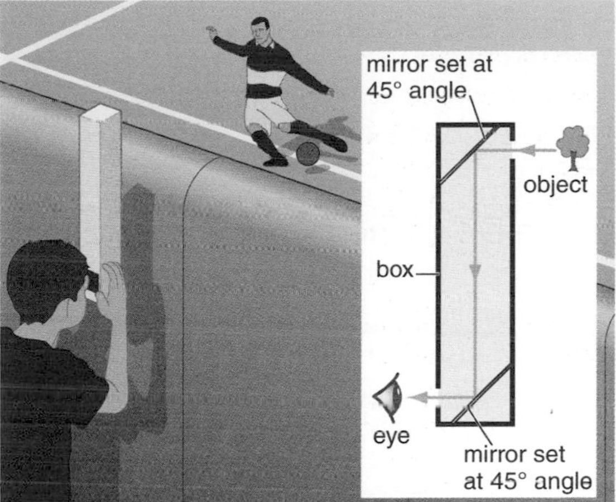

FIGURE 10: Why is a periscope useful at a football match?

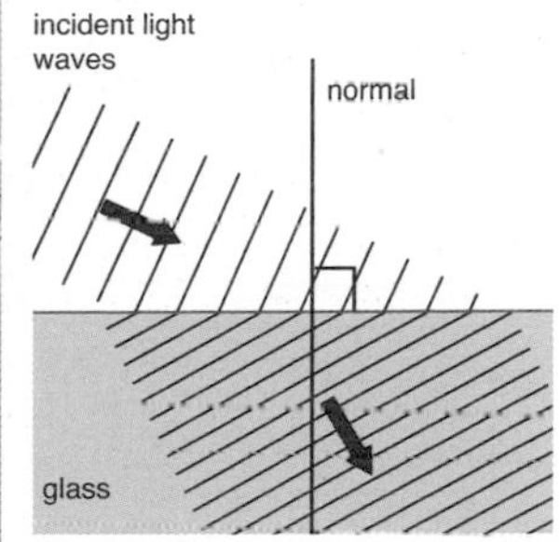

FIGURE 11: When light passes from air into glass, is it refracted towards or away from the normal?

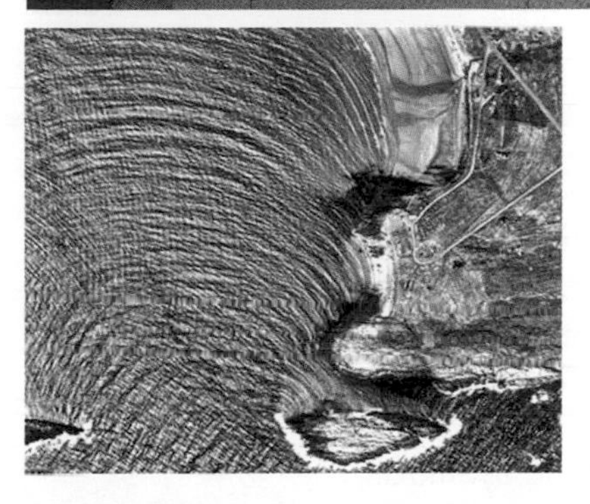

FIGURE 12: Ocean waves diffracted around a coastline. Why can you hear round corners?

Question

13 Why is the aerial on a mobile phone shorter than the aerial on a radio?

Diffraction effects

When water waves pass through a gap of similar size to the wavelength, the waves spread out on the other side of the gap.

If the gap is much larger than the wavelength of the waves, the diffraction effect is not so noticeable.

Most astronomical telescopes contain two mirrors. The secondary mirror is in the path of light from the star or planet being viewed. This means that the light is diffracted around the secondary mirror which reduces the quality of the image.

Diffraction also occurs at the edge of a lens telescope, or microscope, and this gives rise to spikes or rings around the object being viewed.

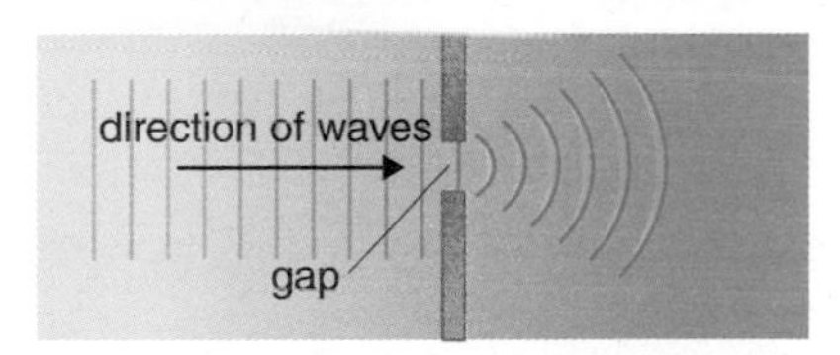

FIGURE 13: For diffraction to occur the wavelength has to be similar to the size of the gap.

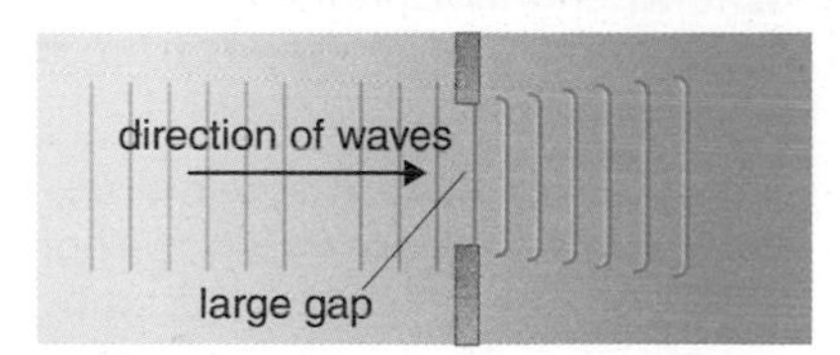

FIGURE 14: Why is diffraction not so noticeable here?

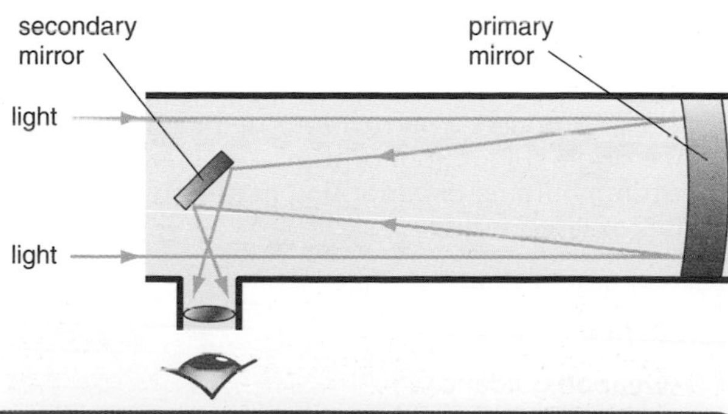

FIGURE 15: What problem does the secondary mirror cause?

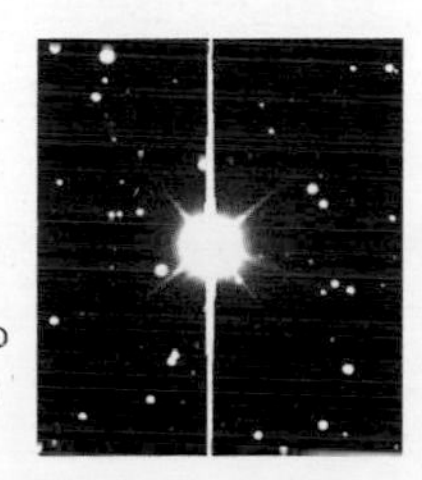

FIGURE 16: Why is it not easy to see distant stars clearly with an optical telescope?

Question

14 In a diffraction experiment, why should the gap be similar in size to the wavelength for diffraction to occur?

Light and lasers

You will find out:

> about what factors affect the choice of how a signal is sent

> how the properties of laser light can be used

'Heroin pouch bursts inside trafficker'

This is a likely headline in a newspaper.

Heroin is an illegal drug. Drug traffickers often put themselves in great danger to import the drug. They wrap the heroin in black plastic tape and swallow it. Sometimes a packet bursts inside them. The trafficker can stop breathing and may die. Doctors can use a special instrument to see inside the body without surgery.

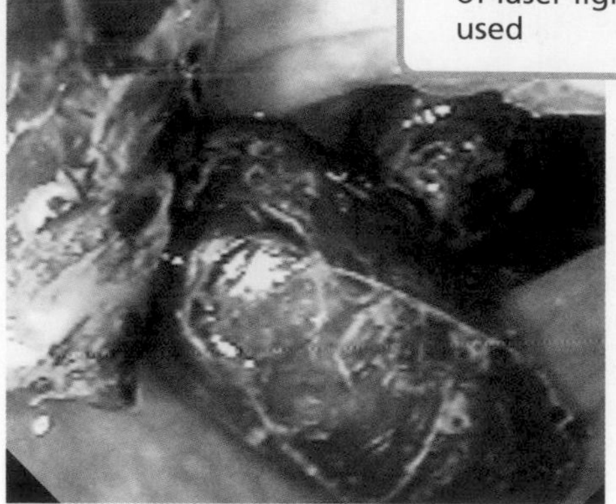

FIGURE 1: This is what a doctor using an endoscope sees inside a person who has swallowed packets of drugs.

Early messages

> Ancient Greek soldiers sent smoke signals; so did the American Indians.

> The British still use chains of beacons to celebrate important events such as the Millennium.

Some messages are too difficult to send by smoke signal.

Sometimes a message has to be sent a great distance. Runners, horses and motorbikes have been used to get messages from one place to another quickly.

Messages can be sent very quickly by a flashing signal lamp. A code is used to represent the different letters. The code used by the lamps was the Morse code and navies used these lamps until 1999.

FIGURE 2: This messenger is riding in wartime and carrying an important message. He is wearing a gas mask to protect him from a gas attack. What else may he need for himself and his horse on his journey?

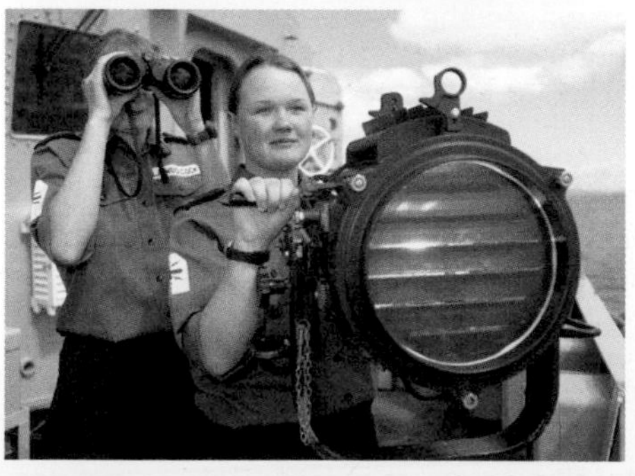

FIGURE 3: Why are two people needed to communicate successfully with another ship in this picture?

Laser

Light **A**mplification by **S**timulated **E**mission of **R**adiation

A **laser** produces a narrow intense beam of light of a single colour (monochromatic). It is capable of cutting, burning or vaporising materials. Other uses of lasers include:

> communication

> dental treatment

> weapon guidance

> surgery

> light shows

> bar code readers.

Questions

1 Why is it dangerous to look into a laser beam?

2 Science fiction films often show a laser beam coming from a laser gun as a special effect. Suggest reasons why we would not see a beam coming from the laser gun.

uses of lasers

Morse code

Mirrors and signalling lamps have been used to send messages using the **Morse code**.

The code is a series of dots and dashes that represent individual letters of the alphabet.

The dots and dashes are a series of on and off signals. This is a digital signal. To cover large distances, the signals are relayed from one signal station to another.

A	B	C	D	E	F	G	H	I
•–	–•••	–•–•	–••	•	••–•	––•	••••	••
J	K	L	M	N	O	P	Q	R
•–––	–•–	•–••	––	–•	–––	•––•	––•–	•–•
S	T	U	V	W	X	Y	Z	
•••	–	••–	•••–	•––	–••–	–•––	––••	

FIGURE 4: The Morse code.

Question

3 Translate this code:

Sending signals

Signals can be sent by light, electricity, radio waves or microwaves. Each type of signal has advantages and disadvantages.

> They are all almost instantaneous.

> Light signals can be easily sent with nothing more sophisticated than a torch, but the signal needs to be coded and is not secure. Anyone can see what is being sent.

> Electrical signals need equipment and wires linking the sender and receiver of the signal. Wires can be cut or damaged. The signal needs to be amplified at regular intervals because of energy loss in the form of heat due to the electric current.

> Radio waves and microwaves can travel large distances and through the atmosphere. They can even travel across space.

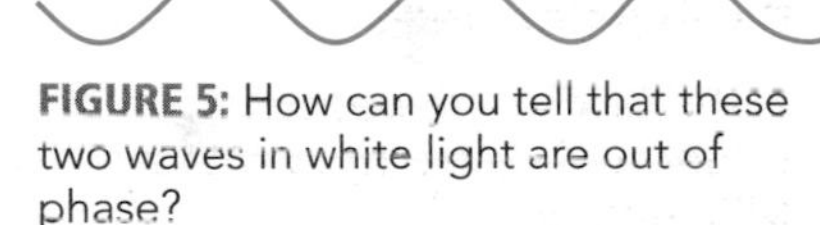

FIGURE 5: How can you tell that these two waves in white light are out of phase?

FIGURE 6: The crests and troughs occur at the same time in these two laser light waves. What term describes this?

Laser light

White light, from a bulb, is made up of many different colours, each of a different frequency and out of **phase**.

The light from a laser is at one frequency and in phase.

We say that the beam is a monochromatic (single colour from one frequency) coherent (in phase) beam. The beam does not diverge very much.

FIGURE 7: What do the fine pits on the surface of a CD represent?

Compact disc

The plastic compact disc (CD) is pressed with a series of fine pits. There are billions of these pits on the bottom shiny surface of the CD. The pits represent a digital signal. The signal is read by a laser.

> The music layer is coated with a fine film of metal that follows the pits exactly.

> The shiny metal film reflects the laser light.

> The metal layer is covered by another layer of transparent plastic.

> The music is sandwiched between two layers of plastic. This stops dust and scratches from affecting the sound.

Questions

4 Suggest why:

a telephone signals between a house and a local exchange are sent by an electric current

b telephone signals between England and America are sent by microwaves.

Trapped light

You will find out:

- about total internal reflection
- about the uses of optical fibres in communication and medicine

Light can stay inside materials such as glass, Perspex or water.

These materials are denser than air. The light is reflected at the boundary between the material and the air.

This is called **total internal reflection**.

Light, infrared radiation or a laser beam can travel along a very thin piece of solid glass called an **optical fibre**. Every time the light meets the boundary with air, it is reflected back into the fibre.

Optical fibres are very flexible.

FIGURE 8: How is light trapped in this fibre optic lamp?

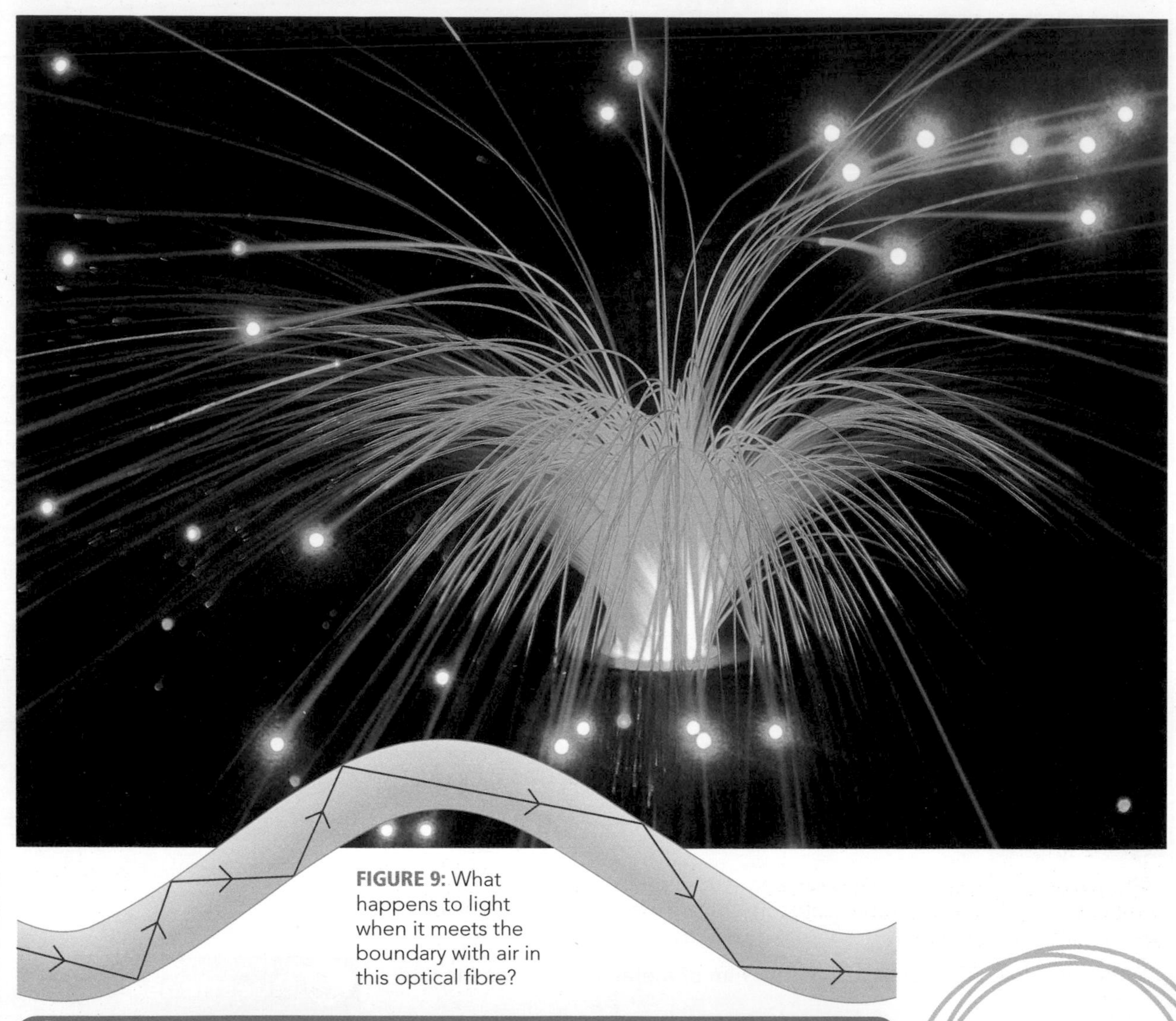

FIGURE 9: What happens to light when it meets the boundary with air in this optical fibre?

Questions

5 What is the relationship between the angle of incidence and the angle of reflection?

6 Using total internal reflection in your answer suggest how a small fish in the sea can see a large fish behind it. You may find it helpful to draw a diagram.

Remember!
An optical fibre is solid, not a hollow tube.

Critical angle

When light is passing from a more dense material such as glass into a less dense material such as air, the angle of refraction is greater than the angle of incidence.

If the angle of incidence increases, it is possible for the angle of refraction to be a right angle (90°). When this happens, the angle of incidence is called the **critical angle**.

Total internal reflection

If the angle of incidence is increased even more, the light is reflected back inside the more dense material. This is total internal reflection.

Optical fibres are used to transmit data at 200 000 km/s which is the speed of light in glass. Telephone conversations and computer data can be transmitted very long distances with little energy loss. The optical fibres work because the angle of incidence is greater than the critical angle. They have a coating around them to improve reflection.

The signals are coded and sent digitally using laser light or infrared.

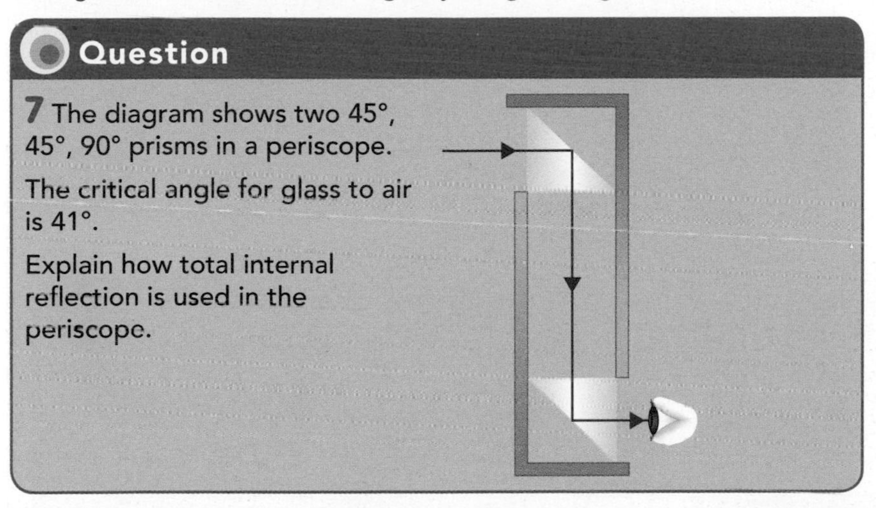

Question

7 The diagram shows two 45°, 45°, 90° prisms in a periscope.

The critical angle for glass to air is 41°.

Explain how total internal reflection is used in the periscope.

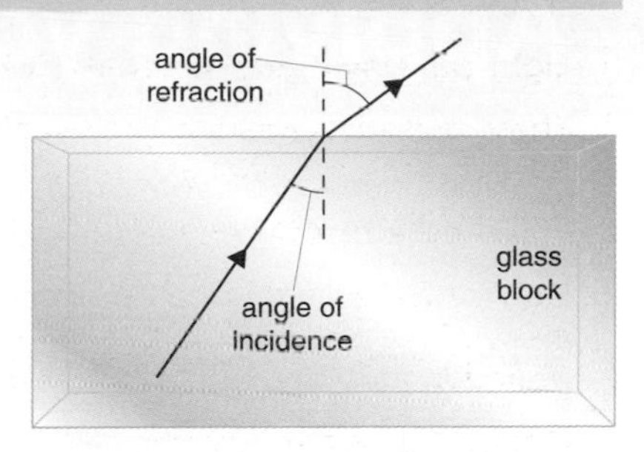

FIGURE 10: Is the angle of incidence more or less than the angle of refraction in this diagram?

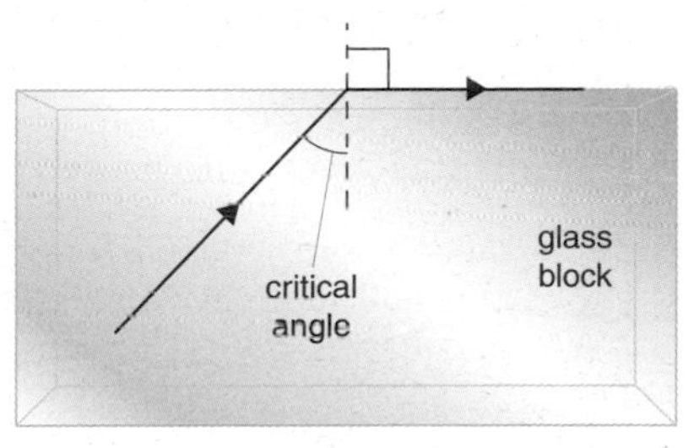

FIGURE 11: When is the angle of incidence called the critical angle?

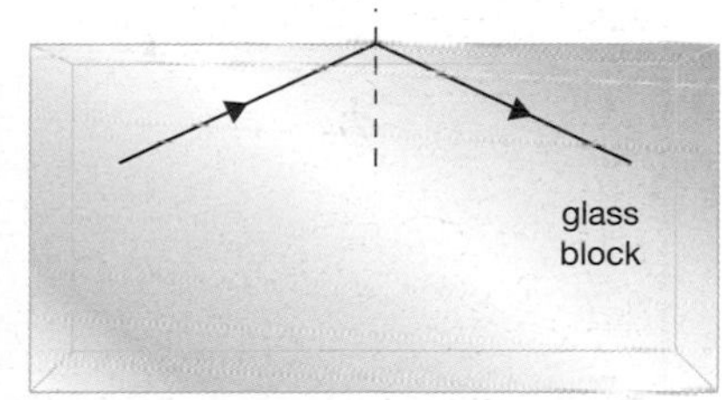

FIGURE 12: What is total internal reflection?

Endoscopy

One very important use of optical fibres is in an **endoscope.** This is an instrument used by doctors that allows them to see inside the body without surgery. It also allows some operations to be performed by keyhole surgery.

Light passes along the outer fibres and lights up the inside of the patient. The reflected light passes back along the inner fibres and the image is viewed through an eyepiece.

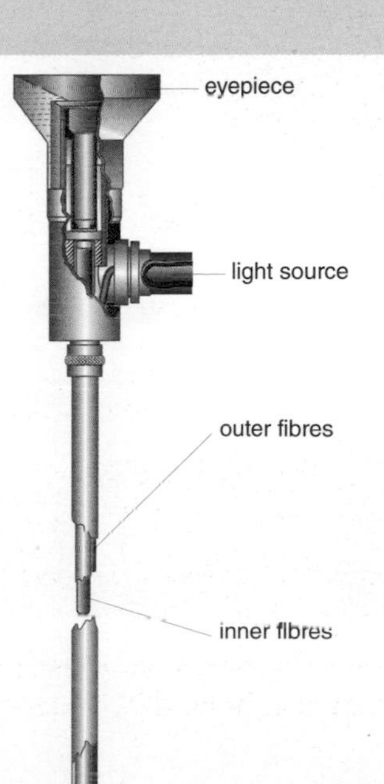

FIGURE 13: How does an endoscope help doctors?

Question

8 In an endoscope the inner fibres must be arranged in the same pattern at both ends, but the outer fibres need not. Suggest why.

Preparing for assessment: Applying your knowledge

To achieve a good grade in science, you not only have to know and understand scientific ideas, but you need to be able to apply them to other situations and investigations. These tasks will support you in developing these skills.

Keeping warm

Jack and Ruby are visiting Butser Ancient Farm in Hampshire, in the south of England. They live on a farm but this one is very different to any other farm that they have seen.

Although Butser is a real working farm it is the only place in Western Europe where you can experience what it might have been like to live on a farm at the time of the Celts and Romans. Based on archaeological excavations of prehistoric sites, the farm has prehistoric livestock, including chickens, and plants such as cereals, as well as Iron Age houses and other buildings.

Jack and Ruby go inside a large roundhouse, about 15 m in diameter. Most roundhouses are smaller than this. The walls of the buildings are mainly woven with willow rods (wattle). Sometimes a mixture of clay, soil and straw is mixed with water to make daub and used to plaster on the wattle walls of the buildings.

The roundhouse has no windows or electric lights so it is very dark inside. Ruby feels cold. She decides fire would be their only way of keeping warm. The Celts and Romans lit fires in the middle of their roundhouses. Jack and Ruby try to imagine what it would be like to live and work there and compare it with their 21st century farm. "I'd rather live on our farm", said Jack. "Oh yes!", replied Ruby. "I think it's time to go home."

Task 1

Look at the picture and imagine that you are in the building. Discuss the ways in which energy is being transferred from the fire and how the inhabitants are getting warm.

Task 2

Describe and explain using words and diagrams how energy is being transferred and where it is being transferred to.

Task 3

Explain why some of the occupants are warmer than others.

Maximise your grade

Answer includes showing that you can...	
	State one way in which heat energy is transferred in the diagram.
	State two ways in which heat energy is transferred in the diagram.
	Recall how one type of transfer works in general.
	Recall how two types of transfer work in general.
	Explain how each type of transfer works in this situation.
	Explain how one type of transfer works with reference to waves or particles (as appropriate).
	Explain how each type of transfer works with reference to waves or particles (as appropriate).
	As above, but with particular clarity and detail.

Cooking and communicating using waves

You will find out:

- > about bodies that give out infrared radiation
- > about heating effects of microwaves
- > how wavelength affects the energy transferred

Microwave energy

There are some things that you should not try to cook in a microwave oven, and a light bulb is one of them – the gas inside (glowing pink in the photograph) will make it explode.

FIGURE 1: A light bulb heated by microwave energy.

Microwaves are part of the electromagnetic spectrum. Radio waves are used to transmit signals to your radio and X-rays are used to make an image of your bone in case doctors think that you have broken it.

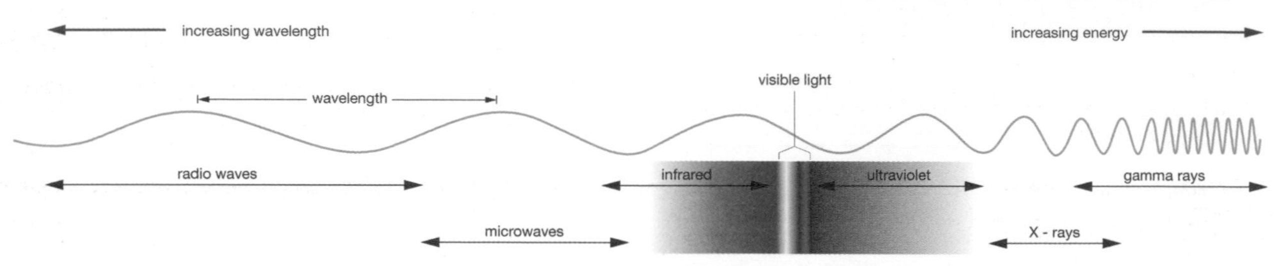

FIGURE 2: The relationship between wavelength and energy.

Infrared radiation

It is not just the flames that are emitting infrared (heat) radiation in Figure 4.

The cat and the newborn kitten are also emitting infrared radiation. All warm and hot objects emit infrared radiation.

> Hotter objects give out more radiation.

> Dull, black objects give out more radiation than shiny, bright objects at the same temperature.

> Infrared radiation is absorbed by the surface of any object. The object warms up. Black, dull surfaces are very good absorbers of radiation.

> White, shiny surfaces are poor absorbers of radiation.

Microwaves

Microwaves are also used for heating. Microwaves in a microwave oven are absorbed by water or fat molecules in food.

Infrared radiation and microwaves are part of a family of waves called the electromagnetic spectrum.

FIGURE 3: How does a microwave oven cook food?

FIGURE 4: Which objects emit the most infrared radiation?

Questions

1 What colour of the visible spectrum is next to infrared radiation in the electromagnetic spectrum?

2 What part of the electromagnetic spectrum has waves with wavelengths longer than microwaves?

Infrared and microwave cookery

> Infrared radiation does not penetrate very far into food.

> It only heats the surface of food.

> Meat that appears to be cooked is often still quite raw on the inside.

FIGURE 5: Why can steak still be pink inside when it appears well done on the outside?

Microwave cookery

> Microwaves penetrate about 1 cm into the outer layers of food.

> Microwaves pass through glass and plastic but, like infrared radiation, are reflected from shiny metal surfaces.

> The door of a microwave oven is made from special glass that reflects microwave radiation.

> Take care: microwaves can cook human flesh as well as the evening meal.

Questions

3 You can warm rolls or defrost a loaf in a microwave oven but you cannot make good toast. Why?

4 Suggest two advantages of using a microwave oven instead of an infrared oven.

Electromagnetic spectrum

Energy is transferred by waves in the electromagnetic spectrum.

The amount of energy associated with microwaves and infrared radiation depends on the wavelength or frequency of the wave.

The shorter the wavelength (or higher the frequency), the more energy is associated with the radiation.

Microwaves have a longer wavelength than infrared radiation and have less energy associated with them.

Infrared cookery

> Energy from infrared radiation is absorbed by the particles on the surface of the food.

> This increases the **kinetic energy** of the particles.

Microwave cookery

> The water or fat molecules in the outer layers vibrate more and their kinetic energy increases.

The energy is transferred to the centre of the food by conduction or convection, depending on the food.

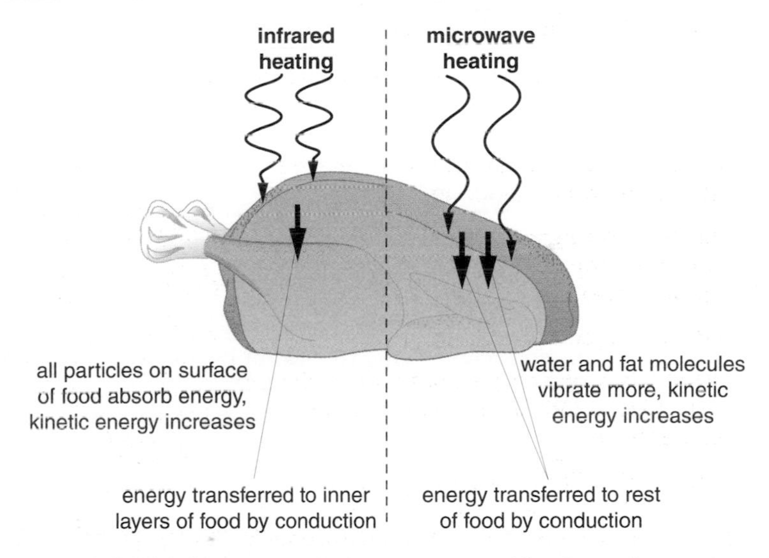

FIGURE 6: What happens to the kinetic energy of food particles in infrared and microwave cookery?

Questions

5 The door of a microwave oven is made from special glass that does not allow microwaves to pass through. Suggest why.

6 Sacha is sitting near to a glowing coal fire. How does the wavelength of infrared radiation from the fire compare to the wavelength of radiation from Sacha?

Microwave communication

You will find out:
- > about other uses of microwaves
- > about possible dangers associated with mobile phones and phone masts

Microwaves are used for communication as well as for cooking.

Recently there have been many reports about the dangers of mobile phones.

Newspaper headlines have read 'Mobiles don't fry the brain, they just warm it'.

As yet there is no scientific proof one way or the other. There is some evidence of warming and rats show signs of stress near mobile phones.

Young people use mobile phones a lot and their brains are still developing.

The Government advises that young people should not use mobile phones for 'non-essential' calls.

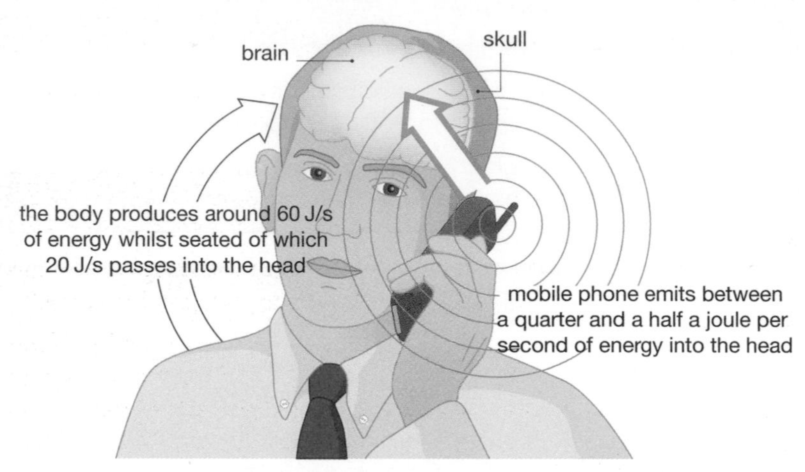

FIGURE 8: The heating effect of mobile phones.

FIGURE 7: Should young people use their mobiles for non-essential calls?

Did you know?

Madley Communications Centre, near Hereford, is the largest satellite station in the world.

It can handle thousands of telephone, fax and television signals at the same time.

Questions

7 How much energy per second enters the head from a typical mobile phone?

8 Why is texting a safer way to communicate than making a call from a mobile phone?

9 Why does the Government advise young people to use mobile phones for essential calls only?

Microwave signals

Long distance communication also uses microwave signals. The **transmitter** and **receiver** have to be in 'line of sight'. Line of sight is lost due to the curvature of the Earth. That is why microwave aerials for sending telephone signals are on high buildings or towers. The Telecom Tower in London dwarfs surrounding tower blocks. It has microwave aerials pointing in every direction.

Some areas of the country do not have good reception. This is because they are not in line of sight with a transmitter.

There may be mountains, hills or tall buildings in the way. Sometimes, weather conditions and large areas of water can affect microwave reception by scattering the signals.

A satellite orbiting Earth is in line of sight with a giant aerial on the ground. The aerial can handle thousands of telephone calls and television pictures at once.

On a smaller scale, a satellite dish on the roof of a house receives microwave signals that deliver satellite television channels.

microwave aerial   how mobile phones work   satellite dish

Many people are worried about where mobile phone masts are situated.

There have been many studies on the effects of the radiation from mobile phones and phone masts. The conclusions have not always been the same. Scientists publish their results and this allows other scientists to see if the results can be repeated.

There are still concerns about the possible risk of cancer due to microwave radiation. Masts should not be sited near schools or playgrounds.

The advice given by experts is that mobile phones should not be used too often or for too long because of possible dangers.

FIGURE 9: What is sending microwaves to this giant aerial?

FIGURE 10: Why have the satellite dishes been fixed on the front walls of the houses?

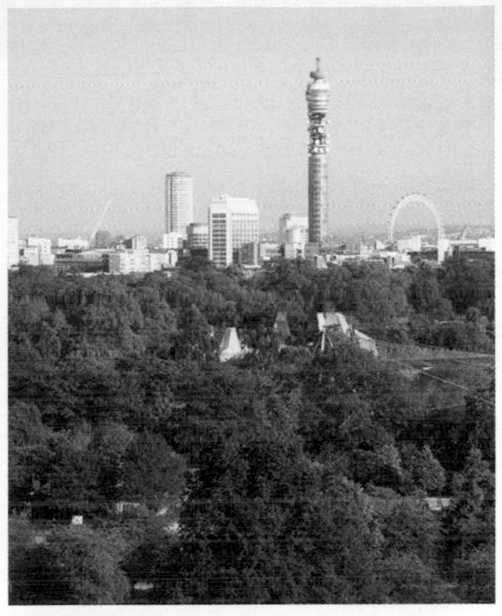

FIGURE 11: Why are microwave aerials on high buildings?

Questions

10 Why are microwave aerials placed on very tall buildings?

11 Suggest why mobile phone masts should not be sited near to a children's play area.

Microwave properties

Even within the band of waves called microwaves, the wavelength can be different. The shortest wavelength for microwaves is 1 mm and the longest is 30 cm.

Microwaves used for communication have a longer wavelength than those used for cooking. This means that there is less energy associated with mobile phones than with microwave ovens.

Microwave signals are refracted as they pass through different densities of atmosphere. There is a horizontal layer in the lower atmosphere where the different densities are such that the microwave signals follow the curvature of the Earth. This is called an atmospheric duct. High frequency microwave signals are attenuated by adverse weather conditions such as rain, snow and fog. They are also reflected by large flat surfaces such as water.

Diffraction

Microwaves do not exhibit very much diffraction around large obstacles, such as hills. This leads to signal loss and explains why the reception on a mobile phone sometimes changes from maximum to zero in a very short distance.

Interference

Signal loss can also be caused when two different signals overlap and **interfere** with each other.

Mobile phones are here to stay. They have many benefits and they may have some negative side effects. It is up to us to decide whether or not the benefits outweigh the negative effects. This decision will affect where mobile phone masts are situated and how much is invested in reducing any harmful radiation from their use.

Question

12 Suggest why aerials used for mobile phone signals are placed close together, particularly in towns.

Data transmission

You will find out:

> how infrared radiation is used
> about analogue and digital signals

Caught on camera

This is an image taken from a TV screen aboard a police helicopter. The police were using infrared thermal imaging, to locate criminals who robbed a bank.

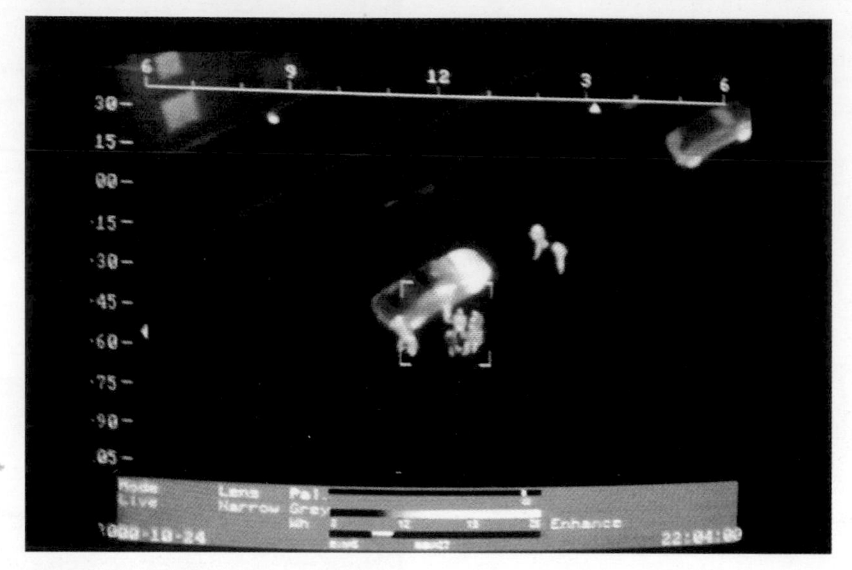

FIGURE 1: In what other situations are infrared cameras useful?

Invisible infrared waves

Uses of infrared waves

> A remote control lets you change the channel on your television from your armchair.

> Motorists can open the garage door when they arrive home on a cold wet night. They use a remote control from inside their car.

Remote controls work by emitting an infrared signal.

Other cordless devices, such as a computer mouse can transmit signals over short distances using infrared radiation. The signal is received by a passive infrared **sensor**.

Infrared sensors are used for security alarms and thermal imaging cameras. They detect low levels of infrared radiation from the body of an animal.

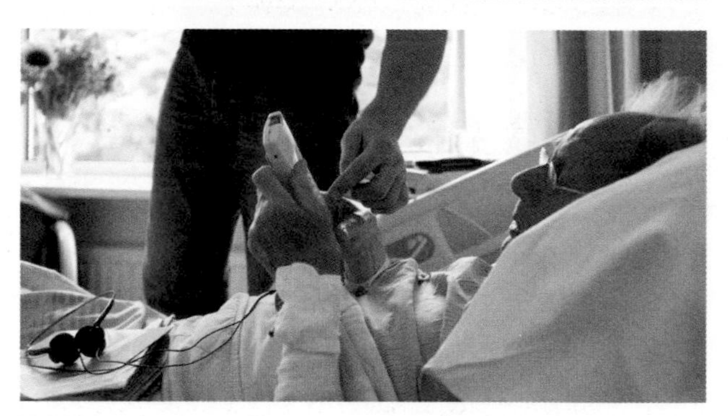

FIGURE 2: Suggest why remote controls are useful.

Digital signals

The signal emitted by a remote control is called a **digital signal**.

A digital signal can only have two values, on or off. On is represented by 1 (or high) and off by 0 (or low). A remote control transmits a coded infrared signal to the device which then changes a TV channel, starts a DVD recorder or increases the volume.

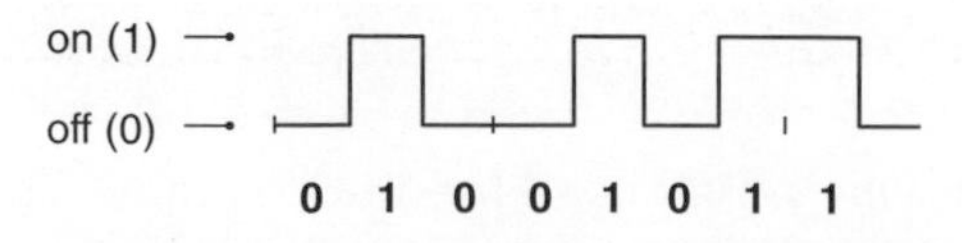

FIGURE 3: How many values can a digital signal have?

analogue signal digital signal

Analogue signals

When a person speaks into a microphone connected to an oscilloscope an analogue wave pattern is seen on the screen. The pattern represents what was said.

Even humming a note produces a complicated wave pattern. The wave pattern is interpreted as follows.

> Height of a wave indicates loudness.

> Distance between waves indicates pitch. High-pitched sounds have waves close together.

An **analogue signal** can have any value. It is continuously variable.

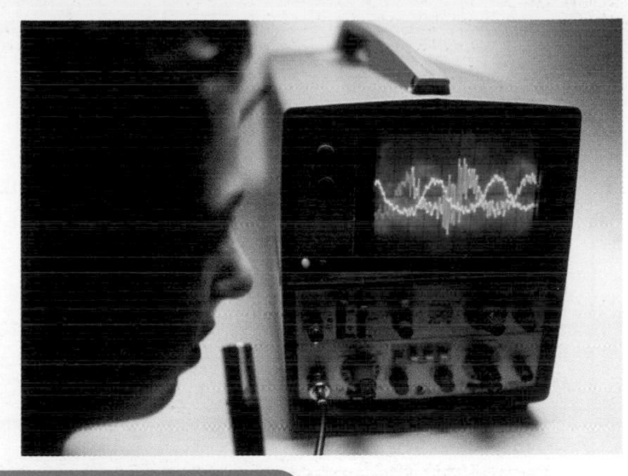

FIGURE 4: What does the analogue wave pattern on the screen represent?

Questions

1 What is the source of the infrared radiation detected by a passive infrared sensor during a burglary?

2 A wristwatch face has hands and numbers 1 to 12.

a Explain why it is described as an analogue watch.

b Make a sketch of a digital watch.

Digital signals

Infrared signals carry information that allows electronic and electrical devices to be controlled.

They do this by sending a coded digital signal from a remote controller to the device. Because digital signals only have two values, any small difference in the signal value (noise) will not be noticed and can easily be removed.

Digital control

When you press a button on the remote control device, this completes the circuit. The circuit sends a coded signal to a **light-emitting diode** (LED) at the front of the remote. The signal looks something like that in Figure 5.

The signal includes a start command, the instruction command (e.g. channel 5), a device code (the particular television) and a stop command. The LED transmits the series of pulses which is received by the device and decoded to allow the television to change channel.

Question

3 Describe the difference between the digital code for 0 and the digital code for 1.

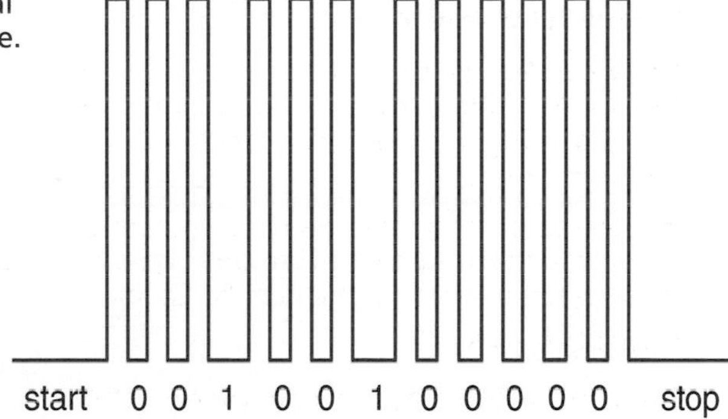

FIGURE 5: Which part of this digital signal encodes the specific instruction and device commands?

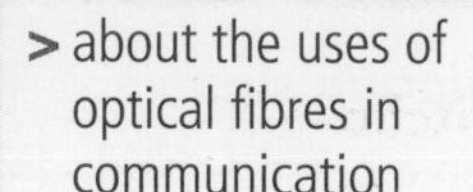

You will find out:
- > about the uses of optical fibres in communication
- > about interference in signals

Optical fibres

As we discovered on page 200, light, infrared radiation or a laser beam can travel along a very thin, flexible piece of solid glass called an optical fibre. Every time the light meets the boundary with air, it is reflected back into the fibre.

Optical fibres are used to transmit data at 200 000 km/s, which is the speed of light in glass. Telephone conversations and computer data can be transmitted very long distances with little energy loss. The optical fibres have a coating around them to improve reflection.

The signals are coded and pulses of light are sent digitally.

FIGURE 6: What happens to light when it meets the boundary with air in this optical fibre?

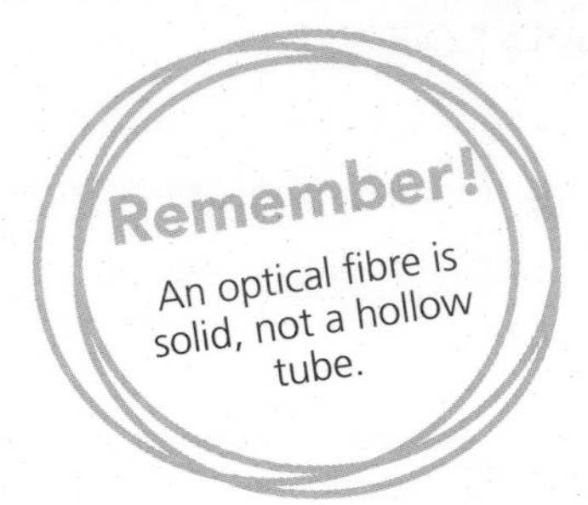

Question

4 What are the advantages of using optical fibres instead of copper wires for data transmission?

Digital advantages

Analogue signals

An analogue speech signal is to be transmitted by radio. A typical frequency for a person speaking is 200 Hz. A typical frequency for a radio **carrier wave** is 200 000 Hz.

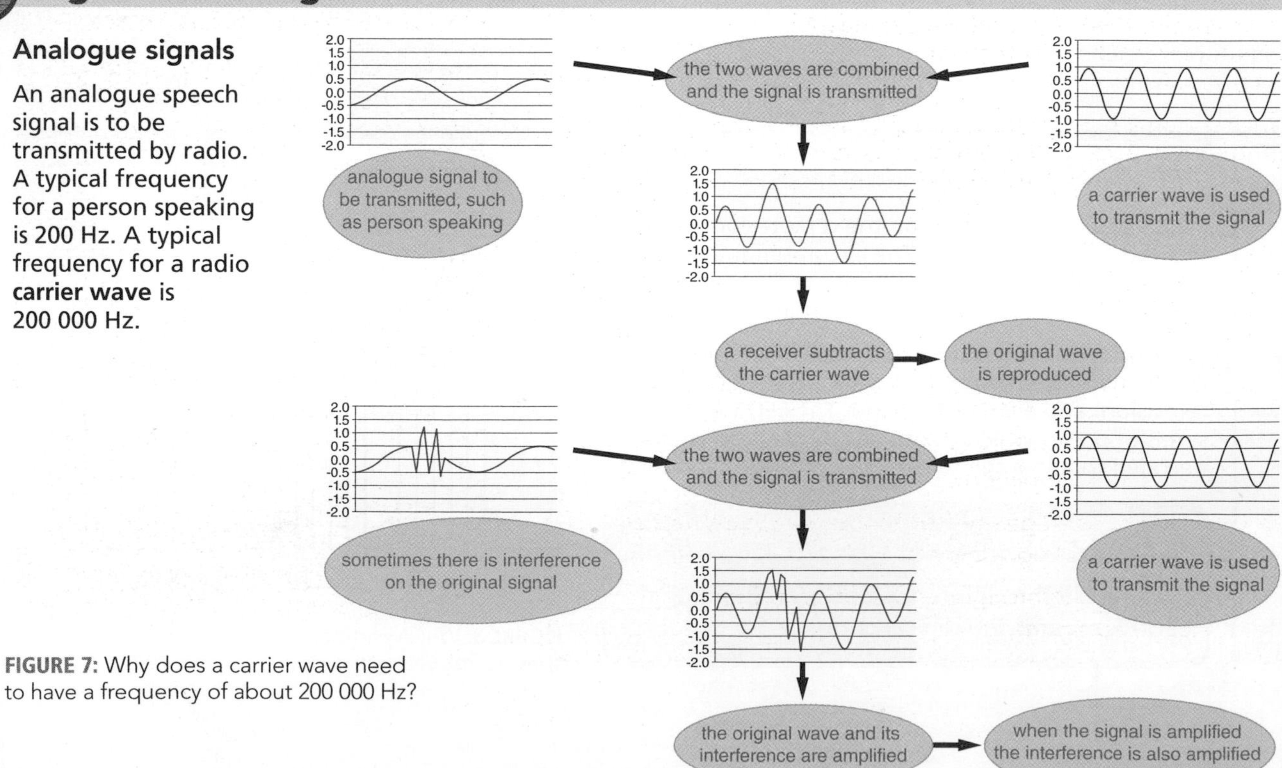

FIGURE 7: Why does a carrier wave need to have a frequency of about 200 000 Hz?

Digital switchover

The properties of digital signals such as multiplexing (see below) contributed to the decision to change from analogue to digital radio and television broadcasting.

From 2009, television signals in the UK started to be changed from analogue to digital. The digital radio switchover is planned for 2015 although that may be delayed until more people buy digital radios.

The switchover for both radio and TV means:

> improved signal quality – both picture and sound

> a greater choice of programmes

> being able to interact with the programme

> information services such as programme guides, and subtitles.

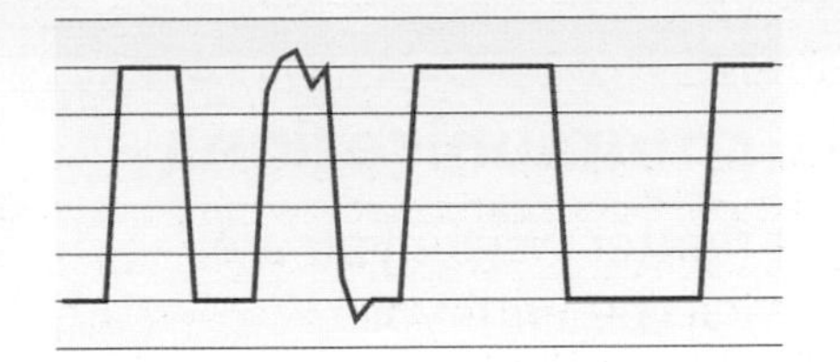

FIGURE 9: Why is the interference easy to eliminate from the final output?

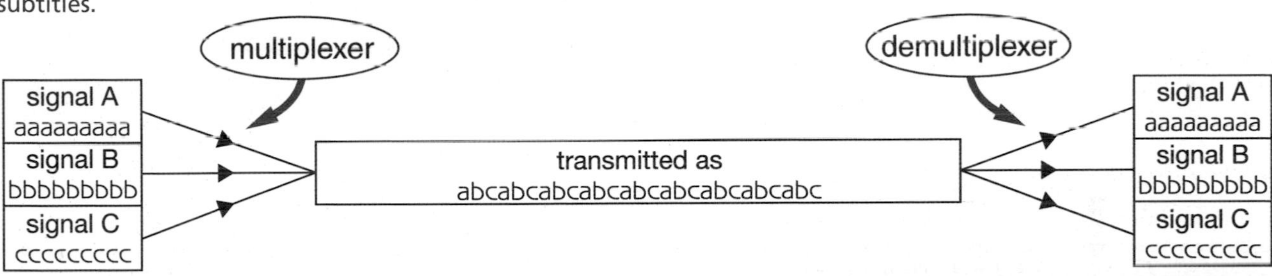

FIGURE 8: What is an advantage of multiplexing?

Multiplexing allows many different digital signals to be transmitted simultaneously. Each digital signal is divided into segments of very short duration. A multiplexer combines the individual signals to be sent and transmits them as a combined signal. A demultiplexer separates them at the receiving end of the transmission.

When a digital signal is transmitted it too can have interference. Because the digital signal is either high or low the interference is not significant in the final output.

Did you know?

As a result of multiplexing, over ten thousand separate telephone conversations can pass along a fibre smaller than the eye of a needle with no interference.

Questions

5 What happens to the interference on an analogue signal when it is amplified?

6 Give two advantages of transmitting a digital signal rather than an analogue signal.

7 What are the advantages of watching a digital television programme?

Wireless signals

Mobile communications

It does not matter where you are, your office can be with you.

The mobile phone and wireless laptop mean that you are available 24 hours a day, 7 days a week.

You can talk to people on the other side of the world. You can access the internet and use e-mail to send and receive documents.

All this is achieved without being connected to a mains power supply or a telephone line, but you do need an aerial.

You will find out:

> about the reflection and refraction of radio signals

> how radio stations interfere with one another

> how wireless technology supports a busy lifestyle

FIGURE 1: What are the advantages of mobile communications?

Wireless technology

Wireless technology uses **radio waves** and microwaves. These waves are part of the electromagnetic spectrum.

Getting the message across

Radio waves behave in the same way as light. They can be reflected off a hill or other obstacles.

The TV aerial on the house in the picture receives two signals. One comes directly from a transmitter. The other is reflected from an obstacle.

The reflected signal:

> travels further

> if analogue, produces a picture shifted to the right.

This is called ghosting.

Sometimes reflected signals allow places not in direct line with the transmitter to receive radio signals.

Digital signals are replacing analogue signals for communication purposes. See pages 208–211 for more information on analogue and digital signals.

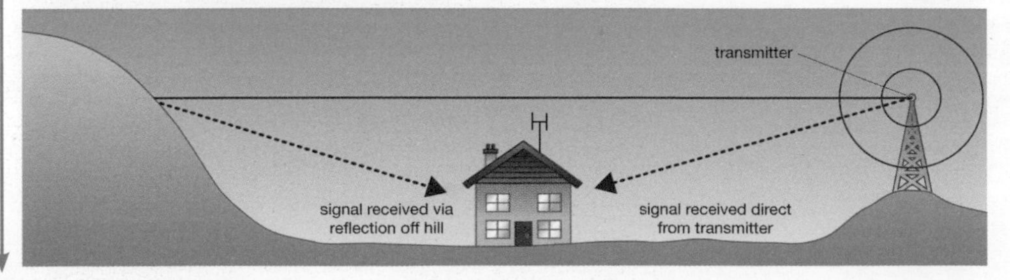

FIGURE 2: What happens to a radio signal when there is a solid object in the way?

microwave communication   radio waves   wireless technology

FIGURE 3: What causes ghosting?

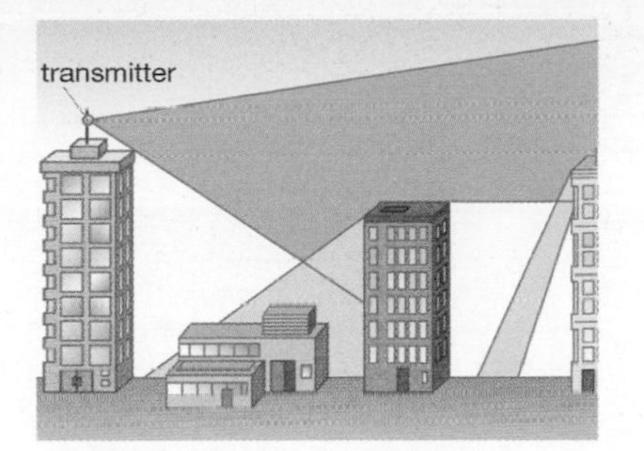

FIGURE 4: Why can radio waves sometimes be received by places in shadow?

Questions

1 Suggest two obstacles, other than a hill, that cause ghosting.

2 A television receives a direct signal and a reflected signal. Describe what you see.

Refraction of radio waves

All electromagnetic waves can be refracted. Radio waves are refracted in the lower layers of the atmosphere.

The amount of refraction depends on the frequency of a wave. Waves with a long wavelength and low frequency show most refraction.

Long-wave radio transmissions show so much refraction that the wave returns to the Earth's surface.

Radio interference

Cirencester is about 100 miles north of Exeter. Radio Devon broadcasts from a transmitter in Exeter at a frequency of 95.8 MHz. Radio Gloucestershire broadcasts from Cirencester at the same frequency. The distance between these stations means that usually only one will be received.

Sometimes weather conditions mean that signals may travel further than normal and there is interference between the two stations. If you are listening to one station, you can hear the other one faintly in the background. Radio stations that are close to one another will broadcast on different frequencies.

The effect of interference between radio stations is reduced if the signals are digital.

Another advantage of Digital Audio Broadcasting (DAB) is the availability of more radio stations. Also instead of having to remember the frequency of each station, you can choose your station by name.

There are disadvantages. There will be some areas of the country where coverage is poor. Compared to FM radio, the audio quality of DAB is not as good although it will improve in the future.

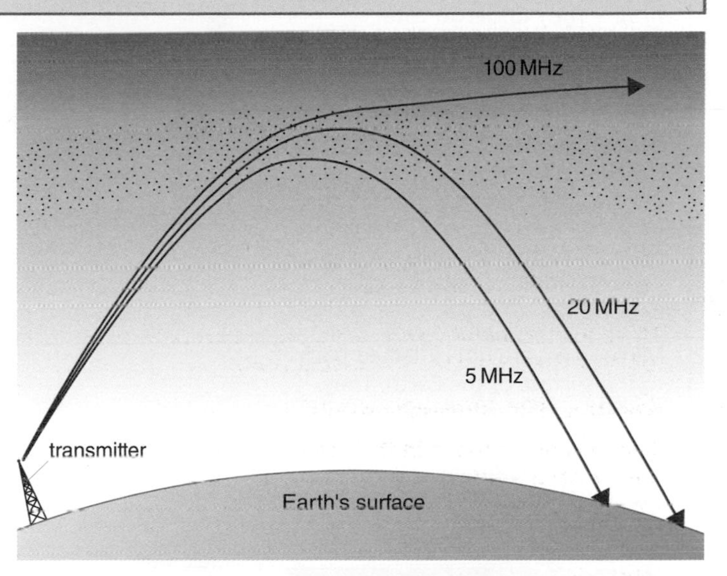

FIGURE 5: What type of wave shows most refraction?

Did you know?

If a signal could follow the curvature of the Earth all the way round, it would take 0.134 seconds to send a message right round the world.

Questions

3 A local radio station broadcasts with a frequency of 103.4 MHz. Why will the signal not be refracted back to Earth from the upper atmosphere?

4 Suggest why people who live in the south of England sometimes hear foreign radio stations in the background when they are listening to their radios.

You will find out:
> how worldwide transmission of radio signals is achieved
> about the diffraction of radio signals

Around the world in 0.134 seconds

One of the layers in the Earth's atmosphere is called the **ionosphere**. Radio waves can be reflected from the ionosphere. Water is also able to reflect radio waves, but land masses are not such good reflectors.

The propagation of radio signals from one place to another because of reflection from the ionosphere is known as sky wave propagation. Radio waves in the ionosphere behave in a similar way to light rays in an optical fibre – they undergo total internal reflection.

As a radio wave travels up in the ionosphere, it passes from a denser to a less dense medium. It continuously bends away from its path until it is totally reflected back to Earth. Coupled with reflection from the oceans, radio signals can be received well beyond line of sight.

Microwaves are not reflected back to the surface of the Earth. They pass through the ionosphere and are received by **satellites** orbiting the Earth. The satellite amplifies the signal and then re-transmits the signal back to Earth.

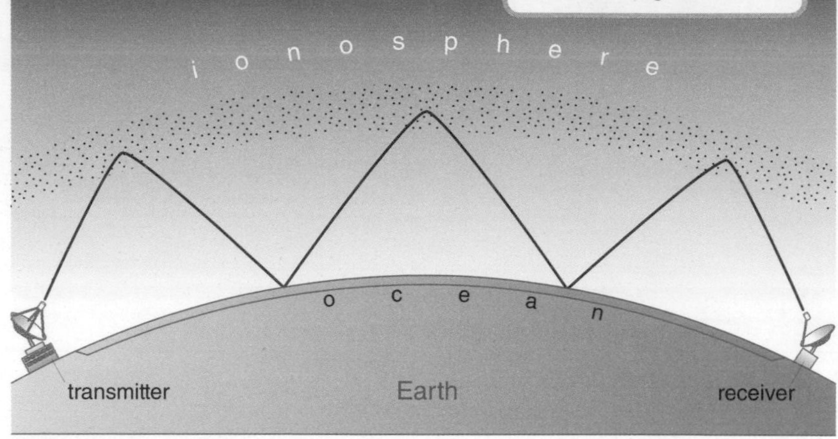

FIGURE 6: How can radio waves be received out of line of sight?

Did you know?

Using a satellite for communication delays a signal by only 0.24–0.28 seconds.

Radio waves are diffracted when they meet an obstruction. The diffraction is only significant if the size of the obstacle is similar to the wavelength (see Figure 7).

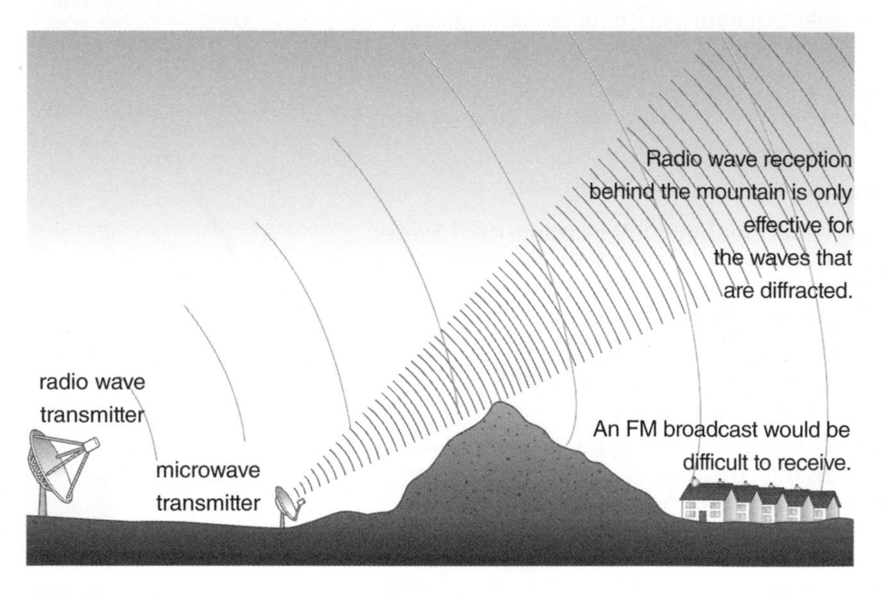

FIGURE 7: Long-wave signals are diffracted by hills but short-wave and microwave signals are not.

Question

5 Jamie lives in a house in a valley. He can listen to long-wave and medium-wave stations on his radio. He cannot get any signal on his mobile phone and VHF (very high frequency used for FM) reception is very poor. Explain these differences.

Communications satellites orbit the Earth every 24 hours at a height of 36 000 km above the equator.

FIGURE 8: What type of signal does a communications satellite receive?

Communication problems

The refraction of radio waves in the atmosphere needs to be taken into account, particularly when transmitting a signal to a communications satellite. The size of the aerial dish on the satellite is not very large and a focused beam of energy needs to be transmitted.

There will be some diffraction at the edges of the dish. The wavelength of the microwaves must be small compared to the dish diameter to reduce this diffraction.

Many radios now receive digital signals and this reduces the amount of interference, especially between radio stations.

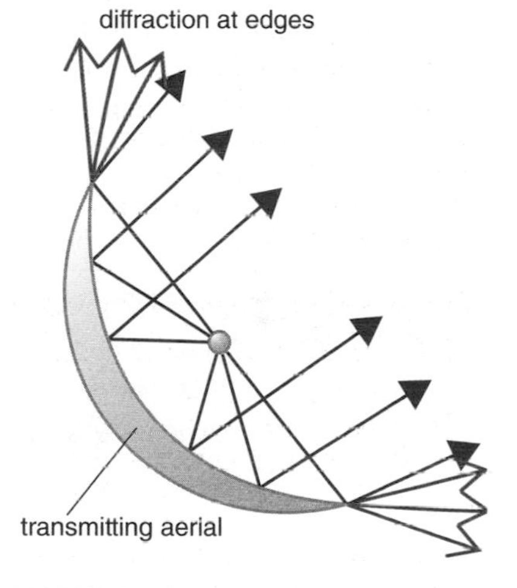

FIGURE 9: Why does a focused beam of energy need to be transmitted from a transmitting aerial?

Questions

6 Suggest why signals to satellites are transmitted as a slightly divergent beam and not as a parallel beam.

7 Why does the size and shape of a transmitting dish aerial dictate the use of microwaves instead of long-wave radio signals?

Stable Earth

You will find out:

> how earthquakes are detected
> about the types of waves that earthquakes produce
> how the properties of earthquake waves allow us to investigate the inside of the Earth

Devastating wave

On Friday 11 March 2011, an undersea earthquake close to Japan was the cause of a tsunami that claimed many lives.

A tsunami is a gigantic wave. Another occurred in the Indian Ocean in December 2004.

With a distance of 500 km between crests and a wave height of 10 m, the wave travelled at a speed of 800 km/h towards islands such as Sri Lanka and The Maldives and also touched the mainland coastline. No one was prepared for the devastating results.

FIGURE 1: What are the longer-term effects of a tsunami?

Earthquakes

An earthquake happens when rocks deep below the surface of the Earth move suddenly at a **fault**.

Shock waves pass through the Earth and travel around its surface. It is the surface waves that cause damage to houses and other structures.

We detect earthquakes using a **seismometer**.

> A heavy weight with a pen attached is suspended above a rotating drum. There is paper on the drum. The base is bolted to solid rock.
> During an earthquake, the base moves but the pen stays still.
> The trace drawn on the paper is a seismograph.

Earthquake description	Average number
very minor	9000 per day
minor	49 000 per year
light	6000 per year
moderate	800 per year
strong	120 per year
major	18 per year
great	1 per year

Questions

1 What force attracts the pen of a seismometer towards the centre of the Earth?

2 Why is the base of the seismometer bolted to solid rock?

Did you know?

Earthquakes happen more often than you think. Thankfully, only a few lead to the sort of disaster seen in Japan in 2011.

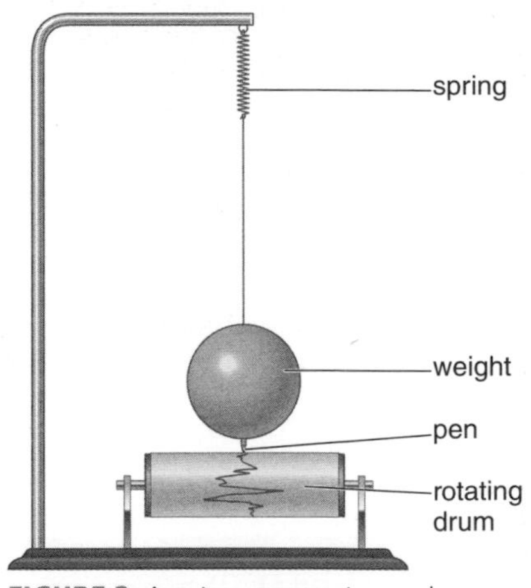

FIGURE 2: A seismometer is used to detect and measure the size of an earthquake.

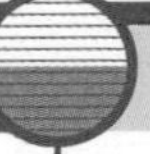

Earthquake waves

An earthquake happens below the Earth's surface at the focus. The **epicentre** is the point on the Earth's surface above the focus.

An **L wave** is a surface wave. It travels out from the epicentre relatively slowly.

Two types of seismic wave created by earthquakes are:

> A **P wave** is a primary (pressure) wave. It is a longitudinal wave and is similar to a sound wave. It travels through the Earth at a speed of between 5 and 8 km/s.

> An **S wave** is a secondary (shear) wave. It is a transverse wave and travels through the Earth at a speed of between 3 and 5.5 km/s. It is slightly slower than a P wave.

The seismograph trace from a seismometer shows the two types of wave.

These waves are important because they help to tell us about the structure of the Earth.

> P waves pass through solids and liquids.

> S waves pass only through solids.

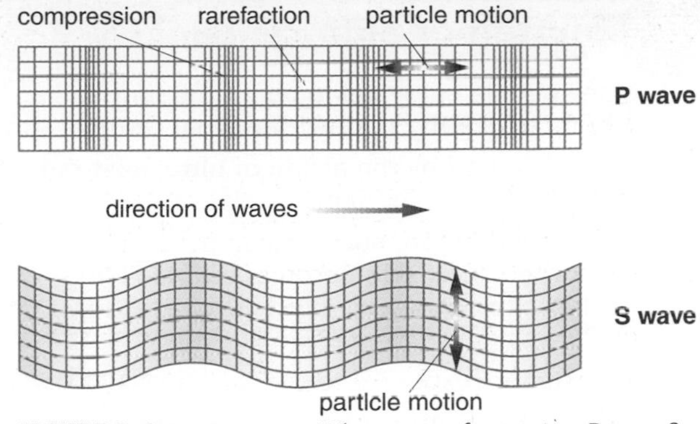

FIGURE 3: Seismic waves. What type of wave is a P wave?

Questions

3 Which type of earthquake wave is similar to a wave in the electromagnetic spectrum?

4 Why does the trace from a seismometer have two separate wave patterns?

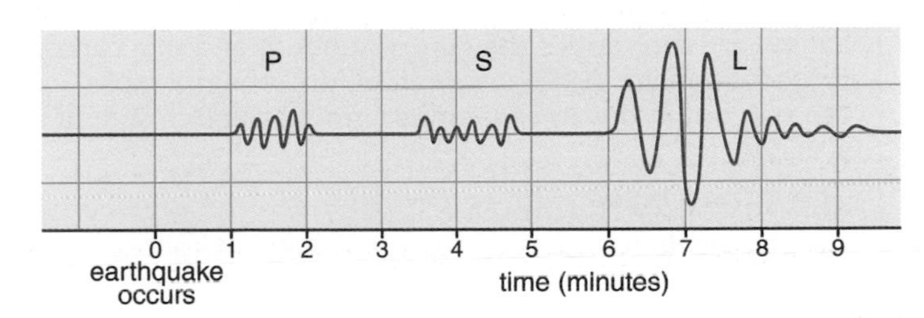

FIGURE 4: Look at this seismograph. Which type of wave travelled fastest after the earthquake?

Inside the Earth

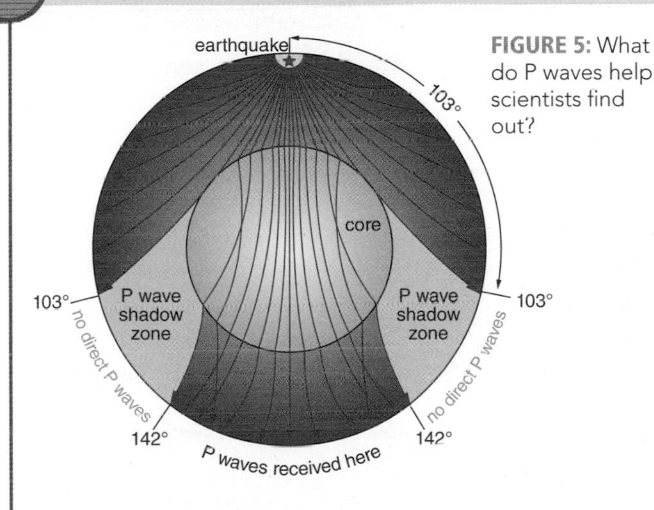

FIGURE 5: What do P waves help scientists find out?

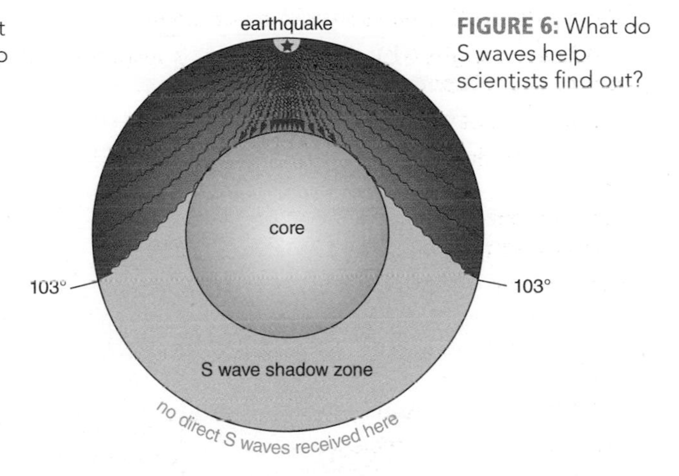

FIGURE 6: What do S waves help scientists find out?

Scientists use the properties of seismic waves to find out more about the Earth's structure.

P waves pass through the Earth and are refracted by the core. The paths taken by P waves help scientists to calculate the size and density of the Earth's layers.

S waves cannot be detected on the opposite side of the Earth to an earthquake. This tells scientists that there is some liquid in their path.

Elsewhere, S waves are detected over the same area as P waves, which indicates that the liquid is found in the Earth's core.

Questions

5 a Draw a trace from a seismometer to show the pattern when all three types of seismic wave are received.

b Draw a second trace to show the pattern from the same earthquake received by a seismometer on the opposite side of the Earth to the earthquake.

P wave S wave L wave

Suntan or skin cancer?

You will find out:

- how to avoid the harmful effects of ultraviolet radiation
- about the ozone layer and how scientists monitor it

Many people enjoy being out in the sunshine and getting a 'healthy suntan', but what causes a suntan?

A tan is caused by the action of **ultraviolet** radiation from the Sun on the skin.

> If you spend too much time in the Sun you can get sunburned. Your skin becomes red, can blister and start to peel. Repeated exposure can cause your skin to shows signs of premature ageing.

> Even more exposure to the Sun can cause skin cancer.

> Exposure to ultraviolet radiation also causes cataracts. This is a clouding of the lens in the eye causing blurred vision.

FIGURE 7: How can you prevent sunburn when you are spending time outside on a sunny day?

Sunscreens can reduce the risks of sunburn and skin cancer. A high sun protection factor (SPF) sunscreen reduces the risk more and means you can stay out in the Sun for longer. Young children should use at least SPF 30.

The ozone layer

The **stratosphere** is part of our atmosphere. It contains a gas called **ozone**. This gas absorbs ultraviolet radiation.

Satellites started to monitor the amount of ozone in the early 1970s. Scientists discovered that the levels of ozone, particularly above Antarctica, were falling faster than expected.

In 1985, the British Antarctic Survey recorded a large drop in atmospheric ozone; they thought their instruments were faulty and replaced all of them. The first explanations were linked to the increased use of fertilizers and later to CFCs in aerosols and refrigerators.

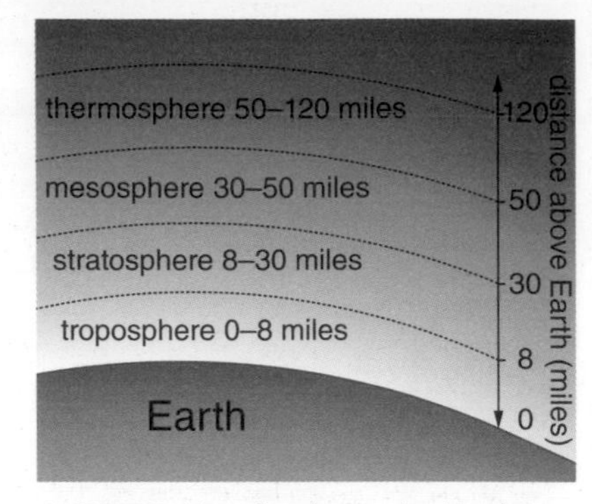

FIGURE 8: How far above the Earth is the stratosphere?

Did you know?

The most common sign of skin cancer is a change on the skin, especially a new growth or a sore that does not heal.

Not all skin cancers look the same. The cancer may start as a small, smooth, shiny, pale or waxy lump. It can appear as a firm red lump. Sometimes, the lump bleeds or develops a crust. It can also start as a flat, red spot that is rough, dry or scaly.

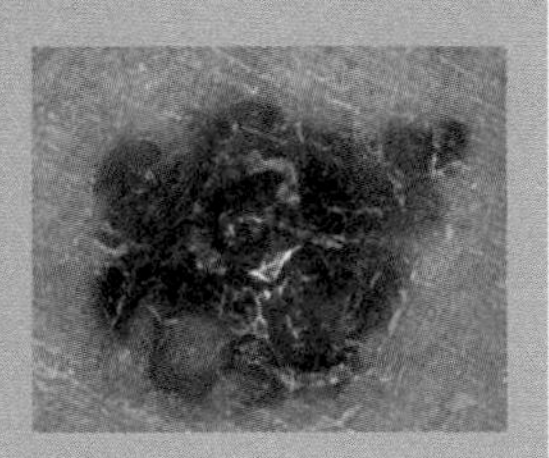

Questions

6 Suggest why young children should use sunscreen with a high SPF.

7 Cricketers often spend a long time in the Sun during a game. Which part of their body is most likely to develop skin cancer if they do not apply enough sunscreen?

Safe sunbathing

A tan is caused by the action of ultraviolet radiation from the Sun on the skin. Inside skin there are cells that produce **melanin**. This is the pigment that causes skin to turn a tan colour. Ultraviolet radiation stimulates melanin production.

People with naturally darker skin are less likely to suffer from skin cancer. Their skin absorbs more of the ultraviolet radiation so less reaches the cells that might become cancerous.

Many scientists are now saying that instead of tanning we should avoid the Sun and keep our natural skin colour. If you do go out in the Sun, you should check the sun index for the day. Many parts of the world publish a daily sun index. The BBC weather centre's sun index gives information about:

> the strength of the Sun

> how long a person can stay in the Sun without burning when not wearing sunscreen.

These risks are for someone with white skin who burns easily. When you use a sunscreen the length that you can safely stay in the Sun can be calculated by:

safe length of time to stay in the Sun with sunscreen on = SPF number × time given in sun index

skin cancer ultraviolet radiation

A sun index of 6 means you can stay in the Sun for 20–30 minutes. If you use a sunscreen with SPF 15, you can stay in the Sun for 15 times longer, which is 5–7.5 hours.

Some people wrongly think that using a sunbed is safer than sunbathing. The amount of ultraviolet radiation from a sunbed is greater than that from the midday Sun. Many cancer research organisations provide warnings about the dangers of exposure to ultraviolet radiation.

Sun index	Risk
1–2	low risk
3–4	avoid being in Sun for more than 1–2 hours
5–6	burns in 30–60 minutes
7–10	severe burns in 20–30 minutes

FIGURE 9: Which SPF will allow you to spend the longest time in the Sun?

Confirming a 'hole'

The scientists who first discovered that the ozone layer was depleting faster than predicted did not believe their results. Their first reaction was to change their equipment because they thought it was faulty. However, the new equipment confirmed their readings. Other scientists also came to the same conclusion. A satellite which was monitoring the ozone layer also recorded lower levels than expected and it is suggested a software error was originally thought to be the cause.

When a number of independent researchers all have the same conclusion, this tends to increase confidence in their conclusions. Scientists often work independently to check each other's findings. They also ask others to read what they are about to publish to make sure the research has met the necessary criteria for being reliable. This is known as peer review.

Questions

8 Dave wants to go out in the Sun. He sees that the sun index for the day is 9. Use the table on the left to answer the following questions.

a How long can he safely stay in the Sun without sunscreen?

b How long can he safely stay in the Sun if he uses sunscreen with SPF 12?

Filling a hole?

Ultraviolet radiation has a range of wavelengths. The shorter the wavelength, the more energy the waves possess and the more harmful they are. The short wavelength ultraviolet radiation breaks apart the molecular bonds in the ozone and the energy is absorbed. This means the most harmful radiation does not reach the Earth.

One of the pollutants discharged into the atmosphere as a result of human activities is CFC gas from aerosols and refrigerators. This gas destroys ozone and so the layer is becoming thinner. This thinning is sometimes called a 'hole'.

Scientists are monitoring this thinning of the ozone layer using satellites. You can see from the satellite image in Figure 10 that the ozone layer is at its thinnest above the South Pole. The ozone hole first appeared over Antarctica because the chemical process that destroys ozone works best in cold conditions. Antarctica is colder than the Arctic.

Governments and the international community were quick to respond to these environmental issues. The Montreal Protocol was signed in 1987 with a target of reducing CFCs by half by 2000. The European Community stopped producing the main CFCs in 1995. The target was met well before 2000.

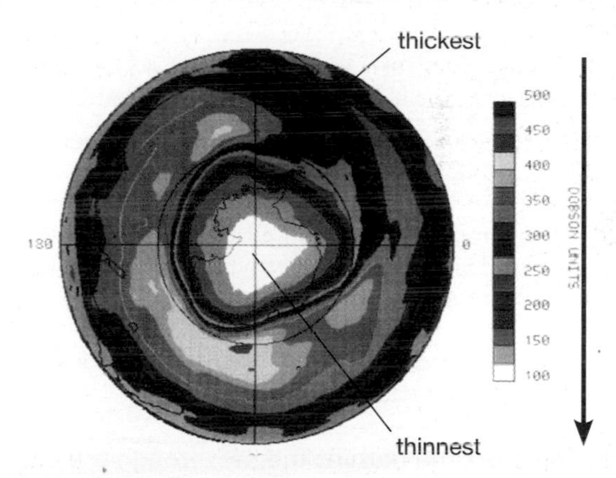

FIGURE 10: Where is the ozone layer at its thinnest?

Questions

9 If the ozone layer becomes thinner what will happen to the following?

a The amount of ultraviolet radiation reaching the Earth.

b The number of people suffering from skin cancer.

Preparing for assessment: Analysis and evaluation

To achieve a good grade in science, you not only have to know and understand scientific ideas, but you need to be able to apply them to other situations and investigations. These tasks will support you in developing these skills.

Tasks

- > Find out if there is a simple relationship between the angle of incidence and angle of refraction.
- > Find out if light always refracts by the same amount no matter what the medium it is passing into.

Context

When light strikes a mirror, it is reflected. It does not matter what material the mirror is made from, the angle of incidence always equals the angle of reflection.

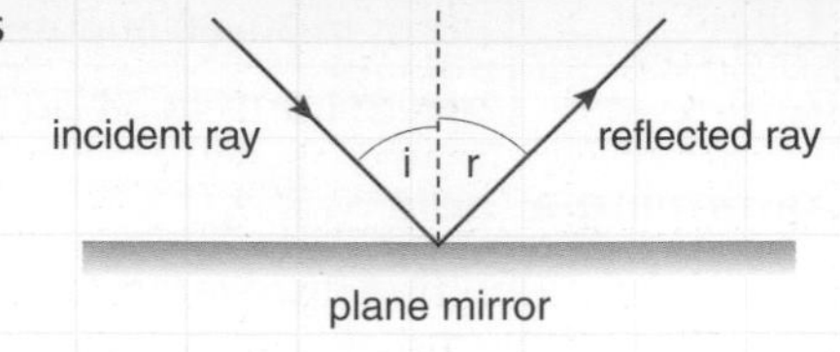

When light passes from one medium into another, it is refracted and changes direction.

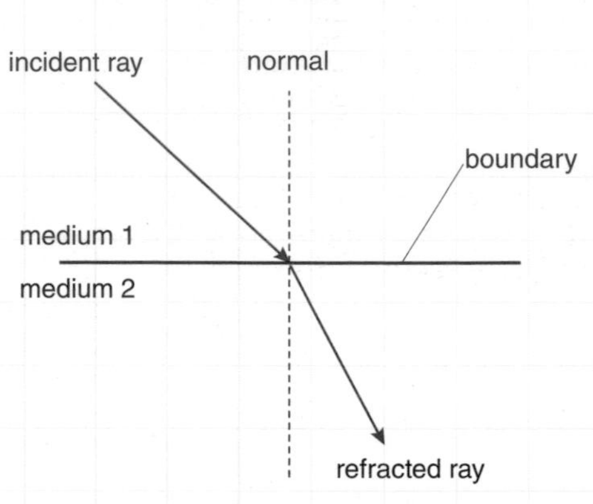

How to do this experiment

1. Place a rectangular glass block onto a sheet of white paper and draw round it.

2. Shine a ray of light from a ray box towards one side of the glass block. Use a pencil to mark the centre of the incident ray and the centre of the emergent ray.

3. Remove the glass block and join the marked points to define the incident ray and the emergent ray.

4. Construct the path of the ray through the block by joining the incident ray to the emergent ray.

5. Measure the angles of incidence in air and refraction in glass.

6. Repeat the experiment four more times using a range of angles.

7. To find out if other materials refract by the same amount, replace the glass block with one made from a different material and do the whole experiment again.

8. A 'water block' can be made by putting the water in a rectangular container made from thin plastic.

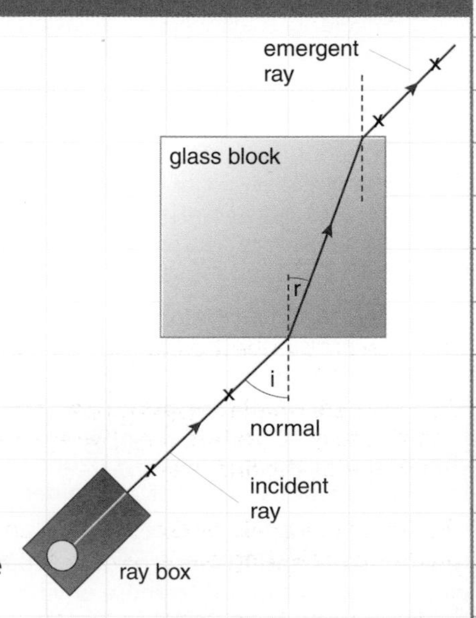

Results, analysis and evaluation

A group of students found these results when they tested glass, water and another transparent material.

GLASS		WATER		TRANSPARENT MATERIAL	
angle of incidence (°)	angle of refraction (°)	angle of incidence (°)	angle of refraction (°)	angle of incidence (°)	angle of refraction (°)
5	3	20	15	15	6
10	7	35	26	30	12
15	10	50	35	45	17
55	33	65	43	60	21
70	38	80	48	75	24

1. Use the same set of axes to plot graphs of angle of refraction (y-axis) against angle of incidence (x-axis) for each material. Use a separate sheet of graph paper.

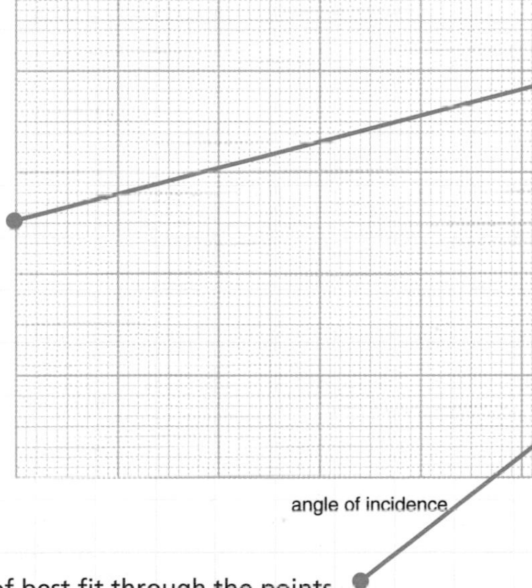

Think about the scale for the axes. Use sensible divisions but make the graph as large as possible.

2. Draw lines of best fit through the points.

Should these be straight lines or smooth curves? Never draw dot to dot.

3. Is there a simple relationship between the angle of incidence and the angle of refraction? Explain your answer by referring to the shape of the line graph.

What shape graph would give a simple relationship?

4. Do all materials refract light by the same amount? Explain your answer by referring to the three graphs.

It is sometimes easier to think about what would happen if the answer is yes.

5. How could the students who found the results for glass improve their technique?

Have a close look at the intervals between their incident angles.

6. How can the students improve the precision of their results?

Do not confuse precision with accuracy; precision is about the repeatability of results.

P1 Checklist

To achieve your forecast grade in the exam you'll need to revise

Use this checklist to see what you can do now. It gives you many of the important points you will need to know. Refer back to the relevant pages in this book if you're not sure and to see if there is anything else you need to know. Look across the three columns to see how you can progress.

Remember you'll need to be able to use these ideas in various ways, such as:

> interpreting pictures, diagrams and graphs
> applying ideas to new situations
> explaining ethical implications
> suggesting some benefits and risks to society
> drawing conclusions from evidence you've been given.

Look at pages 278–299 for more information about exams and how you'll be assessed.

To aim for a grade E	To aim for a grade C	To aim for a grade A
know the difference between heat and temperature **understand** how rate of cooling is affected by temperature **know** that a thermogram indicates temperature **know** what affects energy needed to change temperature or state	**know** the effects of energy flow from hotter to cooler bodies **interpret** data on rate of cooling **explain** how thermograms use colour **understand** specific heat capacity and latent heat **use** specific heat capacity and latent heat equations	**understand** that temperature depends on kinetic energy **know** that heat is measured on an absolute scale
explain why trapped air is a good insulator **describe** examples of energy saving in the home **explain** how trapped air is used to keep homes warm **use** the energy efficiency equation	**explain** how energy transfer can be reduced in homes **interpret** data for different energy saving strategies **use** the energy efficiency equation to complete Sankey diagrams	**describe** how conduction, convection and radiation occur **explain** how design features reduce energy loss **explain** why trapped air reduces energy loss through a cavity wall **use** information on efficiency to draw Sankey diagrams
recognise the features of a transverse wave **know** that electromagnetic waves travel in straight lines **recognise** what happens when reflection or refraction occurs **know** that all electromagnetic waves travel at the same speed **use** the wave equation	**describe** the main features of a transverse wave **describe** how waves diffract at an opening **understand** that refraction occurs due to a change in the wave speed **draw** ray diagrams to illustrate reflection and refraction **arrange** the electromagnetic spectrum in order by wavelength and frequency **manipulate** the wave equation and use standard form	
know that using light increases the speed of communication **recognise** where total internal reflection happens **understand** how light and infrared can travel along an optical fibre **understand** that lasers produce an intense, narrow beam of light **recall** uses of lasers	**explain** advantages and disadvantages of using light, radio and electrical signals for communication **describe** how light behaves at a boundary between two materials **explain** how total internal reflection occurs in optical fibres **know** that laser light is a narrow beam of light of a single colour	**describe** applications of total internal reflection in optical fibres **explain** the term coherent beam of light **explain** how a laser is used in a CD player

To aim for a grade E	To aim for a grade C	To aim for a grade A
interpret information about the electromagnetic spectrum **explain** how the emission and absorption of infrared radiation by an object is affected by its temperature, colour and texture **recognise** that water and fat absorb microwaves **recall** that mobile phones use microwave signals **know** the different views about the risks from mobile phones	**describe** properties of infrared radiation and microwaves **understand** the problems when microwaves transmit information **realise** the evidence of dangers from mobiles is not conclusive	**explain** how microwaves and infrared transfer energy **know** how the energy of microwaves depends on frequency **explain** how to reduce signal loss with microwaves **know** it is not easy to decide on the siting of phone masts
describe uses of infrared radiation **describe** the differences between analogue and digital signals **know** how infrared sensors and thermal imaging cameras work	**describe** how infrared signals control electrical devices **describe** the transmission of light in optical fibres	**explain** how the signal from an infrared remote controls a device **describe** advantages of using digital signals **describe** advantages of using optical fibres **explain** how the properties of digital signals allow us to switch to digital TV and radio
recognise that wireless technology uses electromagnetic radiation **describe** how radiation used for communication can be reflected **describe** the advantages of wireless technology **interpret** information on digital and analogue signals	**recall** how radiation used for communication is refracted and reflected and this can be an advantage or disadvantage for good reception **describe** common uses of wireless technology **describe** advantages and disadvantages of DAB radio	**explain** how long-distance communication uses satellites and the ionosphere **recall** that radio waves (like light) exhibit total internal reflection **explain** the advantages of digital radio
describe the effects of shock waves from an earthquake **recall** effects of exposure to ultraviolet radiation **recognise** that sunscreens can reduce damage to skin **recall** that scientists were surprised to find the ozone hole	**recall** the differences between P waves and S waves **explain** how darker skins have lower cancer risk **interpret** data about sun protection factor **describe** how measurements of ozone reduction were checked	**describe** how seismic waves help to model the Earth's structure **explain** how the ozone layer protects the Earth from ultraviolet radiation **describe** why the ozone layer is depleting and the effect this has **describe** how the discovery of the ozone hole changed attitudes

P1 Exam-style questions

Foundation Tier

1. The speed of radio waves in air is 300 000 km/s.

AO1 **(a)** Write down **two** other types of radiation that travel at this speed. [2]

AO1 **(b)** Radio 1 broadcasts nationally with a **frequency** of 97.3 megahertz. What is meant by the term frequency? [1]

[Total: 3]

AO2 **2 (a)** Katherine sticks aluminium foil onto the wall behind the radiator in her room. This reduces energy loss by radiation. Explain how it does this. [2]

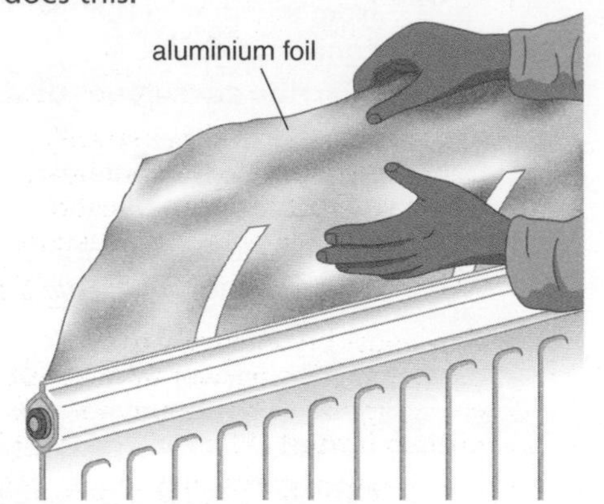

(b) The loft of Katherine's flat has fibreglass wool placed between the joists.

AO2 Explain how fibreglass wool helps to reduce energy loss. [4]

AO1 **(c)** Suggest one other method Katherine could use to reduce energy loss from her flat. [1]

[Total: 7]

3. Infrared radiation is absorbed by the surface of food and the energy is conducted through the food to cook it.

James uses a **microwave** oven to cook his food.

AO1 **(a)** Why does a microwave oven cook food quicker than an oven that cooks by infrared radiation? [1]

AO1 **(b)** The door of a microwave oven has a special grill that stops the microwaves passing through.

What effect will microwaves have on body tissue? [1]

AO1 **(c)** Suggest one other use for microwaves. [1]

[Total: 3]

4. Liz is listening to her radio. Sometimes she can hear another radio station in the background.

AO1 **(a)** What is this called? Choose from this list.

absorption

diffraction

dispersion

interference [1]

AO1 **(b)** Why does this happen? [1]

(c) Anne has a DAB digital radio. When she listens to the same station as Liz, there is no other radio station in the background.

AO1 **(i)** Suggest **one** other advantage of DAB radio. [1]

AO1 **(ii)** Suggest **one** disadvantage of DAB radio. [1]

[Total: 4]

5. Sam and Michelle are on the beach. It is very sunny. They want to get tanned.

AO1 **(a)** What else can happen to their skin if they are exposed to too much ultraviolet radiation? [1]

AO2 **(b)** They are advised that they should spend no more than 15 minutes in the Sun without using sunscreen. How long can they safely stay in the Sun if they use a sunscreen with SPF 20? [1]

[Total: 2]

Worked Example – Foundation Tier

Ian warms a beaker of ice gently with a Bunsen burner.

The graph shows how the temperature changes during the time he is heating it.

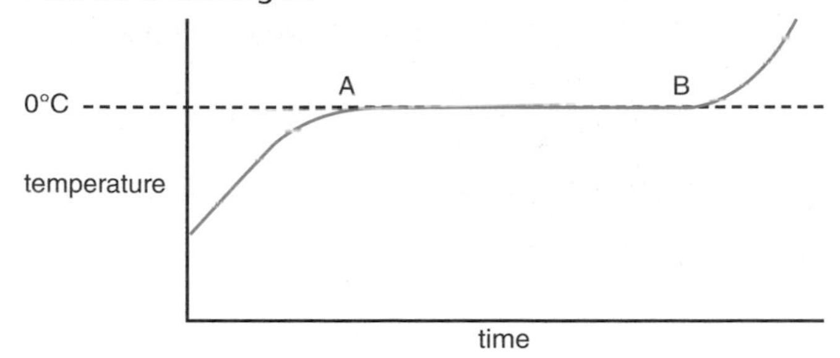

(a) What is happening to the ice between points A and B? [1]

It is changing into water

(b) Ice has a **specific heat capacity** and a **specific latent heat**. What is the difference between specific heat capacity and a specific latent heat? [3]

The specific heat capacity is the amount of heat needed to change the temperature when something gets hotter. Latent heat is the amount of heat needed to change the temperature when something is not getting hotter.

(c) (i) How much energy is needed to change 0.75 kg of ice into water?

Specific latent heat of ice = 340 000 J/kg. [2]

energy needed = mass × specific latent heat

= 340000 × 0.75

453333.33333 J

(ii) The Bunsen burner uses 425 000 J of energy. Calculate the efficiency of the energy transfer process. [2]

efficiency = $\frac{\textit{useful energy output}}{\textit{total energy input}}$

$= \frac{453333}{425000}$

1.07 = 107%

[Total: 8]

How to raise your grade!

Take note of these comments – they will help you to raise your grade.

This answer is correct, but it would be better to use the correct name for the process – melting. 1/1

Specific heat capacity has a precise definition that involves changing the temperature of 1 kg by 1°C.

The question asks about specific latent heat not about latent heat. Students must read the question carefully and answer the question.

Care must be taken not to contradict. The answer talks about changing the temperature but not getting hotter! 1/3

The answer to this question is incorrect. 340 000 has been divided by 0.75 instead of multiplying. But, the working is shown so some credit can be given. Students should always show how they work out an answer to a numerical question. 1/2

The answer is incorrect because it follows on from an incorrect answer to part (i). Marks are normally awarded if an incorrect answer is followed through but the student should have realised that an energy transfer cannot have an efficiency greater than 1. If students find something like this, they should go back and checking what they have already written. 1/2

This student has scored 4 marks out of a possible 8. This is below the standard of Grade C. With a little more care the student could have achieved a Grade C.

P1 Exam-style questions

Higher Tier

1. The diagram represents a transverse wave.

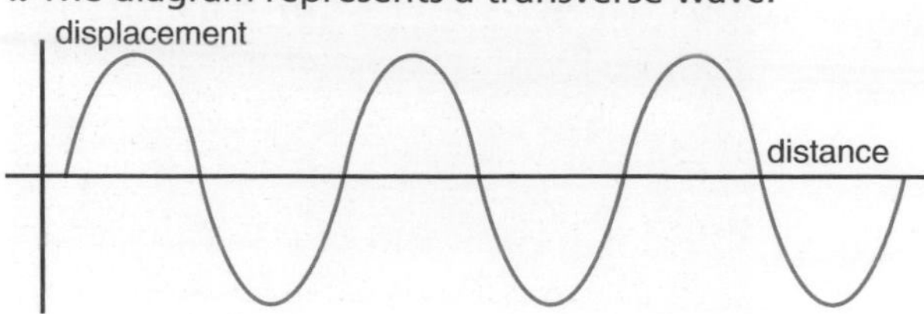

AO1 **(a)** Measure the wavelength of the wave with a ruler. [1]

AO2 **(b)** The wave speed is 300×10^6 m/s. Calculate the frequency of the wave. [2]

AO1 **(c)** Explain what is meant by the term **frequency**. [1]

(d) The wavelength is within the microwave area of the electromagnetic spectrum.

AO1 **(i)** Describe, with the help of a diagram, what happens when microwaves pass through a gap that is much larger than the wavelength. [1]

AO1 **(ii)** Describe, with the help of a diagram, what happens when microwaves pass through a gap which is similar in size to the wavelength. [1]

[Total: 6]

2. Aisha and Debbie are planning to sunbathe. They know that spending too long in the Sun increases the risk of skin cancer.

AO1 **(a)** Explain why Aisha (on the left) is less at risk from skin cancer than Debbie. [1]

(b) The ozone layer, which protects the Earth from ultraviolet radiation, is changing.

AO1 **(i)** What is happening to the ozone layer? [1]

AO1 **(ii)** What is causing the change? [1]

AO3 **(iii)** Why are scientists worried about what is happening to the ozone layer? [1]

[Total: 4]

AO2 **3 (a)** James uses an electrical heater to heat a well-lagged block of lead. The mass of the lead block is 2.4 kg. He finds that 12 480 J of energy are needed to increase its temperature from 16°C to 56°C.

Calculate the specific heat capacity of lead. [2]

(b) When lead melts at 327°C, 23 000 J of energy are needed to melt 1 kg.

AO1 **(i)** Finish the sentence.

23 000 J/kg is the value for the ………………… of lead. [1]

AO1 **(ii)** The temperature does not change when this energy is supplied to lead at 327°C.

What happens to this energy? [1]

[Total: 4]

AO2 **4 (a)** The cavity between the inner and outer wall of this house is filled with insulation material.

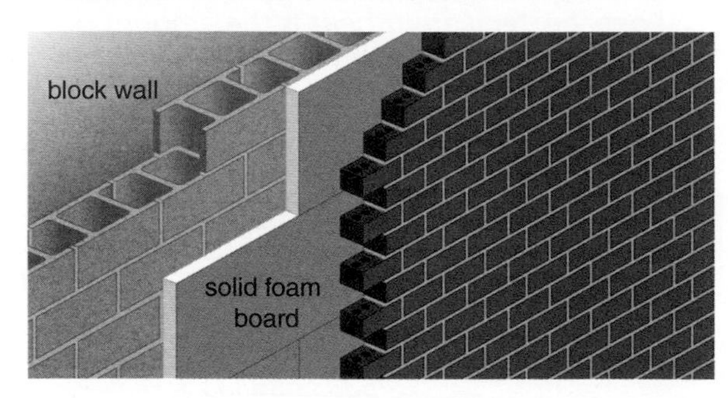

Sometimes, both sides of the insulation material are covered with a shiny foil.

Explain how energy transfer through the walls of a house is reduced. ✐ The quality of written communication will be assessed in your answer to this question. [6]

(b) An older house without cavity wall insulation can have foam injected at a cost of £250. This will save the home owner £125 per year.

AO1 **(i)** What is meant by the term **payback time**? [1]

AO1 **(ii)** Calculate the payback time for cavity wall insulation. [1]

AO2 **(iii)** Double glazing has a payback time of between 5 and 10 years. What advice would you give to someone who says it is more sensible, as far as energy saving is concerned, to install double glazing than cavity wall insulation. [1]

[Total: 9]

AO1 recall the science | AO2 apply your knowledge | AO3 evaluate and analyse the evidence

Worked Example – Higher Tier

Some radios broadcast using an **analogue** signal; some use a **digital** signal.

(a) What is the difference between an analogue signal and a digital signal? [2]

One is continually variable and can take on any value. The other has only two values

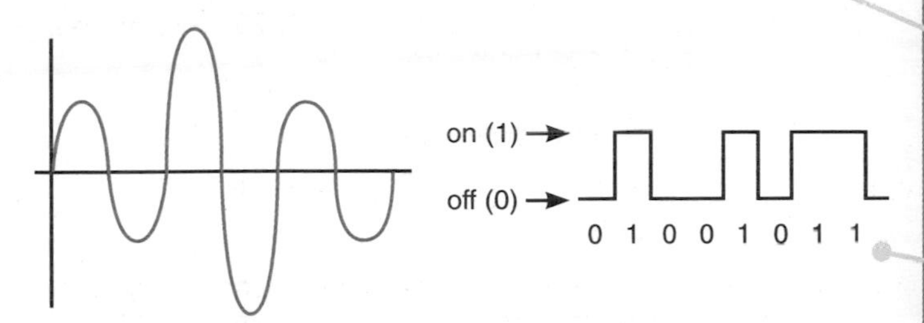

analogue signal *digital signal*

(b) What are the advantages of using digital signals? [2]

There is less interference because the noise is not recognised so is filtered out and not amplified

(c) Digital radio cannot be received in all areas and the quality is often poorer than present reception.

Explain **one** advantage of digital radio. [2]

There will be more stations available

[Total: 6]

How to raise your grade!

Take note of these comments – they will help you to raise your grade.

The answer does not make it clear which is being referred to. Students must always be precise.

Often, there will be space to draw a diagram. A well-labelled diagram can answer the question. 2/2

The question asks for advantages but only one has been given. If the question had asked for an explanation, this would have been worth full marks. The other advantage is the use of multiplexing. 1/2

This is stating one advantage but there is no explanation. The main advantage of digital radio is the lack of interference from other stations. 1/2

This student has scored 4 marks out of a possible 6. This is below the standard of Grade A. With a little more care the student could have achieved a Grade A.

P2 Living for the future (energy resources)

Ideas you've met before

Energy sources

Coal, oil and gas are fossil fuels formed from the remains of plants and animals.

Fossil fuels have taken millions of years to form and cannot be replaced.

The Sun is the ultimate source of energy for the Earth.

Renewable energy sources include water, wind, the Sun and biomass.

 **How is the energy from a fossil fuel released?**

Energy transformations

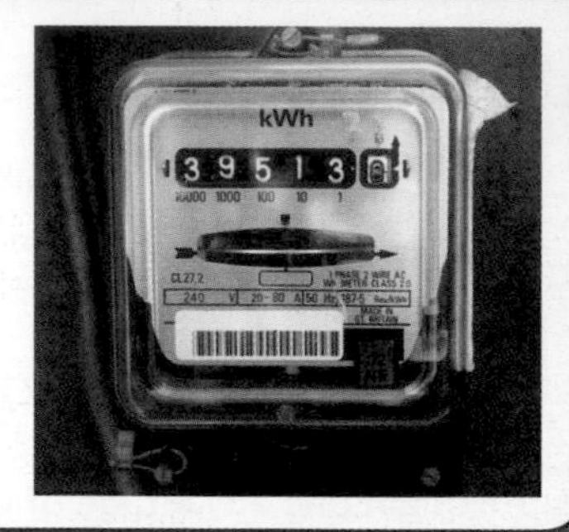

Electricity is a convenient way to transfer energy.

Voltage is measured using a voltmeter.

Fossil fuels and renewable energy sources can be used to produce electricity.

Some electrical appliances transfer more energy than others in a given time.

 **What is the name of the device that produces electricity in a power station?**

Nuclear radiation

Nuclear fuels can be used to produce electricity.

Nuclear fuels are not burned to produce energy.

An element is made of one sort of particle called an atom.

 **Find out the name of a nuclear fuel.**

The Earth in space

The Solar System consists of the Sun with planets and asteroids in orbit around it.

Moons orbit planets.

The Sun and other stars produce their own light.

We can see the planets and moons because they reflect light from the Sun.

Our Solar System is one of many in the Universe.

 **List the planets in our Solar System in order of distance from the Sun.**

In P2 you will find out about...

> how photocells use energy from the Sun to produce electricity

> the advantages and disadvantages of wind technology

> how passive solar heating keeps homes warm

> how the burning of fossil fuels has contributed to global warming

> how a dynamo produces electricity

> how electricity is produced in a power station

> how electricity is transmitted around the country

> the energy lost in a power station

> how to calculate the cost of electricity

> about the different types of nuclear radiation

> about the effects of nuclear waste and how to dispose of it safely

> how radioactive material is handled safely

> the uses of radioactive materials

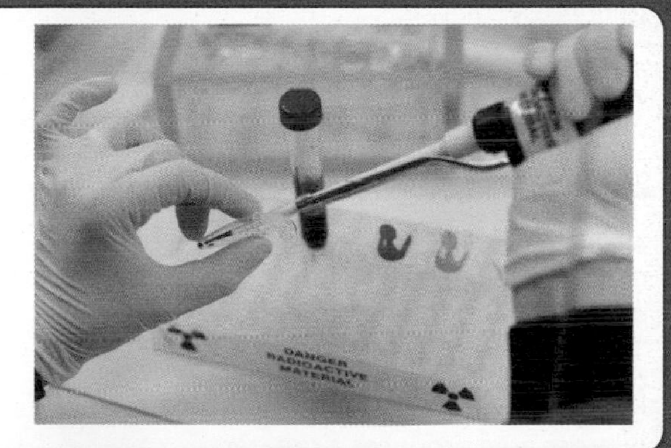

> why planets and moons stay in orbit

> how space is being explored by manned and unmanned spacecraft

> how very large distances in space are measured

> how asteroids have affected Earth in the past

> how scientists think the Universe started

> what happens to stars from their formation to their death

Collecting energy from the Sun

You will find out:

- > how photocells work
- > why photocells are used in particular situations

Total eclipse of the Sun

On 11 August 1999 in Cornwall, the Moon passed over the face of the Sun. The temperature fell by several degrees. Dogs started to bark. Birds flew to their nests. Streetlights suddenly lit up. An eerie silence descended.

It was a total eclipse of the Sun.

FIGURE 1: During a total eclipse of the Sun, the Moon passes between the Earth and the Sun.

Unlimited energy

The Sun is a stable source of energy for the Earth.

Without **solar energy** there would be no life on Earth.

> Light from the Sun allows plants to photosynthesise.

> Heat from the Sun provides the warmth for living things to survive.

Photocells and solar cells

A **photocell** changes light into electricity. A **solar cell** is a photocell that uses light from the Sun. The current (or power) output from a photocell depends on the area exposed to the light. A larger area produces a greater current.

Photocells can be used in places where mains electricity is not easily available.

The large area of photocells on the International Space Station provides a **direct current** (DC) supply to run the spacecraft.

The direct current from a photocell is similar to the current from a battery. It is in the same direction all the time.

FIGURE 2: Would we survive without the Sun's energy?

FIGURE 3: How do photocells help to run the International Space Station?

Question

1 What is the energy source for:

a a photocell?

b a solar cell?

Did you know?

The Sun loses 4 000 000 000 kg of its mass every second as energy.

It still has enough mass left to last for another 5 000 000 000 years.

Electricity from light

Photocells have many advantages over other electrical systems.

> They are very robust and need little maintenance.

> They carry on working for a long time.

> No fuel or lengthy power cables are needed.

> They do not contribute any pollution to the atmosphere in the form of particles or greenhouse gases.

> They use a renewable energy source.

One disadvantage is that they do not produce electricity when it is dark, such as during the night or when it is cloudy.

FIGURE 4: Why do you think there is a photocell on top of the parking meter?

Question

2 The current from a photocell is sometimes used to charge a battery. When would the battery be used?

How do photocells work?

It is not only tropical countries that benefit from using photocells. Even in the United Kingdom some houses have been built with photocells on their roofs, which produce enough electricity for the occupants.

A photocell consists of two special pieces of silicon joined together. Pure silicon is neutrally charged but if a small amount of an impurity such as phosphorus is added, there are more free electrons which make this a better conductor than pure silicon. This is known as **n-type** (negative) **silicon**.

If a different impurity, such as boron, is added there is an absence of free electrons. The absence of an electron is known as a hole and the silicon is known as **p-type silicon**.

FIGURE 5: What do these photocells produce?

When these two types of silicon are placed together it is called a **p–n junction**. An electric field is created between the two layers.

Sunlight is made up of tiny packets of energy called **photons**. When photons are absorbed in a photocell the energy causes electrons to become free. They move in the electric field and leave the junction to flow in an external circuit. The current (or power) produced by a photocell depends on:

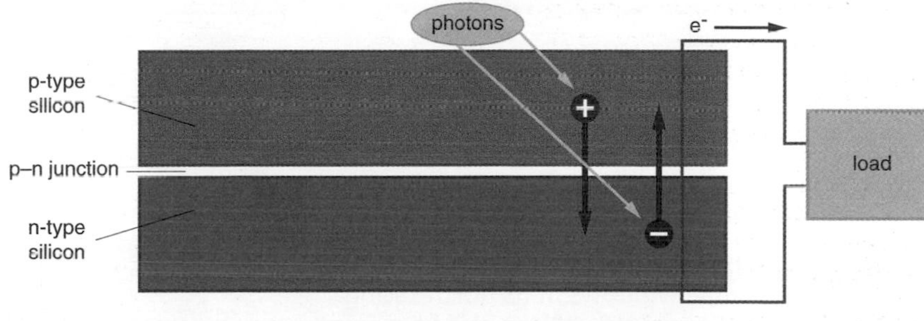

FIGURE 6: How is an electric field set up in a photocell?

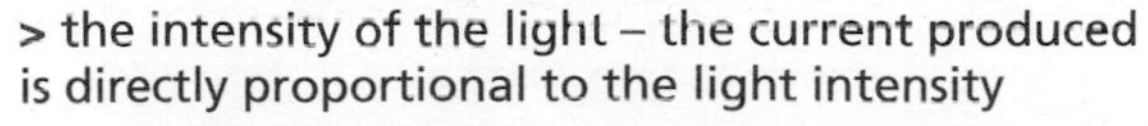

> the intensity of the light – the current produced is directly proportional to the light intensity

> the surface area exposed – the current produced is directly proportional to the surface area

> the distance from the light source – the current is inversely proportional to the square of the distance from the light source.

Questions

3 Why are photocells placed on south-facing roofs in England but on north-facing roofs in Australia?

4 a In a photocell, what is the charge of a hole?

b Suggest what the 'p' in p-type silicon stands for.

Renewable energy

You will find out:
- how passive solar heating can reduce demand on other energy resources
- about the advantages and disadvantages of wind technology

The Sun is a **renewable** source of energy.

As well as being used to produce electricity from a photocell, the energy from the Sun can be used for heating.

A solar water heater on the roof of a building is made from rectangular collectors. Inside each collector is a series of small tubes. These tubes pass over a black plate. The black plate absorbs **radiation** from the Sun and warms the water passing through the tubes.

Warmed water rises to the storage tanks as it is heated. Colder water sinks down into the collector.

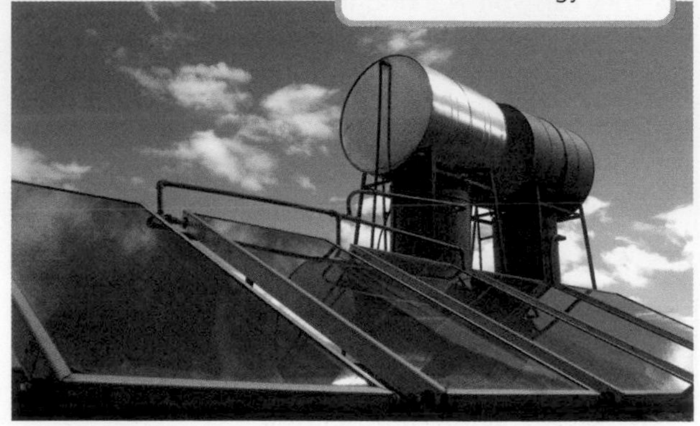

FIGURE 7: What is the large shiny cylinder above the solar panels?

Wind

The temperature difference between land surfaces causes **convection currents**.

This movement of air is usually called wind. Wind is used to turn a wind turbine and produce electricity.

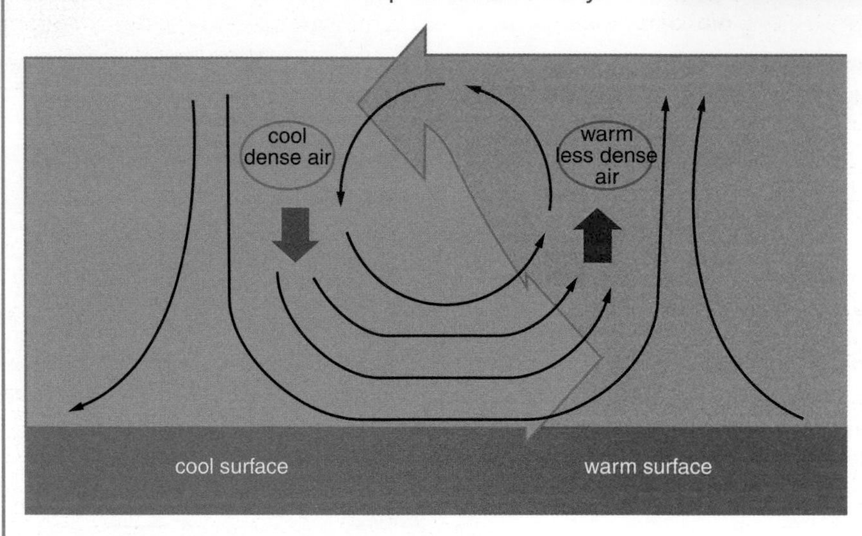

FIGURE 8: What causes convection currents?

FIGURE 9: A wind farm. Why do you think the turbines are placed high up on hills or out at sea?

Passive solar heating

A home that uses **passive solar heating** makes full use of direct sunlight for heating purposes.

The home has large windows in its south-facing walls and small windows in its north-facing walls. This allows natural light and heat from the Sun to be used and reduces the need for other heating.

During the day, energy from the Sun warms the walls and floor of a room.

At night the energy is **radiated** back into the room from the warm walls and floor.

Curved solar reflectors focus energy from the Sun and can be used to directly heat water, for example, when camping.

Questions

5 What is the best colour for absorbing radiation?

6 Which process (conduction, convection or radiation) describes the transfer of energy by the water in a solar water heater?

7 Why should south-facing windows in homes in the UK be large?

8 Why are the reflecting surfaces of solar collectors curved?

passive solar heating   wind farm benefits

a

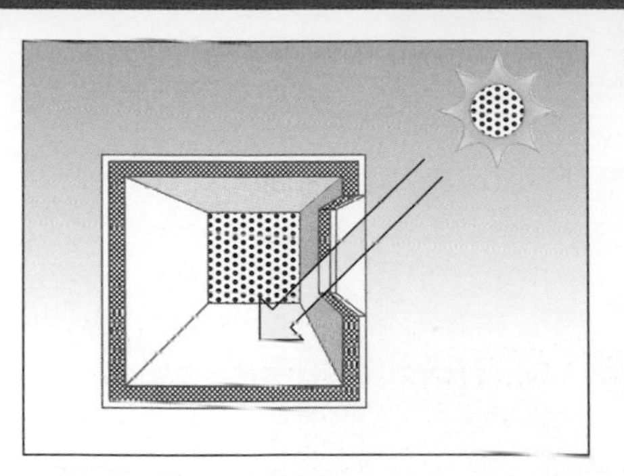

b

FIGURE 10: How does passive solar heating warm a home: **a** during the day? **b** at night?

Did you know?

A typical wind turbine on a wind farm provides enough electricity for 1500 homes.

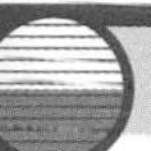

Wind farms

Wind farm – friend or foe?

There is some controversy surrounding wind farms.

Few people dispute the fact that they do not discharge polluting gases into the atmosphere and do not contribute to **global warming**.

There are, however, major concerns about their visual impact on the countryside and their noise. They do not work if there is little wind and must be shut down if wind speeds are greater than 88 km/h

Individual wind turbines do not take up much space, but a wind farm will occupy quite a large area. Some are being sited out at sea and others occupy land which can be also grazed by cattle.

Question

9 Some farmers do not mind their fields being used for wind turbines. Suggest why.

More on solar heating

The temperature inside a greenhouse is higher than the temperature on the outside.

This is why plants that normally only grow in tropical climates can grow at the Eden Project in Cornwall.

The Sun is very hot and produces infrared radiation with a short **wavelength**. Glass is transparent to this short wavelength radiation. The ground and plants inside the greenhouse absorb this radiation, warm up and re-radiate infrared radiation. The plants are not as hot as the Sun and the wavelength radiated is therefore longer. This radiation will not pass through glass but is reflected back inside.

It is not just greenhouses that are warmed in this way. The temperature inside your home rises for the same reason.

To make sure that a solar collector works as efficiently as possible, a computer moves the dish to face the Sun throughout the day.

Question

10 Explain why it is cooler outside a greenhouse than it is inside.

FIGURE 11: How does the wavelength of infrared radiation from the Sun compare with the wavelength emitted by plants in a greenhouse?

FIGURE 12: Why are these solar collectors all pointing the same way?

Generating electricity

You will find out:

> how a dynamo generates electricity

Lighting up

Rajab goes for a bicycle ride at dusk. His bicycle is fitted with a dynamo. When Rajab pushes the pedals round, the movement is changed to electricity and is used to light up his bicycle light.

This means he does not need a direct current battery for his light.

FIGURE 1: You must have lights on your bike when riding after dark.

Mini generator

A bicycle **dynamo** is a small electrical **generator**.

A magnet rotates inside a coil of wire and an **alternating current** (AC) is produced.

FIGURE 2: How does this dynamo produce an alternating current?

If a magnet is moved near to a coil of wire a current is produced in the wire. If a wire is moved near to a magnet a current is produced in the wire.

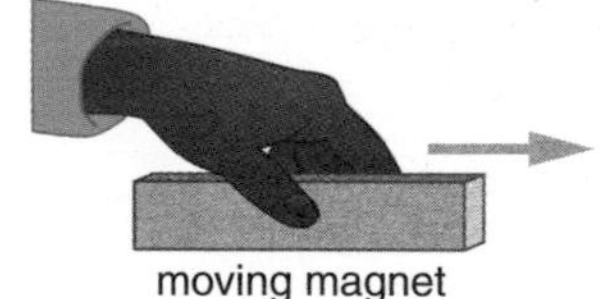

FIGURE 3: What happens when the magnet is moved closer to the coil of wire?

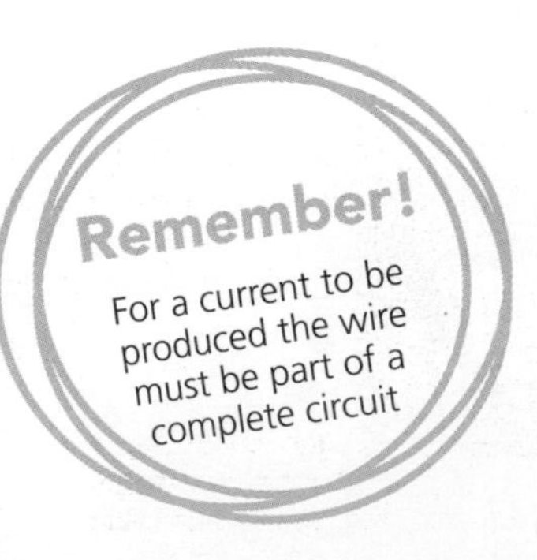

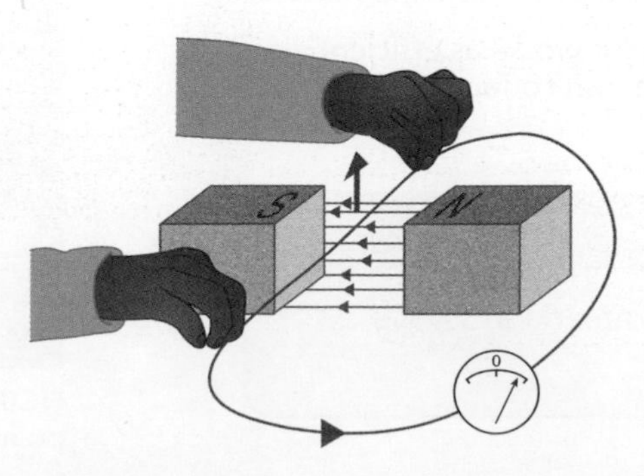

FIGURE 4: What happens when the wire is moved in the magnetic field between the two magnets?

power stations renewable fuels

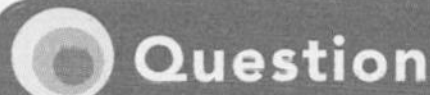

Question

1 What is the difference between a current from a battery and a current from a dynamo?

FIGURE 5: When a new fossil fuel power station is to be built what do you think it should be near to?

Larger currents and voltages

There are three ways to increase the current from a dynamo:

> use a stronger magnet

> increase the number of turns of wire on the coil

> rotate the magnet faster – on a bicycle this means pedalling faster.

The dynamo produces a changing **voltage** and current. An **oscilloscope** shows how the alternating voltage changes with time. The maximum voltage and the time for one cycle (the period) can be found from the trace.

To calculate the frequency, the following equation is used:

frequency = 1 ÷ period

The unit is hertz, Hz.

Questions

2 Describe the brightness of the bulb in Rajab's bicycle light connected to his dynamo when he has stopped at traffic lights.

3 What happens to the brightness of the bulb in Rajab's bicycle light connected to his dynamo when he pedals very fast downhill?

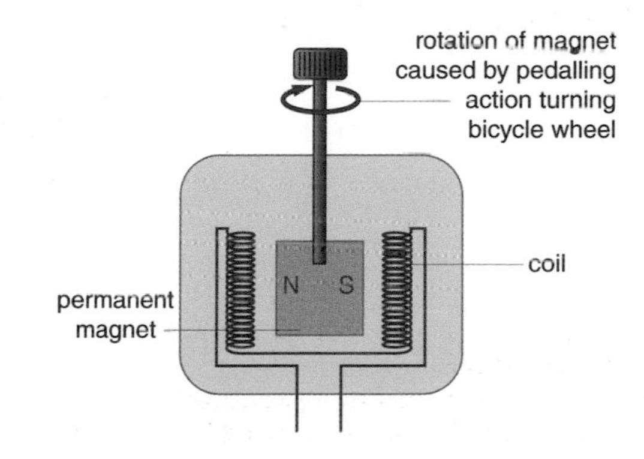

FIGURE 6: What changes could you make to this dynamo to increase the current output?

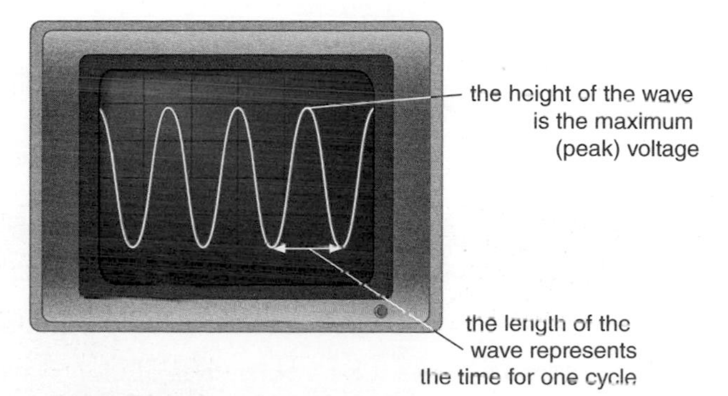

FIGURE 7: What can you measure from an oscilloscope trace?

Carrying electricity

You will find out:
- how electricity is carried
- how electricity is produced
- about energy efficiency

Power stations use an energy source such as coal to generate electricity.

> Electricity is transmitted round the country through a network of power cables called the **National Grid** to **consumers**.

> Consumers are homes, farms, offices, schools, shops, factories and hospitals.

> Some of the energy from the fuel is not transferred to electricity. It is lost to the environment, usually as heat.

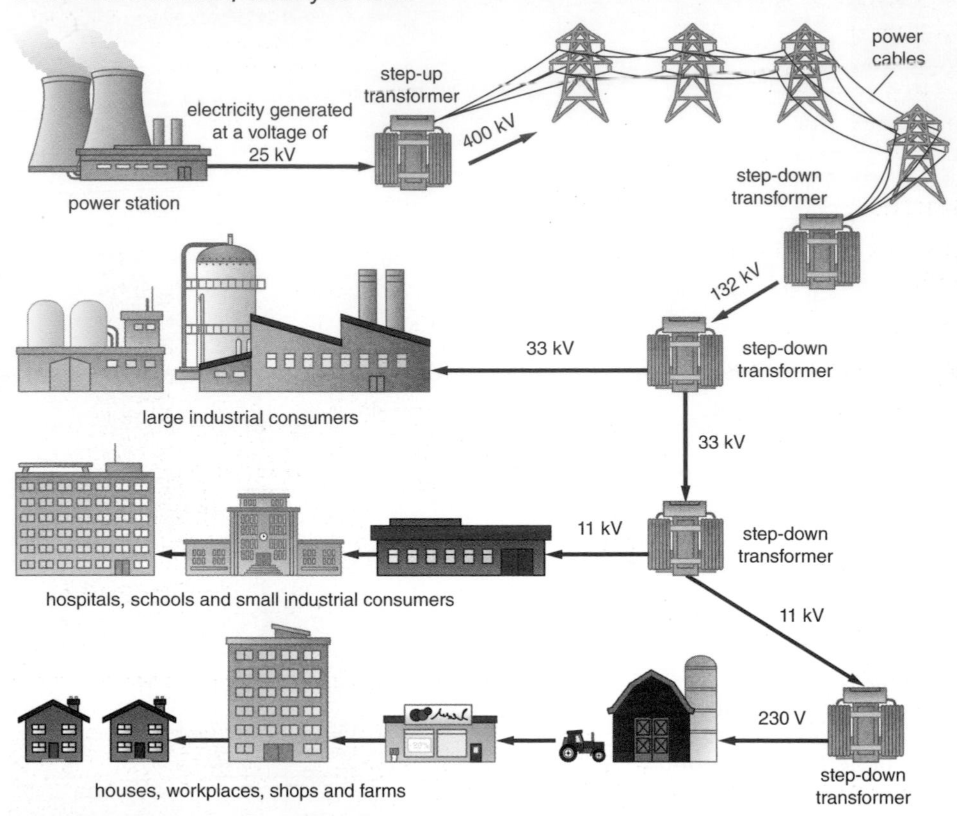

FIGURE 8: What do transformers do to the size of voltages?

Did you know?

In power stations around half the energy stored in coal is lost as heat from cooling towers.

Did you know?

The dense white clouds you see coming from a power station are not smoke.

They are steam coming from the cooling towers.

Questions

4 What substance is cooled in cooling towers in a power station?

5 Explain what is meant by a consumer of electricity.

6 A power station uses coal as its fuel. For every 1000 J of energy stored in the coal, it produces 330 J of useful energy output. Calculate the efficiency of the power station.

how power stations work

Simple alternating current generator

Like a dynamo, a generator in a power station works on the principle that a magnetic field rotating relative to a conducting coil generates a voltage in the coil. In a simple alternating current generator:

> a coil of wire is free to rotate between the poles of a magnet

> as the coil rotates it cuts the magnetic field and a current passes through the coil.

The same thing would happen if the coil was stationary and the magnet rotated. It is the relative movement between the two that is important.

Generators in power stations

> A fuel such as coal is burned in a boiler room to heat water and produce steam.

> The steam drives a **turbine**.

> The turbine turns a generator which produces electricity.

> A lot of energy from the fuel is wasted and not transferred into electricity.

Energy efficiency

Energy **efficiency** is a measure of how well a device transfers energy.

$$\text{efficiency} = \frac{\text{useful energy output } (\times 100\%)}{\text{total energy input}}$$

A power station is not very efficient. For example:

> for every 100 J of energy stored in coal, 15 J is wasted in the boiler

>a further 45 J is wasted in the cooling towers and 5 J in the generator.

The remaining energy is converted to electricity.

fuel energy input	=	waste energy output	+	electrical energy output
100 J		65 J		35 J

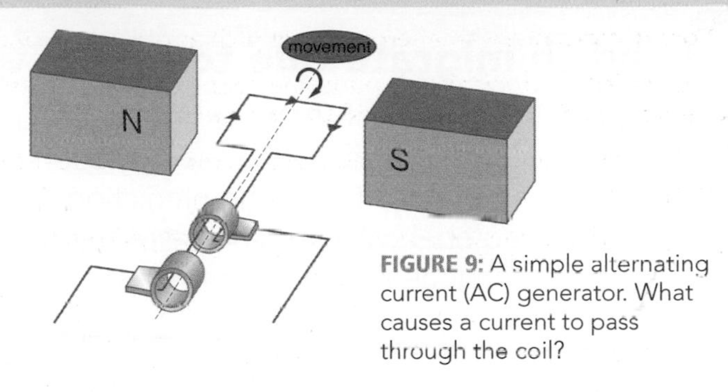

FIGURE 9: A simple alternating current (AC) generator. What causes a current to pass through the coil?

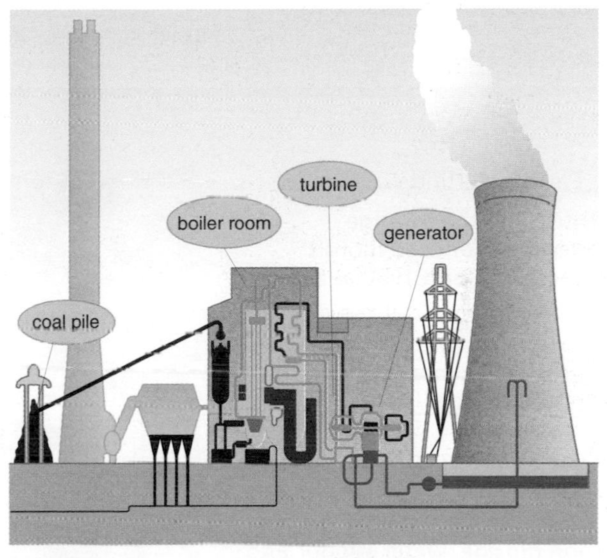

FIGURE 10: What is the role of a boiler room in a power station?

Questions

7 Other than from the cooling towers, where is energy lost in a coal-fired power station?

8 What happens to the steam after it has turned a turbine in a power station?

9 Calculate the energy efficiency of a power station where 67 per cent of available energy is lost to the environment.

Energy efficiency

Question

10 A certain power station is 35 per cent efficient. It produces 2 million joules of electrical energy each second. How many joules of energy are lost to the environment each second?

Remember!

A question may require you to change the subject of an equation. Practise rearranging equations until you are really confident.

Global warming

You will find out:

- about the gases in the Earth's atmosphere that prevent heat radiating into space
- about the natural and man-made sources of these gases

Jellyfish migrate due to global warming

Highly poisonous jellyfish that are normally found in the Mediterranean and Caribbean Seas may soon be found in the waters around Britain. This is the prediction of scientists who think that, as global warming increases sea temperatures, these jellyfish will be able to spread from their traditional areas. There is certainly evidence of greater numbers of large and poisonous jellyfish being found in waters where they had never been found before.

FIGURE 1: Why are jellyfish moving their habitat?

Global warming

The Earth is getting warmer.

Scientists use the average temperature for each month between 1951 and 1980 as a basis for comparison.

The picture shows that the temperature in most parts of the Earth was warmer in July 2010 than this average. Some places were nearly 6°C warmer.

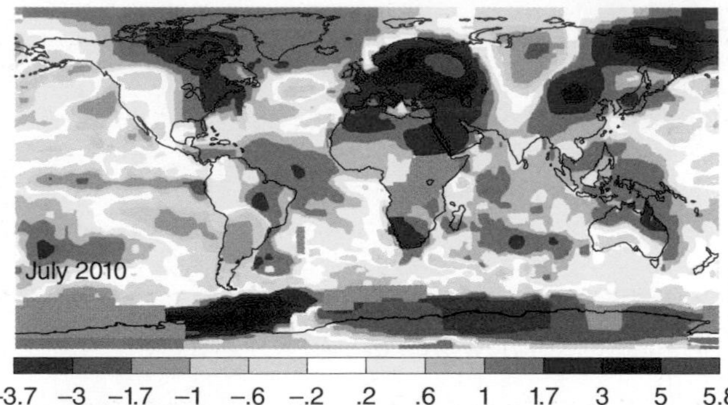

FIGURE 2: How much warmer than the 1951-80 average was it in England in July 2010?

Greenhouse gases

Carbon dioxide, water vapour and methane are gases which can be found in the Earth's atmosphere.

They stop heat from the Earth radiating into space. This is known as the greenhouse effect and the gases are three examples of **greenhouse gases**.

Questions

1 What will happen to the polar ice caps and sea levels if the temperature of the Earth continues to rise?

2 Why is carbon dioxide called a greenhouse gas?

Where do the greenhouse gases come from?

Carbon dioxide occurs naturally in the atmosphere. It is released during natural forest fires, volcanic eruptions, decay of dead plant and animal matter, evaporation from the oceans and respiration. The amount of natural carbon dioxide in the atmosphere has changed significantly during the Earth's history. When the Earth was first formed, most of the atmosphere was probably carbon dioxide. During the last few hundred million years, the amount of carbon dioxide has been declining. Today, natural sources release about 150 billion tonnes each year.

Since the industrial revolution in the 18th century, the amount of man-made carbon dioxide has increased as we have burned fossil fuels. The UK produces about 150 million tonnes each year and the global total is 7 billion tonnes from burning fossil fuels, waste incineration, deforestation and cement manufacture.

Plants absorb carbon dioxide during **photosynthesis** and the oceans also absorb carbon dioxide. If all carbon dioxide emissions stopped today, it would take hundreds of years for plants and oceans to absorb enough to return to the levels that existed before the industrial revolution.

Water vapour is the most significant greenhouse gas and almost all of the water vapour occurs naturally. A mere 0.001% comes from human activity. Half of the greenhouse effect is due to water vapour and a further quarter is due to clouds (water droplets).

greenhouse gases

Methane is produced when organic matter decomposes in an environment lacking oxygen. Natural sources include wetlands, termites, and oceans. Man-made sources include the mining and burning of fossil fuels, the digestive processes in farmed animals such as cattle, rice paddies and the burying of waste in landfills. Most of the 500 million tonnes annual emissions of methane come from man-made sources. Methane is removed from the atmosphere by chemical breakdown but this cannot keep pace with increased emissions, so concentrations are increasing.

Most infrared radiation can pass through the Earth's atmosphere, but longer wavelengths are absorbed by greenhouse gases, warming the atmosphere.

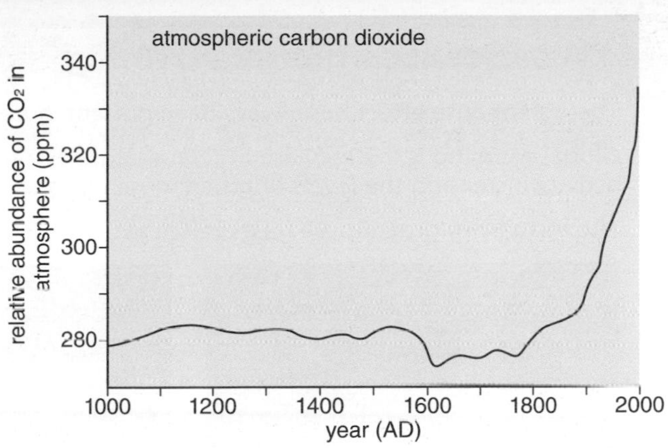

FIGURE 3: When was there a sudden rise in CO_2 emissions?

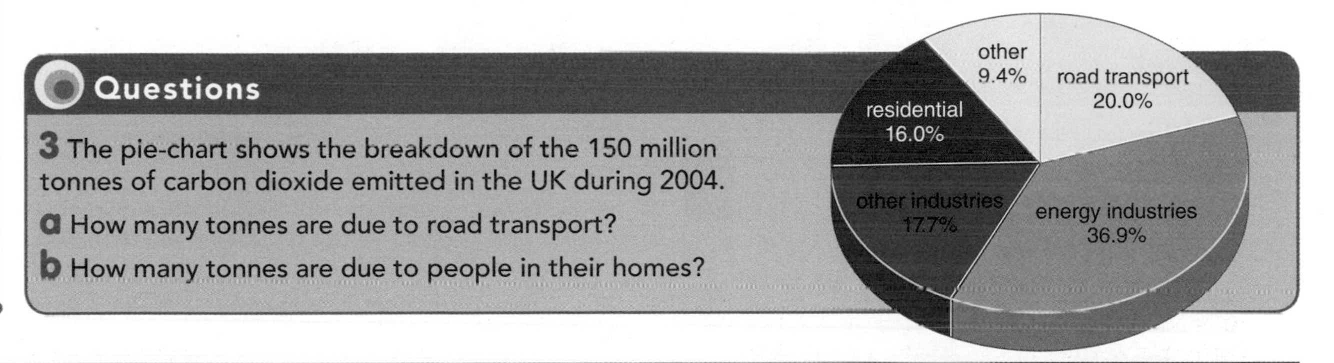

Questions

3 The pie-chart shows the breakdown of the 150 million tonnes of carbon dioxide emitted in the UK during 2004.

a How many tonnes are due to road transport?

b How many tonnes are due to people in their homes?

The greenhouse effect

Much of the infrared radiation from the Sun has a relatively short wavelength. This radiation is absorbed by and warms the Earth. The Earth then re-radiates the energy as infrared radiation with a longer wavelength. This longer wavelength radiation is absorbed by the greenhouse gases which warms the atmosphere.

Question

4 The table provides data from the United States Department of Energy showing the concentrations of greenhouse gases in the atmosphere. The data shows the amounts prior to the industrial revolution and the changes since then. Water is usually omitted from such tables. Methane has an effect 21 times greater than carbon dioxide. This means that, compared to carbon dioxide, the *relative* pre-industrial baseline is 17 808 ppb (21 x 848).

Greenhouse gas	Concentration in parts per billion (ppb)				
	pre-industrial baseline	natural additions	man-made additions	total concentration	percent of total greenhouse gases
carbon dioxide CO_2	288 000	68 520	11 880	368 400	99.438
methane CH_4	848	577	320	1745	0.471
other	310	12	17	339	0.091
total	289 158	69 109	12 217	370 484	100.000

a Compared to carbon dioxide, what is the *relative* total concentration of methane?

b The relative total concentration of other gases is 104 011ppb. Calculate the relative *percentage* contributions from carbon dioxide, methane and other gases.

You will find out:

- > the causes of global warming
- > why there are differences of opinion about how to deal with global warming

What causes global warming

The greenhouse effect has always been present. It keeps the Earth warm.

Global warming is the increase in temperature as a result of human activity increasing the levels of greenhouse gases.

The increase in greenhouse gases is due to:

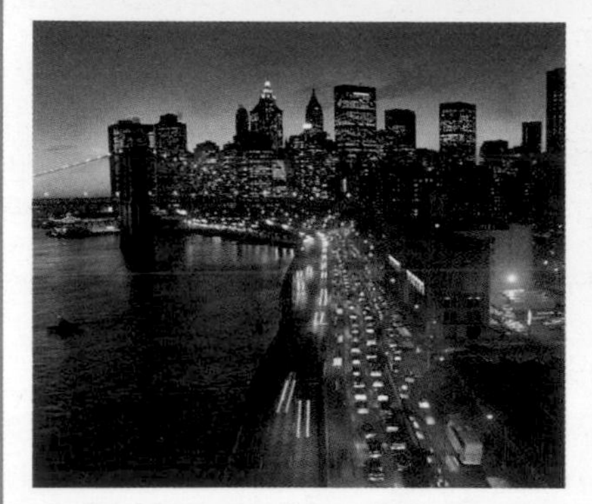

FIGURE 4: Increased energy use

FIGURE 5: Increased carbon dioxide and methane emissions

FIGURE 6: Deforestation

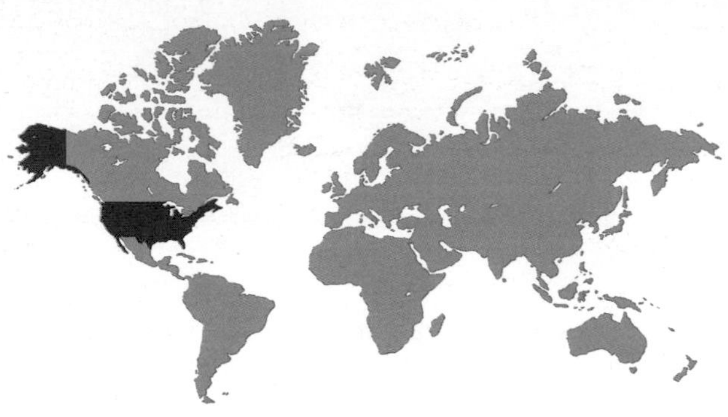

FIGURE 7: By 2010, which country had not agreed to reduce greenhouse gas emissions?

By 2010, most countries in the world (shown in green on the map) had agreed to reduce carbon dioxide and methane emissions.

The aim was to reduce emissions by 5.2% (compared to 1990) before 2012.

Scientists will need to monitor concentrations of greenhouse gases around the world and share their findings. This is the only way we will be able to obtain a full picture of what is happening. It is not easy to measure global warming because the temperature of the Earth naturally fluctuates year on year.

Question

5 Why do governments want to reduce greenhouse gas emissions?

Dust warms, dust cools

Dust in the atmosphere can have opposite effects. The smoke from the factories reflects radiation from the town back to Earth. The temperature rises as a result.

The ash cloud from the volcano reflects radiation from the Sun back into space. The temperature falls as a result.

FIGURE 8: How does smoke affect temperatures on Earth?

FIGURE 9: How does a volcanic ash cloud affect temperatures on Earth?

It is important that decisions are made on the basis of scientific evidence, not on the basis of unsubstantiated opinions. There are many myths surrounding global warming and some research reports may choose to ignore evidence which does not support the researcher's preconceived idea.

Many newspapers, and the Vice-President of the United States, used a photograph of polar bears which had been reproduced in a report by the Canadian Wildlife Agency. The bears cling to the top of what is left of the ice floe. Their plight was seen as a perfect symbol to show the dangers of global warming. In reality, the photograph was taken during the summer when the ice normally melts; the bears are quite capable of swimming 100 miles or more so were in no danger.

FIGURE 10: Are these polar bears in danger?

What should we do?

The vast majority of scientists agree that the evidence supports climate change. In March 2010, the British Meteorological Office led a review which examined more evidence than ever before. Dr Peter Stott, from the Met Office, said all the new evidence on melting sea ice in the Arctic and rising global temperatures point to man-made global warming.

Other research shows contradictory evidence. In 1936, part of the United States had 49 consecutive days when the temperature was above 30°C. A similar thing happened in 1955. But in 1992 there were only five consecutive days and in 2002 only three. Does this point to global cooling?

Question

6 Discuss whether or not it is acceptable to use evidence, such as the polar bear photograph, out of context to illustrate a point.

Manipulating data

Using the same data from the United States Department of Energy on page 239 and including water vapour as a greenhouse gas, a very different picture emerges.

greenhouse gas	% of greenhouse effect	% natural	% man-made
water vapour	95.000	94.999	0.001
carbon dioxide CO_2	3.618	3.502	0.117
methane CH_4	0.360	0.294	0.066
other	1.022	0.928	0.094
total	100.000	99.723	0.278

This shows the contribution from human activity is less than 0.3% and from carbon dioxide just over 0.1%.

Those who deny that human activity is a contributory factor claim that these percentages are insignificant and within the normal variation of the Earth.

Other scientists point to the continued increasing temperatures since the industrial revolution, the melting of the polar ice caps and the rise in sea levels as evidence that we need to take action now.

Remember!

You may be asked to explain how scientists can agree about the greenhouse effect, but have very different views on the contribution of human activity to global warming.

causes of global warming Kyoto protocol

Fuels for power

You will find out:

> about power

> about the cost of using electricity

How much does it really cost?

The electricity bill has arrived and Holly's dad is asking her not to leave the television and light on in her room.

But compared to the energy used in other parts of the house, such as the kitchen, the amount Holly's things use in her room is very small.

Most televisions and lights have a power rating of less than 100 watts.

A typical toaster or kettle has a power rating of at least 2000 watts (2 kilowatts).

Appliance	Power in W
	60
	180
	750
	1000
	2250

FIGURE 1: Which appliance costs most per hour to use?

FIGURE 2: "Who's used all this electricity?"

Power

Power is a measure of the rate at which energy is used. The unit is **watt** (W) or **kilowatt** (kW).

The cost of using an electrical appliance depends on:

> its power rating

> how long it is used for.

Question

1 A current of 2 A passes through the bulb in a car headlamp. The car has a 12 V battery.
power = voltage × current
Calculate the power rating of the bulb.

Measuring power

power = voltage × current

Most electrical appliances show their power rating. This is the rate at which they transfer energy.

Electrical consumption is the total amount of energy that has been used in a period of time.

The amount of energy used is measured on a **kilowatt hour** meter. If you look carefully at an electricity meter you may see a rotating disc. The faster the disc

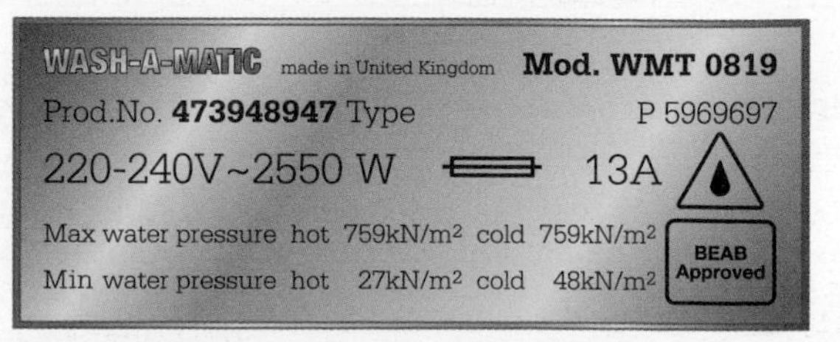

FIGURE 3: Most electrical appliances show their power rating. What is the power of this electrical appliance?

rotates the more electrical energy is being used. When a shower or cooker is being used the disc goes round very quickly. If only a light is switched on the disc hardly moves.

energy supplied = power × time

The unit is kilowatt hour (kWh).

The cost of using an electrical appliance depends on the energy used.

cost = energy × cost per kilowatt hour

The unit is pence (p).

FIGURE 4: What is the reading on this electricity meter?

Questions

2 A 7000 W shower is used for 3 hours a week.

a How much energy is used each week?

b Electricity costs 12p per kWh. How much does it cost to use the shower each week?

3 A microwave oven is designed to operate at 250 V. A current of 3 A passes.

a Calculate the power rating of the microwave.

b It takes 12 minutes to cook a meal. Electricity costs 12p per kWh. How much does it cost to cook the meal?

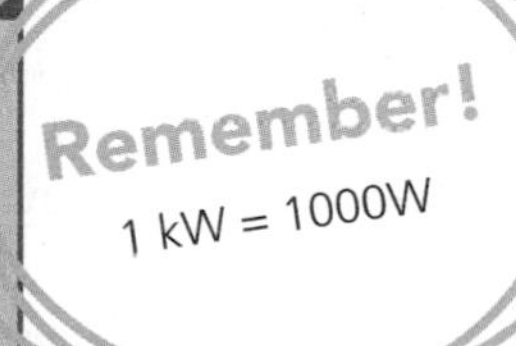

Cheaper electricity

Some electricity supply companies have a reduced price for "off-peak" electricity. If electricity is used at night it is cheaper. This is because electricity still has to be produced but most consumers do not need to use it.

> Night storage heaters use off-peak electricity to heat up concrete blocks. These release heat the following day.

> Washing machines can be programmed to wash clothes overnight.

The electricity producer benefits by maintaining a more stable output and avoiding start-up costs.

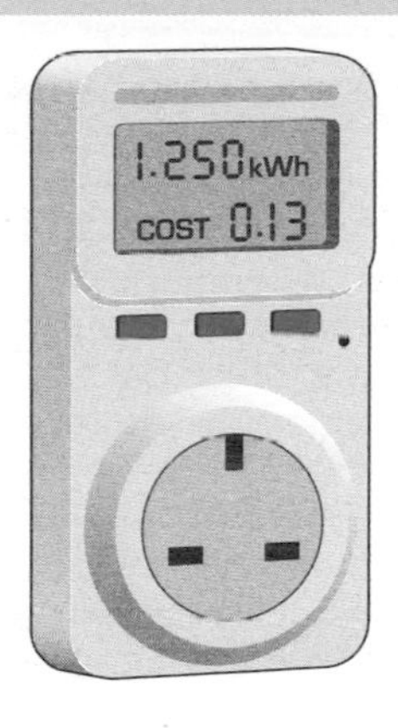

FIGURE 5: A home energy monitor.

Questions

4 What are the benefits to the producer of selling off-peak electricity?

5 A toaster has a power rating of 2.3 kW. Mains voltage is 230 V. Calculate the current in the toaster.

6 Charlie plugs a 3 kW kettle into an energy monitor. At the end of the day, the monitor reads 1.25 kWh. How long was the kettle used for during the day?

average monthly cost for electricity

Power station changes

You will find out:
- how electricity is moved around the country
- about the types of fuels used in power stations

Power station fuels

Fossil fuels are the most common energy source for power stations. They release energy as heat. Fossil fuels are:

> coal   > natural gas   > oil.

Fossil fuels are **non-renewable** energy resources. This means that eventually they will all run out.

Some power stations are using **biomass** as a fuel.

Biomass is a renewable energy resource. Biomass fuels are:

> wood   > straw   > manure.

Renewable energy resources will not run out. For every tree that is cut down a new one can be planted.

A different type of fuel that is used as the energy resource in **nuclear** power stations is **uranium** and sometimes **plutonium**.

In electricity transmission networks a **transformer** increases or decreases the size of an alternating voltage.

FIGURE 6: Suggest why there are a lot of transformers in this electricity substation.

Remember!
A transformer does not change alternating current into direct current.

Did you know?

The voltages from a transformer at a substation are very high.

As soon as you see a sign like this, do not go any further. It does mean what it says.

Questions

7 Why is it dangerous to go into an electricity substation?

8 Why are coal, oil and natural gas called fossil fuels?

9 What is the advantage of using a renewable energy resource?

Energy sources

Some energy sources are more appropriate than others in a particular situation. It is difficult to imagine a British train being powered by solar cells! Each energy source has its own advantages and disadvantages. Some factors are relative, for example, both coal and oil reserves are running out but coal will last longer. The choice of energy source will depend on its availability – even renewable sources that will last for ever are not always available. Some sources have greater risks associated with them. Deep mine coal is difficult and dangerous to extract; uranium has health risks associated with exposure to it. The effect on the environment cannot be ignored. Burning fossil fuels contributes to climate change, but wind turbines may be considered noisy and unsightly.

FIGURE 7: What is the voltage carried through the National Grid?

The National Grid

The **National Grid** distributes electricity around the United Kingdom at voltages as high as 400 000 V. The high voltage leads to:

> reduced energy loss

> reduced distribution costs

> cheaper electricity for consumers.

Transformers in the National Grid step down (reduce) or step up (increase) the voltage. The voltage at our homes is 230 V.

what are transformers used for?

Transformers are not only used when distributing electricity.

Many household items such as chargers and adaptors contain transformers to reduce the voltage.

Your mobile phone battery cannot be recharged directly from the mains at 230 V.

FIGURE 8: What voltage does a mobile phone work on?

Question

10 Suggest why there are step-up and step-down transformers in the National Grid.

Transmission loss

When a current passes through a wire the wire gets hot. The greater the current, the hotter the wire.

When a transformer increases the voltage the current is reduced. This means there is less heating effect and therefore less energy lost to the environment.

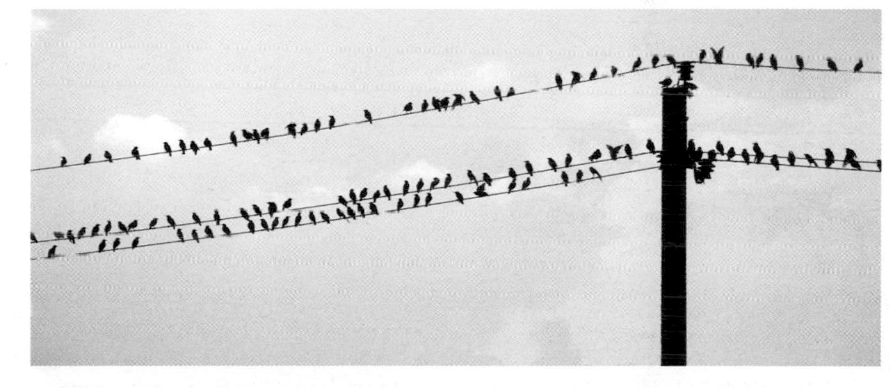

FIGURE 9: Why do birds like to sit on overhead power lines?

power = voltage × current
The power of transmission remains constant so increasing the voltage reduces the current

Question

11 What is the advantage of transmitting electricity around the country at high voltage?

step-up transformer   step-down transformer

Preparing for assessment: Applying your knowledge

To achieve a good grade in science, you not only have to know and understand scientific ideas, but you need to be able to apply them to other situations and investigations. These tasks will support you in developing these skills.

The risks from radon (see pages 248–251)

Chloe lives in a house in Cornwall, in the southwest of England, with her parents and brothers. The house is quite a new one, but it is built in an area where there is a lot of granite. In Science lessons at Chloe's school her teacher had explained about a gas called radon.

Radon is a gas that forms naturally due to the decay of uranium in the ground. It forms in larger amounts in areas such as Devon and Cornwall where there is a large amount of granite in the ground. Radon decays and gives out radioactive particles, which remain suspended in the air. Normally this is not a problem, but if the particles are in air that is inside a building, the levels can rise further.

People inhaling air that contains these radioactive particles take in alpha radiation and are at a greater risk of developing lung cancer. This is a particular problem in houses with well-fitting doors and windows, as the air does not circulate as easily.

Chloe told her parents about what she had learnt and her mother said, 'I've been meaning to get something done about this. Diane, over the road, got some detector device to put in the house to see if the radon was a problem. We should do that as well.'

Her dad said that he would find out where they could get the detectors from. 'I don't know if we have to pay for them,' he said, 'but we should get them anyway. I just don't like the idea of there being any of that radiation here in our house.'

Chloe laughed. 'Don't be daft,' she said, 'There's radiation around wherever you are. It's what the radiation does to you that matters.'

Task 1

Think about Chloe's reaction to her dad, when he said that he did not want any radiation in their house. What might she have said to him to explain her ideas?

Task 2

What are the dangers of being exposed to radiation – why can inhaling alpha radiation be more dangerous than exposure to beta or gamma radiation?

Task 3

In fact, Chloe's Science teacher had been explaining to them about ionising radiation. Explain, using diagrams if it helps, what the word 'ionising' means.

Task 4

Chloe has to do Science homework. She has to explain how the ionisation effects of radiation involve electrons being transferred. Using words or diagrams, suggest what she might write.

Maximise your grade

Answer includes showing that you can...	
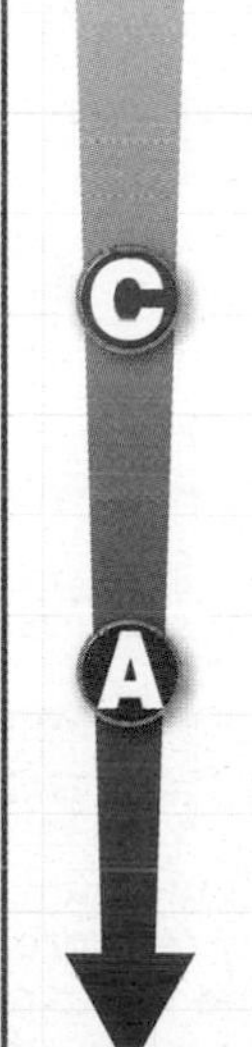	Suggest that alpha radiation may be harmful.
	State what the harmful effect of alpha radiation could be.
	Describe alpha radiation as being not very penetrating.
	Explain what part of the body absorbs alpha radiation and how this might affect the lungs.
	Explain the meaning of ionisation.
	Explain ionisation in terms of electron transfer.
	Explain ionisation in terms of removal of electrons from particles gain of electrons by particles.
	As above, but with particular clarity and detail.

Nuclear radiations

You will find out:

> about the effects of different types of ionising radiation

> about the disposal of nuclear waste

> about the properties of ionising radiation

Cancer clusters near nuclear plants – double the risk

In 2008, Britain was deciding whether or not to build new nuclear power stations. In the same year, German scientists published a report that claimed children under five, who live close to a nuclear power plant, have twice the risk of suffering from leukaemia.

This is a blood cancer. They looked at the records of 6300 children over a period of 23 years who lived within 5 km of a nuclear power plant. Statistically, there should have been 17 leukaemia cases in children under five. They found 37.

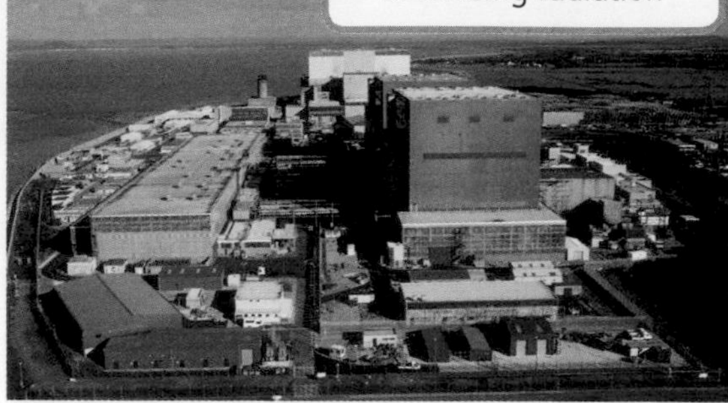

FIGURE 1: Do you think nuclear power plants, like this one at Hinkley Point, should be built near to people's homes?

Radiation around us

Nuclear power stations do not produce smoke or carbon dioxide. Carbon dioxide is a greenhouse gas that causes global warming.

Nuclear power stations do produce some waste. The waste material is **radioactive** and can be harmful. Waste heat is carried away by cooling water often obtained from the sea, rivers or lakes.

Radioactive materials give out nuclear radiation. There are three main types of nuclear radiation, alpha (α), beta (β) and gamma (γ).

Radiation from radioactive sources causes **ionisation**. In ionisation the radiation changes the structure of any atom exposed to the radiation.

The cells in our bodies are made up of many atoms, so body cells can be changed by radiation. One important chemical in a cell is DNA. If DNA changes due to ionisation the cell behaves differently. This is called **mutation**.

Sometimes when a cell mutates it divides in an uncontrolled way. This can lead to cancer.

FIGURE 2: This map shows the positions of nuclear power stations in the UK.

Questions

1 Why are so many of the nuclear power stations in the UK located on the coast?

2 Describe what can happen to a body cell if it is exposed to radiation.

Nuclear fuel, waste and radiation

The fuel used in a nuclear reactor is uranium. Uranium is a non-renewable energy resource. Once the uranium has been used the waste remains radioactive for thousands of years.

One of the waste products from a nuclear power station is **plutonium**. Plutonium can be used to make nuclear weapons. A nuclear explosion causes death and destruction over a very large area. The affected area remains contaminated with radioactive material and is unusable for many years.

Disposal of radioactive waste

> Low-level radioactive waste is buried in landfill sites.

nuclear waste management

> High-level radioactive waste is encased in glass inside stainless steel containers and buried deep underground.

> Some radioactive waste can be reprocessed into new and useful radioactive material.

Properties of ionising radiation

Alpha, beta and gamma radiation come from the **nucleus** of a radioactive atom.

Nuclear radiation causes **ionisation** of materials that it passes through. Alpha radiation causes most ionisation and gamma radiation causes the least.

Positive **ions** are produced when atoms lose electrons. Negative ions are produced when atoms gain electrons.

FIGURE 3: A nuclear explosion. Why is the damage caused by this type of explosion so devastating?

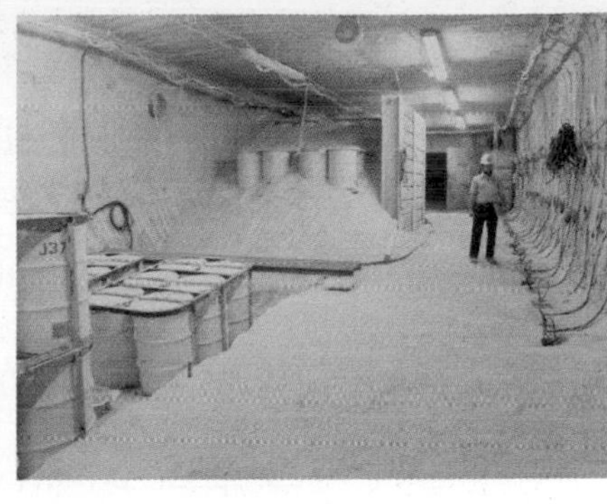

FIGURE 4: An underground burial site for nuclear waste. Suggest why these sites should not be built near to earthquake zones.

Identifying radiation

Alpha radiation only travels a few centimetres in air. It is not very penetrating and is absorbed by a sheet of paper or skin.

Beta radiation has a range of about 1 metre in air. It will pass through paper but is absorbed by a few millimetres of aluminium.

The range of gamma radiation in air is, in theory, infinite. In practice the amount of radiation decreases until it cannot be distinguished from natural background radiation. Gamma radiation is very penetrating and, although a few centimetres of lead will stop most of the radiation, some can pass through several metres of lead or concrete.

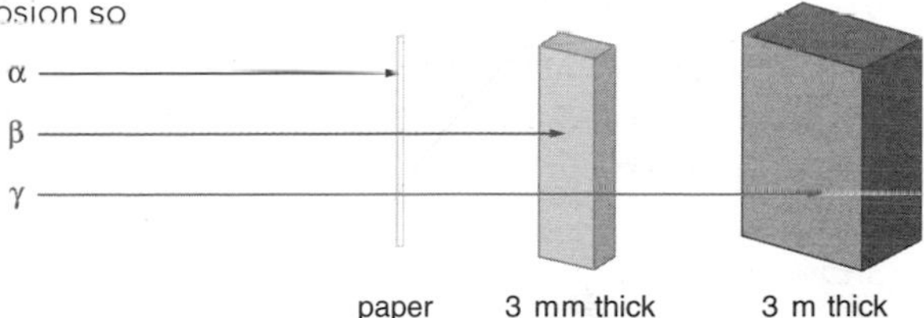

FIGURE 5: Different penetrations of the three types of radiation. Which two radiations are absorbed by the material shown and which type of radiation penetrates the material?

Questions

3 Suggest why radioactive waste must be stored in steel drums deep in the ground.

4 Suggest two ways that radioactive waste might prove harmful to fish.

Why choose nuclear power?

Despite the possible dangers from nuclear power and nuclear waste there are advantages in using nuclear power stations:

> fossil fuel reserves are not used

> no pollution or greenhouse gases are discharged into the atmosphere.

The disadvantages are:

> very high cost of maintenance and of dismantling old nuclear power stations

> the risk of incidents similar to the one at Fukushima, Japan

> the need for careful disposal of radioactive waste.

Ionisation effects

The formation of ions can cause chemical reactions. Such reactions may disrupt the normal behaviour of molecules inside the body. They may cause strands of DNA to break or change; protein molecules may change their shape. These effects are potentially harmful.

Remember!

Being exposed to radioactivity does not make you radioactive.

Did you know?

Nuclear power provides 18 per cent of the UK's energy needs. In France it provides nearly 70 per cent.

Questions

5 Most present nuclear power stations in the United Kingdom are to be closed down by 2020. Suggest how the energy needs of this country could be met in the future.

6 A radioactive source emits alpha and gamma radiation. How can you show this using a radiation detector and different materials?

Safe handling of radioactive material

You will find out:

> how radioactive material is handled safely

> how radioactivity is used in the home, industry and medicine

> about the risks of radioactive material

Radiation from radioactive materials damages living cells.

The amount of damage depends on the type of radiation and how much radiation the cells have been exposed to.

To reduce exposure, people who handle radioactive material:

> sometimes wear special protective clothing

> make sure that the distance between the radioactive substance and themselves is as large as possible

> use shielding to absorb radiation

> use the material for the shortest amount of time possible and return it to a secure, labelled container

> always use tongs or remote handling techniques when moving radioactive substances.

In school your teacher uses much lower levels of radioactive material than those used in industry, but will take similar precautions when handling radioactive material.

FIGURE 6: What precautions have these workers taken before handling radioactive material?

Uses of ionising radiation

The fact that ionising radiation damages living cells means that exposure to radiation can be a cause of cancer.

It also means that the radiation can be used to kill cells and living organisms. Cancer cells within the body can be destroyed by **radiotherapy**.

Cobalt-60 is a radioactive material used to treat cancers.

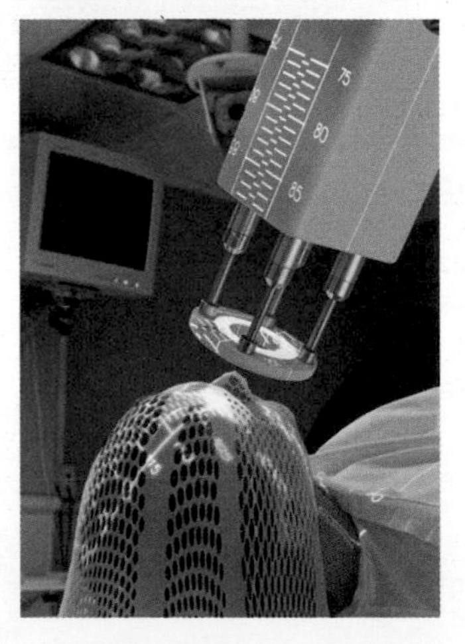

FIGURE 7: A patient undergoing radiotherapy in the treatment of cancer. Why is ionising radiation effective in treating cancer?

Question

7 In schools only teachers are allowed to use radioactive sources. Why is it too dangerous to let students use them?

Using nuclear radiation

Smoke alarms contain a radioactive source that emits alpha radiation. The radiation ionises the oxygen and nitrogen atoms in air. This causes a very small electric current that is detected.

When smoke fills the detector in the alarm during a fire, the air is not so ionised. The current is less and the alarm sounds.

Measuring thicknesses in industry

Radioactive substances are used to measure and control the thickness of metals and paper as they are manufactured. Beta radiation is used for paper in a rolling mill, but gamma radiation is used for metal sheets.

As the sheet passes between the rollers it is pressed into the required thickness. If the pressure on the rollers is reduced the sheet becomes thicker. The amount of radiation passing through the sheet decreases. The radiation detector senses this and transmits a signal back to the rollers to increase the pressure.

FIGURE 8: How does a smoke detector work?

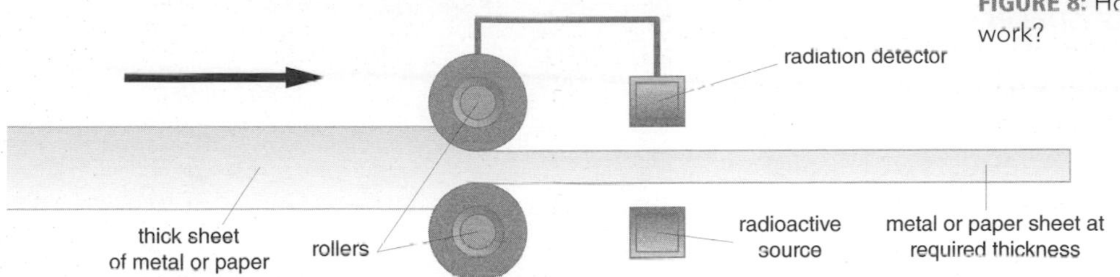

FIGURE 9: Using a radiation detector to keep the thickness of metal or paper constant. What happens if the pressure on the rollers increases?

Uses in medicine

Gamma radiation (and sometimes beta radiation) is used in medical screening to trace the passage of blood and other substances around the body. A radioactive liquid called a **tracer** is injected into the patient and after a time a special picture is taken.

Many cancers form deep inside the body. For this reason, penetrating gamma radiation is most commonly used to attack tumours in cancer treatment.

The instruments that doctors use are **sterilised** by gamma radiation. The radiation kills microbes and bacteria that could lead to infections in patients.

Non-destructive testing

Gamma radiation can be used to test such things as welds or castings. A source of gamma radiation is placed on one side of the object to be tested and a detector on the other. Any increase in the amount of radiation detected will indicate a fault in the weld or casting.

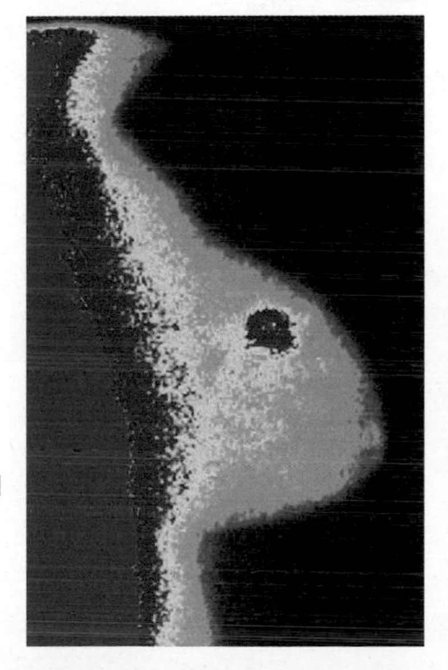

FIGURE 9: A patient's breast after a tracer has been given. The red part of the breast shows a high concentration of tracer due to an increased blood supply which indicates a cancer.

Questions

8 Suggest two reasons why an alpha radiation sourcè is suitable for use in a smoke detector.

9 What type of radiation is used to check the thickness of paper in a paper mill? Why?

10 What type of radiation is used to check the thickness of aluminium in a rolling mill? Why?

Problems of dealing with radioactive waste

Radioactive waste can remain radioactive for thousands of years so it has to be stored safely. The waste must be stored so that it cannot get into the natural underground water systems and hence into lakes and rivers.

Although the waste is generally not suitable for making nuclear weapons there is a risk that terrorists could use waste to contaminate water supplies or areas of land both in towns and countryside.

As scientists discover more about the effects of exposure to nuclear radiation, the acceptable level to which we can be exposed may change in the future.

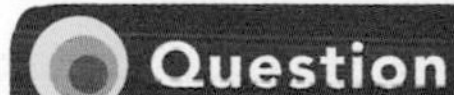

Question

11 Suggest areas in the world where it is not safe to store radioactive waste. Explain your reasons.

uses of gamma radiation

Exploring our Solar System

You will find out:

- about the bodies in space that make up the Universe
- about the planets in our Solar System
- why planets and moons stay in orbits

'Houston, we've had a problem here'

Imagine being a long way from home when your car breaks down and there is no one nearby to help. Now imagine being over 300 000 km from home when there is an explosion in your spacecraft.

You have no light, electricity or water and your oxygen supply is running out. You have to rely totally on scientists at Mission Control Centre in Houston, in the United States of America, to get you home.

Thankfully, the spacecraft was guided home successfully and the crew of Apollo 13 landed safely.

FIGURE 1: What happens at Mission Control, Houston?

Our place in the Universe

Scientists have been studying the **Universe** for a long time.

Stonehenge is an ancient monument. It was built over 4000 years ago. The people who built it knew where to place the massive stones.

The Universe contains billions of galaxies. Our **galaxy** is called the Milky Way.

Each galaxy contains billions of **stars**. Our star is called the Sun. Earth is just one of the planets that orbit the Sun in our **Solar System**.

The stars we see are in our own galaxy. Stars are very hot and produce their own light.

FIGURE 2: What can you see very early on Midsummer's Day between two special stones at Stonehenge?

The Moon can be seen as it orbits the Earth. Sometimes, other planets, orbiting satellites, **meteors** and **comets** can be seen in the night sky. Meteorites are large rocks that do not burn up as they fall to Earth.

Some things, such as a **black hole** at the centre of a galaxy, can never be directly seen.

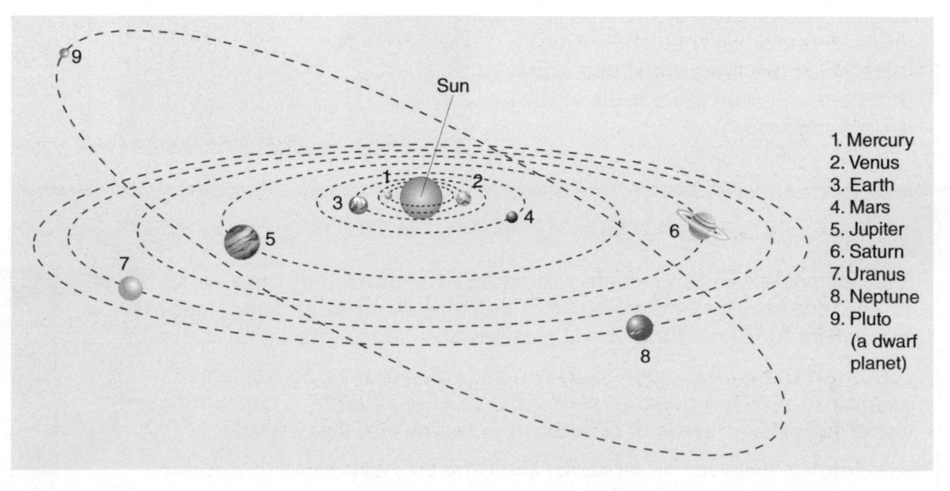

FIGURE 3: Suggest why some pictures of our Solar System, like this one, show Pluto and others do not.

is Pluto a planet?

Remember!
To help you remember the order of the planets you can use the mnemonic:
My **V**ery **E**asy **M**ethod **J**ust **S**et **U**p **N**aming planets

Questions

1 Arrange the following in order of size, starting with the smallest.
- **comet** · **galaxy** · **moon** · **planet**
- **Solar System** · **star** · **Universe**

2 Which of the following produces its own light?
- **comet** · **moon** · **planet** · **star**

What's out there?

It is now accepted that there are eight planets in our Solar System, orbiting the Sun. Gravitational forces determine how planets move.

Since its discovery in 1930, Pluto was considered to be the ninth planet in our Solar System. But in August 2006, a group of astronomers voted to strip Pluto of its planetary status. They said it is too small to be considered a planet. It is smaller than several moons in the Solar System so has been reclassified as a dwarf planet.

Planet	Diameter in km	Average distance from Sun in million km	Time to orbit Sun in Earth units
Mercury	4800	57	88 days
Venus	12 200	108	225 days
Earth	12 800	150	1 year
Mars	6800	228	1.9 years
Jupiter	143 000	778	11.9 years
Saturn	120 000	1429	29.5 years
Uranus	51 000	2870	84 years
Neptune	50 000	4500	165 years
Pluto	*4000*	*5900*	*248 years*

> Comets have very elongated orbits. They pass inside the orbit of Mercury and then reach way past Pluto. Comets are made of ice and dust. Most are smaller than 10 km in diameter.

> Each star you see in the night sky is one of millions of stars that form a galaxy. Stars vary in size from being not much larger than Jupiter to 10000 times larger than Earth. A star is a mass of glowing gas with temperatures measured in thousands of degrees celsius. Planets orbit stars and, with the help of space telescopes, planets that orbit other stars are being discovered.

> A meteor is a small rock made from grains of dust that get very hot as they come into contact with Earth's atmosphere. They burn up and heat the air around them. The air glows and a streak of light, called a 'shooting star', is seen. A meteorite is a larger rock that does not burn up but it falls to Earth.

> A black hole occurs where a large star has collapsed in on itself under gravity. It cannot be seen because light cannot escape from it. A black hole has a very large mass but a very small size.

Questions

3 Pluto was only discovered in 1930. Suggest why it took so long to discover Pluto.

4 Suggest why Mars is more likely to have a surface capable of supporting life than any planet other than Earth.

More on the Solar System

Around and around

Everything moves in a straight line unless a force acts on it. A force that acts towards the centre of a circle is called a **centripetal** force.

Question

5 What provides the centripetal force to keep the Moon in orbit around Earth?

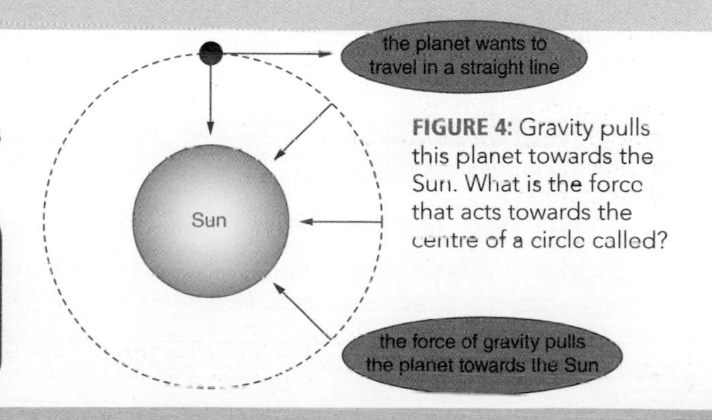

FIGURE 4: Gravity pulls this planet towards the Sun. What is the force that acts towards the centre of a circle called?

Is anybody out there?

You will find out:

- about the distances involved in space travel
- about manned and unmanned space flights
- how very large distances are measured in space

On 12 December 1901, Marconi sent the first radio message from England to the United States.

On 16 November 1974, scientists sent a coded message towards a star system near the edge of our galaxy. The message gave information about life on Earth.

Because of the large distances involved the scientists do not expect a reply for at least 40 000 years.

FIGURE 5: The Marconi radio transmitter was used to send the first radio message across the Atlantic Ocean.

Scientists have also sent unmanned spacecraft into space. Some of them carry pictures showing a man and a woman and details of our Solar System and will never return to Earth.

Other spacecraft contain recordings of sounds from Earth and greetings in many languages. Unlike spacecraft containing astronauts, unmanned spacecraft do not need to carry food, water or oxygen.

For the past 50 years, unmanned spacecraft (**probes**) have been collecting information from outer space. Some have landed on the Moon and some have landed on Mars. Probes have passed by or orbited the other planets, or have been sent to investigate comets, sending back data.

A probe may measure:

- temperature
- gravitational forces
- radiation
- magnetic fields
- a planet's atmosphere and surroundings.

Remote vehicles have driven over the surface of Mars. These robots take photographs, and analyse rocks and the atmosphere. One day they will collect rocks to bring back to the Earth.

The Hubble Space Telescope orbits Earth collecting information from the furthest galaxies.

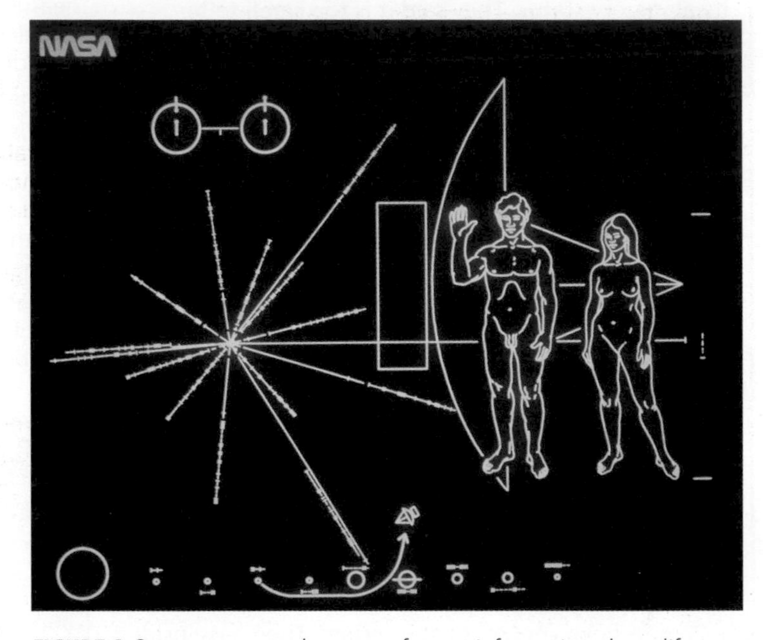

FIGURE 6: Some unmanned spacecraft carry information about life on Earth. Can you suggest why?

Questions

6 When did scientists first send a radio message into space?

7 Suggest three advantages of sending an unmanned rather than a manned spacecraft into space.

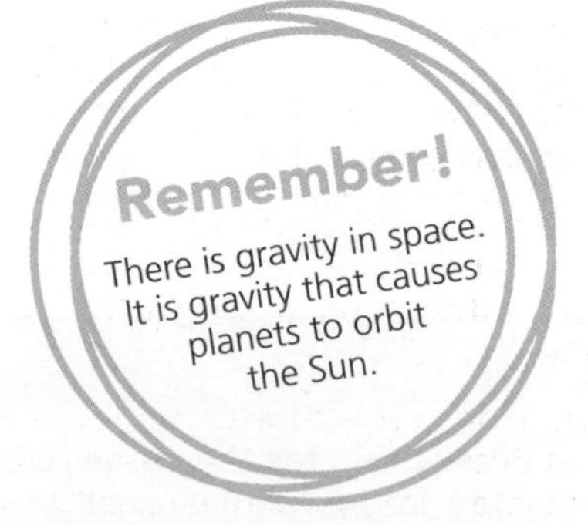

NASA abandon plans for Moon base   ISS sightings

Exploring the planets

Manned spacecraft

Unmanned space probes can go where conditions are deadly for humans. Although astronauts wear normal clothing inside their spacecraft, when they go outside for a spacewalk, or for a walk on the Moon, they need to wear a special **spacesuit**.

> Without an atmosphere to filter the sunlight the surface of the suit facing the Sun might reach a temperature of 120 °C. The other side of the suit may be as cold as −160 °C. The spacesuit keeps the astronaut's body at a normal Earth temperature.

> A dark visor stops the astronaut being blinded.

> The suit is pressurised and has an oxygen supply for breathing.

If astronauts do visit other planets they will be away from home for a very long time. The spacecraft will need to carry large amounts of fuel as well as food, water and oxygen for the astronauts. The environment in the spacecraft needs to be kept as similar as possible to that on Earth.

Throughout their mission the astronauts will be subject to lower gravitational forces than they are used to.

Earth is 150 000 000 km away from the Sun. That might seem a long way, but compared to some of the distances measured in the Universe it is not.

Light travels 300 000 km each second. So light from the Sun takes about eight minutes to reach Earth. Very large distances in space are measured in **light years**. This is the distance light travels in one year.

FIGURE 7: What measurements does this remote vehicle take on the surface of Mars?

FIGURE 8: Astronauts on a spacewalk. What do you think has to be taken into account when planning how many supplies to load on to a spacecraft?

Questions

8 Why can an astronaut wear normal clothes inside a spacecraft?

9 The gravitational pull of the Moon is only one-sixth of the gravitational pull of Earth. How does this affect the ability of an astronaut to move around on the Moon?

A long way to go!

Is it all worth it?

In 2004, President Bush of the United States announced that **NASA** was working towards a permanent Moon base within 20 years and sending astronauts to Mars sometime after 2020. Experts have put the cost at as much as £400 billion.

Five years later, plans for the Moon base were abandoned and efforts are now being concentrated on landing astronauts on Mars.

Unmanned spacecrafts cost less and do not put lives at risk, but they have to be very reliable because there is usually no way of repairing them when they break down.

Did you know?

A light year represents 9 500 000 000 000 000 km. That's a long way.

The star nearest to Earth other than the Sun is called Proxima Centauri and is 4.22 light years away.

Some stars are billions of light years away, so imagine converting that to kilometres.

Question

10 How long does sunlight take to reach Pluto? Give your answer in hours.

Threats to Earth

You will find out:
> how the Moon was formed
> about the properties of asteroids
> how asteroids have affected Earth in the past

Fatal collision

About 65 million years ago Earth was struck by a large asteroid 10 km in diameter. A crater called the Chixulub Basin in Mexico, measuring 300 km across, is thought by scientists to have been caused by the collision.

This event is believed to have caused great climatic changes and so conditions were no longer suitable for dinosaurs to stay alive and they became extinct.

FIGURE 1: What caused the climate changes that made life for the dinosaurs impossible?

Earth's remnant

Both the Moon and Earth are estimated to be 4.6 billion years old. There is evidence to suggest that they were formed after two planets collided. Earth was formed as a result of the collision and the remaining debris collapsed together to form the Moon.

The Moon is a natural **satellite** that orbits Earth.

FIGURE 2: 'Earthrise.' How are Earth and the Moon thought to have formed?

What happens when an asteroid hits Earth?

Asteroids are large rocks in orbit around the Sun. If an asteroid leaves its orbit and approaches the Earth we see it as a meteor. If it hits Earth as a meteorite it leaves a large crater. The Barringer Crater in Arizona was formed over 50 000 years ago by the impact of an iron-containing meteorite colliding with Earth. The meteorite is thought to have had a mass of more than 10 000 tonnes. The crater is 183 m deep and has a diameter of 1200 m.

FIGURE 3: What caused this crater in Arizona?

The damage caused by the giant asteroid that hit Earth 65 million years ago was much worse.

> An enormous crater was made.

> Hot rocks rained down.

> Fires were widespread.

> **Tsunamis** flooded large areas.

> A cloud of dust and water vapour was thrown up into the upper atmosphere and spread around the globe.

> Sunlight could not penetrate the dust cloud and temperatures on Earth fell.

It is estimated that 70 per cent of all species on Earth, including the dinosaurs, became extinct as the Earth's climate changed.

Questions

1 How old is Earth?

2 Why did most plants become extinct 65 million years ago?

3 There is not a large lump of iron in the centre of the Barringer Crater. Suggest what happened to the iron in the asteroid as it hit Earth.

The origin of the Moon

During the first million years after its formation, there was a major disturbance of the Solar System. There was a collision between Earth and another planet. The force of the collision was sufficient to almost destroy both planets.

The dense molten iron became concentrated into Earth's core and the less dense rocks started to orbit the new Earth in a ring. These rocks clumped together to form the Moon.

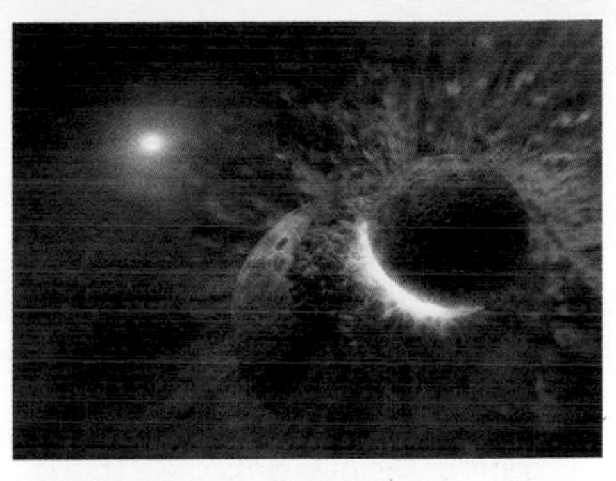

FIGURE 4: Earth collided with a smaller planet that broke up. The pieces joined together. What did they form?

What is an asteroid?

An asteroid is a mini-planet or 'planetoid' that orbits the Sun. Most asteroids are in a 'belt' between Mars and Jupiter. It is a lump of rock and is left over from when the Solar System was formed.

What evidence is there for asteroid collisions with Earth?

Scientists know that asteroids have collided with Earth in the past because of the evidence they have found. As well as large craters, geological evidence supports collisions from asteroids in the past.

In 1980, geologists were analysing rock samples near to where an asteroid was thought to have struck. They found low levels of the metal iridium. Iridium is not normally found in the rocks of Earth's crust but is quite common in meteorites. Asteroids contain many of the elements found in the interior of planets.

Fossil evidence

Fossil evidence also supports asteroids having collided with Earth. Many fossils are found below the iridium layer but few are found above it.

The 150 m high tsunami that followed the impact also disrupted the fossil layers. It carried debris that contained a variety of fossil fragments as far as 300 km inland.

Questions

4 Most asteroids are in orbit between Mars and Jupiter. Eros has an orbit time of 1.76 years. Where does this asteroid orbit?

5 Describe how the motion of the Moon has changed since it was formed.

Evidence for the origin of the Moon

The debris blown out of both the Earth and the planet it collided with came from their iron-depleted outer layers. The iron core of the other planet melted on collision and merged with Earth's iron core.

The evidence supports the model.

> The average density of Earth is 5500 kg/m^3 while that of the Moon is only 3300 kg/m^3.

> There is no iron in the Moon.

> The Moon has exactly the same oxygen composition as the Earth, but rocks on Mars and meteorites from other parts of the Solar System have different oxygen compositions. This shows that the Moon formed in the same vicinity as Earth.

Why do asteroids not form planets?

All bodies in space, including planets, were formed when clouds of gas and dust collapsed together as a result of gravitational forces.

The size of gravitational forces depends on the mass of the object.

Jupiter has a very large mass compared to the mass of an asteroid, so its gravitational force prevents an asteroid joining with other asteroids to form a planet. The asteroids remain in a belt and orbit around the Sun.

Questions

6 One earlier theory was that the Moon was formed at the same time as Earth as a 'double planet'. How would the Moon be different if this theory was correct?

7 Why does Jupiter have a larger effect on asteroids than any other planet?

8 Geologists have recently found fossil remains of animals that did not live in the same environment close together. They have also found 70-million-year-old fossils above those that are 65 million years old.

Explain these observations.

dinosaur extinction

A comet's tail

You will find out:
- > about the properties of comets
- > how scientists are monitoring the danger from near-Earth asteroids and comets

A comet is a chunk of ice filled with dust and rock.

When a comet comes near to the Sun's heat, its ice core warms up and a glowing tail is thrown out. The tail is formed from very small pieces of **debris**.

Near-Earth objects

Some asteroids and comets have orbit paths that pass close to the orbit of Earth. They are called **near-Earth objects** or **NEOs**. A slight movement from their orbit may mean that they collide with Earth.

Scientists look for near Earth asteroids and comets using a **telescope**.

Did you know?

In 1682, British astronomer Edmund Halley saw a comet and worked out it would next be seen in 76 to 77 years.

In 1759, right on time, the comet was seen again.

The next time Halley's comet is due to pass close to Earth is in 2062.

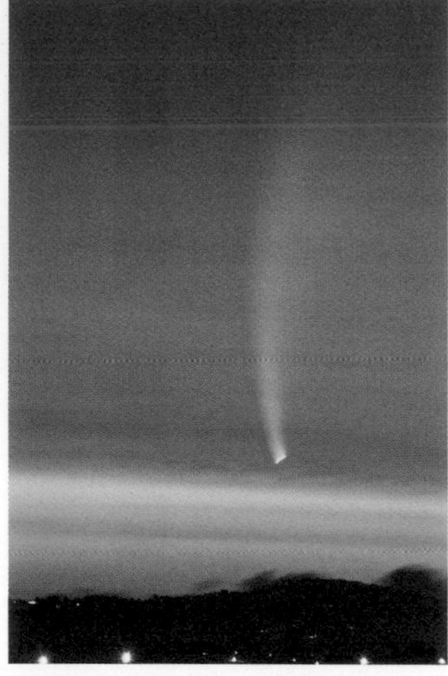

FIGURE 5: A comet's tail can be seen from Earth. What is a comet's tail made from?

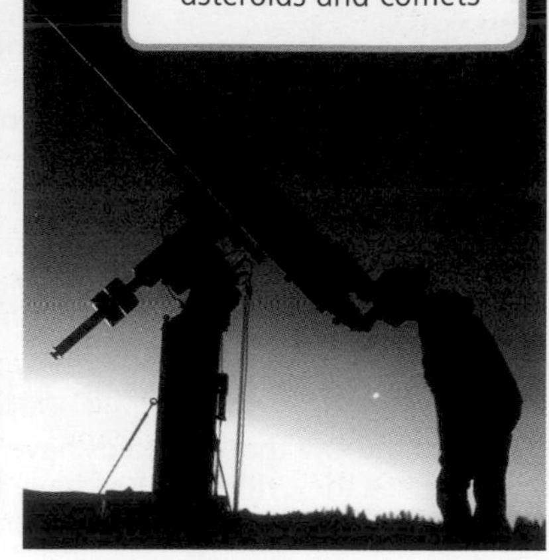

FIGURE 6: A telescope is used to look at objects in the night sky. What are two examples of NEOs that can be seen through a telescope?

Questions

9 You can only see a comet's tail when it is near the Sun. Suggest why.

10 Why can you not see asteroids in the night sky with the naked eye?

A comet's orbit

Most planets have circular-shaped orbits.

The orbit of a comet is very **elliptical**. Most comets pass close to the Sun, inside the orbit of Mercury. They then pass well outside the orbit of Pluto. Their speed increases as they get closer to the Sun.

A comet is made of dust and ice, just like a dirty snowball. As a comet passes near to the Sun, the ice melts. A glowing cloud of gas and dust forms around the comet. Solar wind from the Sun blows the gas and dust into the comet's tail. This always points away from the Sun, so a comet travels tail first as it goes away from the Sun. The tail may be hundreds of millions of kilometres in length.

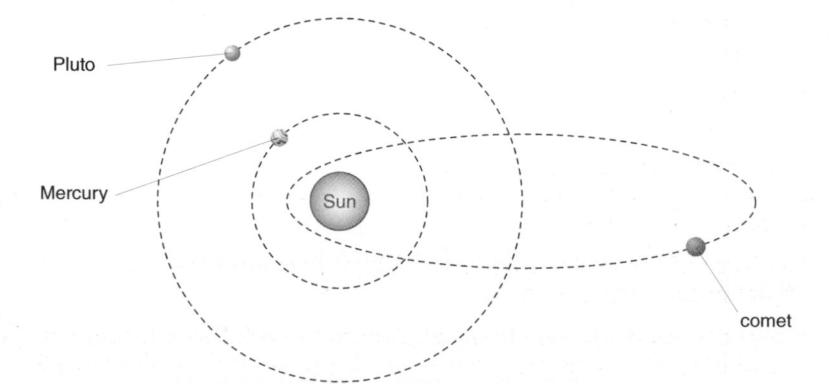

FIGURE 7: A comet has an elliptical orbit. What happens as it approaches the Sun?

comets and asteroids

Predicting a collision

Imagine you saw an object at point **A** yesterday.

Today it is at point **B**.

If it carries on moving as it is, tomorrow it will be at point **C**.

But, if it speeds up, it may be at point **D**.

Or if it changes direction, **E** may be a likely place, then it may be at **F** the next day.

The longer we study the path of an asteroid or comet, the more information we have to predict its path. The orbit pattern of Earth is well known. The orbit of the asteroid or comet is plotted and continually updated to see if there is any risk of a collision.

It is not easy to track near-Earth objects. Many are very small, less than a few metres across, so are very difficult to spot. A large number spend much of their time in an orbit between us and the Sun. It is not easy to see anything when you are looking into the Sun.

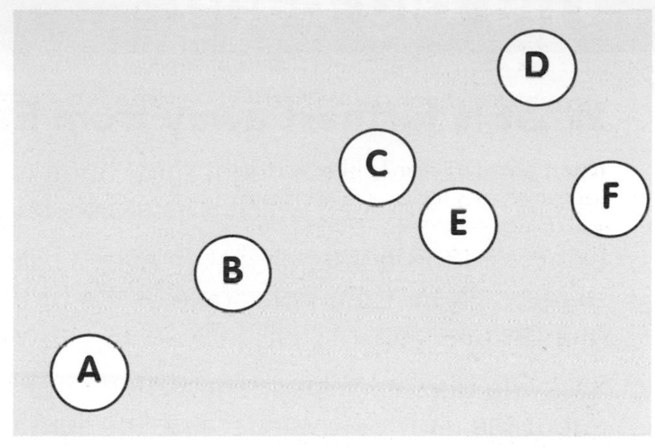

FIGURE 8: Predicting the path of a near-Earth object is not easy.

Question

11 Astronomers have not been studying near-Earth objects for very long. Suggest why they sometimes change their minds about whether a particular asteroid is likely to collide with Earth in the future.

A comet's speed

The gravitational attraction from the Sun causes a comet's speed to increase as it approaches the Sun.

The further away from the Sun a comet is the slower it travels.

What do we do if a NEO approaches?

The positions of NEOs are constantly monitored by telescopes and satellites.

If an object was found to be on a collision course with Earth, there are several options.

One option would be to launch a rocket containing a very large explosive. This would be detonated a long way from Earth, near to the object.

The force of the explosion may be enough to change the path of the object so that it misses Earth.

If that did not work and the object did collide with Earth, it could be the end of life on Earth as we know it, just like it was for the dinosaurs 65 million years ago.

Did you know?

New NEOs are being discovered almost every day.

About 100 NEOs pass close enough to Earth to be classed as a threat each year.

Question

12 Why would an explosion to change the course of an asteroid on a collision course with Earth have to be a long way away?

NASA near-Earth objects

The Big Bang

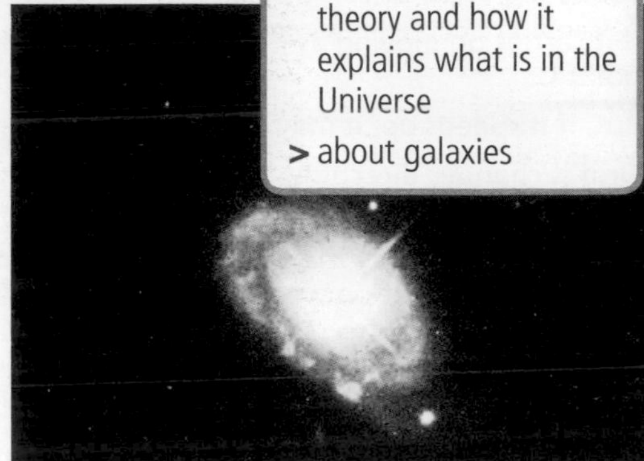

You will find out:
- > about how the Universe started
- > about the Big Bang theory and how it explains what is in the Universe
- > about galaxies

What is furthest away from Earth?

It might just look like a bright spot, but it is very special. It is a quasar, which is incredibly bright.

Light from this quasar has taken over 2 billion years to reach Earth. Light from some of the furthest quasars takes over 12 billion years to get here.

Studying quasars helps scientists find out more about the early Universe because the light from a quasar left it so long ago.

FIGURE 1: Quasars are the most distant objects in the Universe.

Universal theories

There have been many theories about the Universe, how it began and how it has changed over time.

One of the earliest models of the Universe came from Claudius Ptolemaeus (Ptolemy) over 2100 years ago. He suggested the Universe was like an onion with the Earth at the centre. The Moon, Sun and other planets were in layers around the Earth. The outer skin contained the stars.

In 1514, Copernicus was the first to suggest that planets orbited the Sun. Less than 100 years later, Galileo supported Copernicus.

Today, most scientists accept the Big Bang theory for the formation and structure of the Universe.

The first few seconds

About 15 billion years ago all of the matter in the Universe was in a single point. The temperature was 1000 million million million million °C.

Scientists do not know what caused the Universe to explode but in the smallest fraction of a second it did. There was a massive fireball of particles and radiation.

The Universe expanded rapidly and cooled down. Scientists call this the **Big Bang**. After one-ten-millionth of a second the Universe had cooled to below 1 million million °C. Within a few seconds, particles such as electrons, protons and neutrons were formed. Hydrogen and helium were the first elements formed within a few minutes.

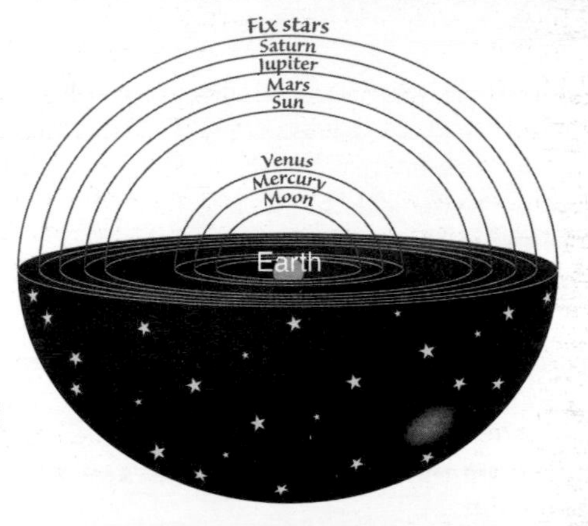

FIGURE 2: Who suggested this model of the Universe?

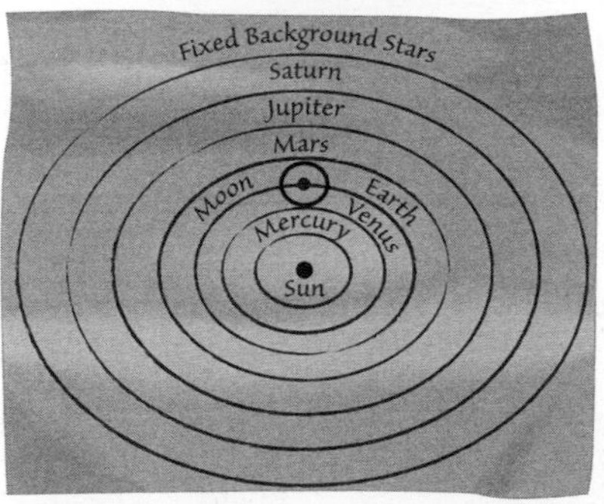

FIGURE 3: When was this model first suggested?

Did you know?

Scientists have estimated that the mass of the Universe is about 30 000 000 000 000 000 000 000 000 000 000 000 000 000 000 000 000 000 kg.

If scientists are correct about the mass of the Universe then it will continue to expand.

Questions

1 Name three of the earliest particles formed in the Universe

2 What were the first elements in the Universe to be formed?

history of the Universe Big Bang space size of Universe

What's wrong with Ptolemy?

Copernicus realised that Ptolemy's model did not explain the periodically observed backward motion of Mars, Jupiter and Saturn. It also failed to account for the fact that Mercury and Venus never moved more than a certain distance from the Sun.

With the help of the newly invented telescope, Galileo observed four moons orbiting Jupiter. This confirmed that not everything orbited the Earth and supported Copernicus. The Ptolemaic model no longer provided a full explanation.

The expanding Universe

Scientists have found that, with only a few exceptions, all of the galaxies are moving away from Earth and apart from each other. The furthest galaxies are moving fastest. The Universe is expanding.

This behaviour of the Universe can be modelled very simply using a balloon. Dots are drawn onto the balloon. These represent the galaxies. As the balloon is blown up, the dots move apart from each other. The bigger the balloon gets, the further the dots move apart.

Microwave signals are constantly reaching Earth from all parts of the Universe. This is believed to be a 'remnant' of the Big Bang.

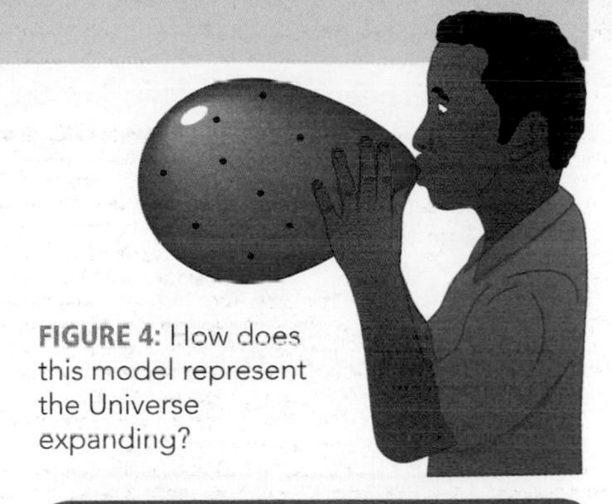

FIGURE 4: How does this model represent the Universe expanding?

Question

3 The Centaurus galaxy is moving more slowly than the Sombrero galaxy.
Explain how you know which one is furthest away from Earth.

Why not adopt the Copernican/Galilean model?

In the 16th century the Roman Catholic Church was very influential and despite Copernicus's ideas continued to support Ptolemy's model that the Earth was at the centre of the Universe. Ptolemy's model was also simpler to understand. There are allegations that Galileo was threatened with torture and life imprisonment to change his mind but these have not been substantiated. It was 350 years before the Church finally cleared Galileo.

In the 17th century, Newton was working on his theory of universal gravitation. This suggested that all bodies attract one another. A result of this would be that the Universe would collapse as the stars all mutually attracted. To counter this, he proposed an infinite Universe with an infinite number of stars. Today, we believe that gravitational collapse is prevented because the Universe is constantly expanding as a result of the Big Bang.

Red shift

When a source of light is moving away, the wavelength appears to increase. This shifts the light towards the red end of the visible spectrum.

When scientists look at the spectrum of light from the Sun it has lines on it. When scientists look at light from a distant star, there is the same pattern of lines in the spectrum but the lines have moved closer together and towards the red end of the spectrum.

This is known as **red shift** and it shows that the star is moving away from Earth. The faster the star is travelling, the more the amount of red shift is observed. Scientists use this information to calculate the age and starting point of the Universe.

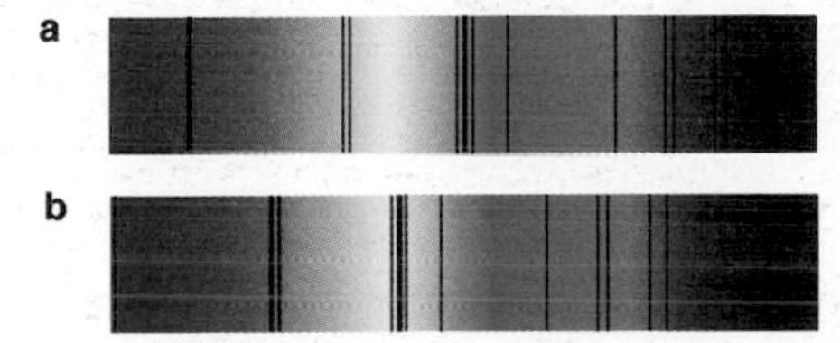

FIGURE 5: a The spectrum of white light from the Sun (the black lines show that helium is present). **b** The spectrum of light from a distant star showing red shift.

Questions

4 Blue shift occurs when a galaxy is moving towards Earth.

a Describe what a scientist sees when she looks at light from a galaxy moving slowly towards Earth.

b What will she see if she looks at light from a galaxy moving very quickly towards Earth?

what is red shift?

You will find out:
- about the life cycle of stars

A star is born

Man has been studying the stars for a very long time. Star maps are not new. The **constellations** have been drawn for centuries. The arrangement of stars in a constellation is just the way they appear from Earth.

But the stars in the sky have not always been there. New stars are forming all the time. They have different sizes and masses.

A star starts its life as a swirling mass of gas and dust in a huge cloud of material in space.

All stars, including the Sun, will one day die. Very large stars become **black holes**. Not even light can escape from a black hole because of the strong gravitational forces. The 'escape speed' needed for anything to leave is greater than the speed of light.

FIGURE 6: This picture of Leo is based on a map published in 1690.

Did you know?

A star's lifetime may be millions of years.

The smallest stars are neutron stars. They are only 20 km across.

In 1054, an exploding star produced a supernova remnant called the Crab Nebula.

Chinese astronomers recorded that they could see the Crab Nebula in daylight for the next 3 weeks.

Questions

5 How does a star start its life?

6 Constellations have names such as Pisces, Capricorn, Scorpio and Libra. What is a constellation?

Gone – but not forgotten

What happens to a star at the end of its life depends on its size. When a star has used up all of the hydrogen 'fuel' in its core, the core contracts and no more energy is produced.

The Sun

As the core of a medium-sized star such as the Sun contracts, the outer part expands. It cools and changes colour from yellow to red. The star becomes a **red giant**.

The Sun will become so big that it will swallow up Mercury and Venus and reach Earth. While the star is in its red giant phase, shells of gas are thrown out. These are called **planetary nebula**. The nebula from the Sun will stretch to the edge of our Solar System.

The core of the original star shrinks until it is about the same size as Earth. It is very hot and shines brightly as a **white dwarf**. It is not making energy so eventually cools, changing colour from white, through yellow to red and becomes a **black dwarf**.

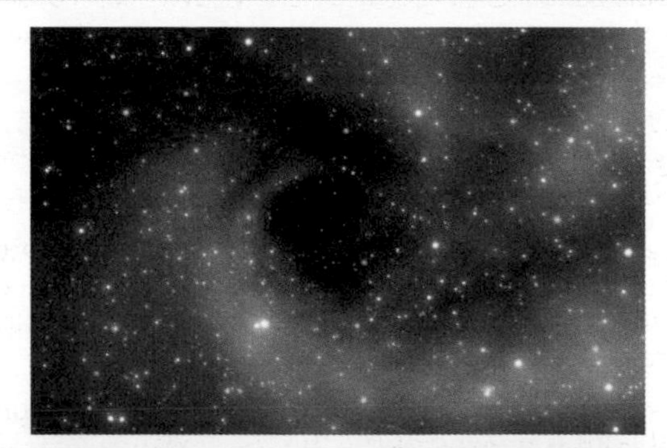

FIGURE 7: The birth of a star. Why do stars look so small in the night sky?

FIGURE 8: A planetary nebula. What causes these?

life cycle of a star telescope black hole

More massive stars

A massive star only shines for a few million years before it uses up all its hydrogen. The core starts to contract and the outer part expands as a red **supergiant**. Suddenly, in less than a second, the core collapses and the whole star explodes and is thrown outwards. This explosion is called a **supernova** and the small remaining core is a **neutron star** (or a black hole if the star is very massive). Neutron stars are very dense. A neutron star contains mainly neutrons and just one teaspoonful has a mass of 100 000 000 tonnes!

The material thrown out as the core explodes, collides with gas and dust in space and forms a glowing cloud of gas called a supernova remnant. The Crab Nebula is an example of a supernova remnant.

Over a period of time, the supernova remnant merges with other dust and gas in space. As gravitational forces act, so a new star is formed.

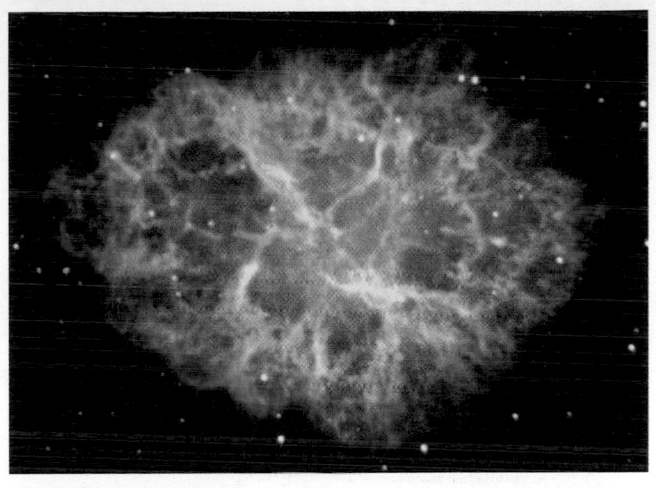

FIGURE 9: Crab Nebula is a supernova remnant formed in 1054 that is still visible today.

Questions

7 Our Sun is sometimes referred to as a 'second generation star'. Suggest why.

8 Atoms have a nucleus that contains protons and neutrons. The nucleus is surrounded by a cloud of electrons. Use your knowledge of the structure of an atom to explain why a neutron star is so dense.

A star's life history

A star starts its life as a cloud of gas and dust. Gravity can cause this cloud, or **nebula**, to contract into a spinning ball of gas. The gas ball is so tightly packed that it gets hot and starts to glow.

This **protostar** is shining, but not brightly and it cannot be seen because of the gas and dust surrounding it. As gravity causes the protostar to become even smaller, it gets hotter.

After millions of years, the temperature in the core of the protostar is high enough for **thermonuclear fusion** to take place. This is the joining together of hydrogen nuclei to form helium nuclei. Huge amounts of energy are released.

The star is now visible and remains visible while there is enough hydrogen. The life of a star depends on its size. Small stars live longer than large stars. Although large stars have more hydrogen, they use it at a faster rate. A medium-sized star, like the Sun, may shine for 10 billion years. A large star may only shine for a few million years.

What happens to a star at the end of its life depends on its mass (see opposite and above).

The core of a neutron star could continue to collapse even more. It has a large mass and small volume. As a result it becomes so dense and gravitational forces are so large that not even light can escape. This is a black hole.

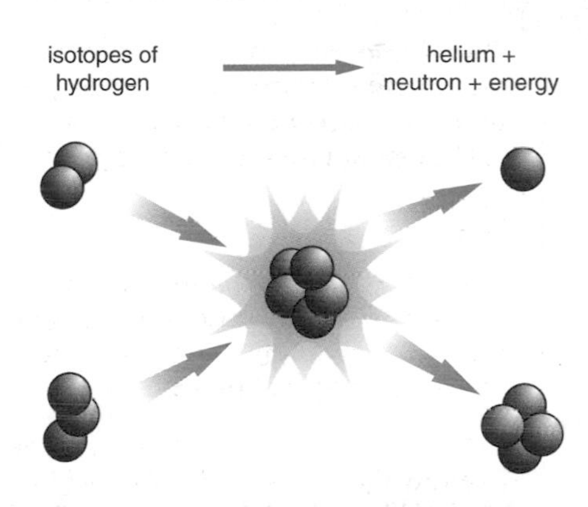

FIGURE 10: What has to happen to a protostar before thermonuclear fusion can take place?

Questions

9 What causes dust and gas to collect together to form a star?

10 What happens during thermonuclear fusion?

Preparing for assessment: Planning and collecting primary data

To achieve a good grade in science, you not only have to know and understand scientific ideas, but you need to be able to apply them to other situations and investigations. These tasks will support you in developing these skills.

Tasks

- Plan an investigation to see how the area of a photocell exposed to the light affects the current produced.
- Once your plan has been approved, perform the investigation, record your results and write a simple conclusion.

Context

Photocells use light to produce electricity. Solar cells are photocells that use the light from the Sun to produce electricity.

To obtain large amounts of electricity, a large area of solar cells is needed.

Solar cells are also used to produce small amounts of electricity, for example to charge a battery. This solar cell is under some trees which means it does not receive as much light from the Sun.

Planning your investigation

You can measure the voltage from a photocell using a digital voltmeter.

These are the things you will need to consider when planning your investigation.

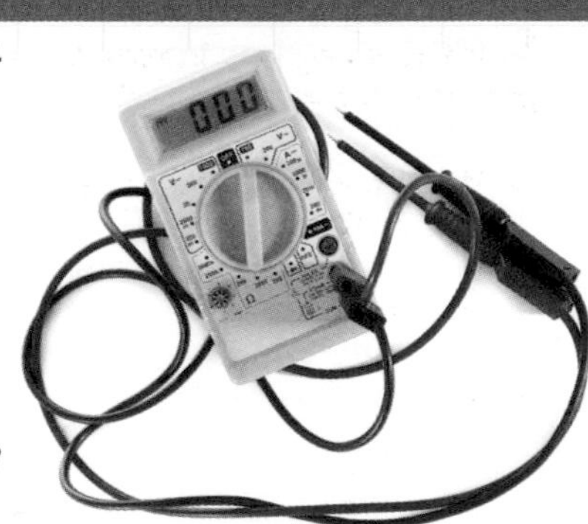

1. You can develop your plan in groups of two or three if you wish.

2. How will you measure the current from the photocell? You cannot just replace the voltmeter with an ammeter.

3. How will you change the area of the photocell exposed to the light?

4. What do you need to keep the same to make it a fair test?

5. How many different areas will you need before you can identify a trend?

6. Will you need to repeat your readings? If so, how many times?

7. You should carry out a risk assessment before you start the investigation. What precautions should you take?

8. Write the plan for the investigation. Try to write the plan in a logical order and ask yourself if someone can perform the investigation following just your plan.

Performing the investigation

Once your plan has been approved you can perform the investigation.

1. Identify all of the different photocell areas you used.

2. If you repeated any readings, all of these will need to be recorded as well as the average result.

3. Record your results in a table like this. You may need to add extra rows or columns.

area of photocell in cm^2	current in A	

4. If you were to complete this as a GCSE Controlled Assessment, you would go on to plot a graph and evaluate the investigation.

5. Is there any way in which you could have improved on how you performed the investigation?

6. What have you found out about how the area of the photocell affects the current?

How many sets of readings will you need to take to identify a trend?

Do you need to repeat readings? If so, how many times? Why do scientists repeat readings?

What graph would you draw? What would the labels be on the axes? How would you use the graph to decide on the answer to the task?

Think about accuracy and precision. How are they different?

P2 Checklist

To achieve your forecast grade in the exam you'll need to revise

Use this checklist to see what you can do now. It gives you many of the important points you will need to know. Refer back to the relevant pages in this book if you're not sure and to see if there is anything else you need to know. Look across the three columns to see how you can progress.

Remember you'll need to be able to use these ideas in various ways, such as:

> interpreting pictures, diagrams and graphs
> applying ideas to new situations
> explaining ethical implications
> suggesting some benefits and risks to society
> drawing conclusions from evidence you've been given.

Look at pages 278–299 for more information about exams and how you'll be assessed.

To aim for a grade E	To aim for a grade C	To aim for a grade A
understand that photocells convert light into DC electricity **describe** how the Sun's energy can be harnessed	**describe** advantages and disadvantages of photocells **describe** advantages and disadvantages of wind turbines	**describe** how light produces electricity in a photocell **understand** how a photocell's current depends on light intensity **explain** why passive solar heating works **recall** that solar collectors track the position of the Sun
describe the dynamo effect **recall** that a generator produces AC and a battery produces DC **describe** the stages in electricity production and distribution **realise** there is a lot of energy wasted in a power station **use** the equation in the context of a power station: efficiency = $\frac{\text{useful energy output}}{\text{total energy input}}$ (× 100%)	**describe** how a dynamo can be changed to produce a greater current **describe** how simple AC generators work **describe** how electricity is generated at a conventional power station	
understand what causes the greenhouse effect **recall** examples of greenhouse gases **describe** how climate change is linked to global warming **describe** difficulties in measuring global warming **explain** why scientists should share their data	**describe** how different wavelengths of radiation behave **name** natural and man-made sources of greenhouse gases **explain** how both human and natural activity affect the weather **list** evidence for and against man-made global warming **distinguish** between opinion and evidence-based statements	**explain** the greenhouse effect in terms of infrared wavelengths **interpret** data on abundance and impact of greenhouse gases **interpret** data about global warming and climate change **explain** how scientists can agree about the greenhouse effect but disagree about the effects of human activity
use the equation: power = voltage × current **recall** that the unit of power is the watt or kilowatt **use** data relating cost to power of appliance and time used **recall** that transformers can increase or decrease voltage **recall** that fuels release energy in heat	**use** the equation: energy supplied = power × time **state** that the unit of electrical energy is the kilowatt hour **calculate** the cost of energy supplied **explain** why transformers are used in the National Grid **describe** and evaluate advantages and disadvantages of different energy sources	**use** the kilowatt hour as a measure of the energy supplied **list** points for and against using off-peak electricity **explain** how increasing voltage leads to less energy waste

To aim for a grade E	To aim for a grade C	To aim for a grade A
recognise where radiation can be beneficial or harmful **recall** the three types of ionising nuclear radiation **describe** how to handle radioactive materials safely **recall** that nuclear power does not cause global warming **describe** nuclear waste as radioactive and harmful	**describe** examples of beneficial uses of radiation **describe** the relative penetrating power of alpha, beta and gamma radiation **describe** how to dispose of radioactive waste **recall** that uranium is non-renewable **recall** that plutonium is a waste product from nuclear reactors	**describe** how alpha, beta and gamma radiation can be identified **explain** problems of dealing with radioactive waste **describe** the advantages and disadvantages of nuclear power
identify the relative positions of the Earth, the Sun and planets **explain** why stars give off their own light and can be seen **recall** that radio signals take a long time to travel through space **compare** the resources needed by different types of spacecraft **describe** why unmanned spacecraft are sent into space	**recall** the relative nature and sizes of bodies in space **describe** a light year as the distance light travels in a year **describe** difficulties of man travelling between planets **recall** that humans would die where unmanned spacecraft go	**understand** that centripetal force keeps bodies moving in orbit **explain** why a light-year is used to measure distance in space **explain** points for and against using unmanned spacecraft
recall that the Moon may be the result of two planets colliding **describe** the makeup of a comet and of an asteroid **describe** consequences of an asteroid colliding with Earth **describe** that a near-Earth object may collide with Earth **describe** how near-Earth objects may be seen	**describe** how planets colliding could have formed the Earth and Moon **describe** the position and orbits of asteroids and comets **describe** evidence for past asteroid collisions **describe** how observing NEOs can predict their future paths **explain** why NEOs are difficult to observe	**discuss** evidence for a collision forming the Earth–Moon system **explain** why the asteroid belt is between Mars and Jupiter **explain** why the speed of a comet changes as it approaches a star **suggest** actions to avoid the threat of a NEO collision
describe ideas about the Big Bang theory **recall** that stars have a finite 'life' and are different sizes **recognise** that models of the Universe have changed over time **recognise** the Ptolemaic, Copernican and Galilean models **understand** why not even light can escape from black holes	**recall** that most galaxies are moving away; the furthest move fastest **describe** the end of the 'life cycle' of a small and a large star **explain** how better technology has led to new models **describe** why Copernicus and Galileo developed new models	**explain** the meaning of red shift and what we can find out from it **describe** the life history of a star **explain** why the Copernican and Galilean models were controversial and were not widely adopted for a long time

P2 Exam-style questions

Foundation Tier

1 A coastal footpath is lit by a 'streetlight' the electricity for which is obtained from batteries. The batteries are charged by electricity obtained from solar cells.

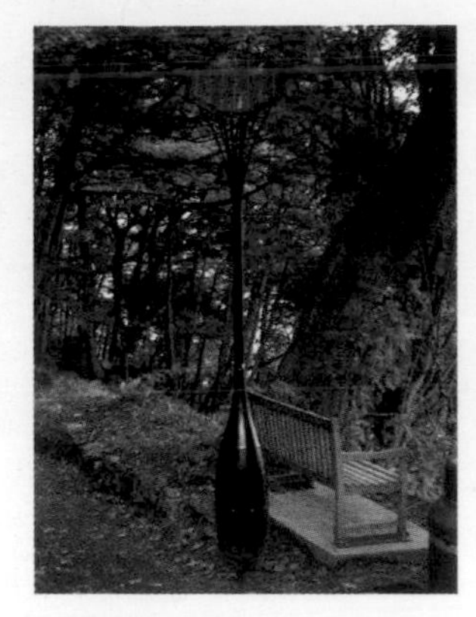

AO1 **(a)** Write down four advantages of using solar cells to produce electricity. [4]

AO1 **(b)** Describe one disadvantage of using solar cells. [1]

[Total: 5]

2 Ella attaches a dynamo to her bicycle lamp to produce the electricity needed for the lamp.

AO1 **(a)** What can Ella do to make the lamp shine more brightly? [1]

AO1 **(a)**
AO2 **(b)** The output voltage from the dynamo is 12 V. The current in the lamp is 1.5 A when it is operating normally. Calculate the power rating of the lamp. [2]

[Total: 3]

AO1 **3** There are three main types of ionising radiation. Alpha and gamma are two of the types.

(a) Write down the name of the other main type of ionising radiation. [1]

(b) Smoke detectors use a source of alpha radiation. Describe how a smoke detector works. Use ideas about ionisation. [3]

(c) Describe **three** safety precautions a teacher should take in a school laboratory when handling a source of gamma radiation. [3]

[Total: 7]

4 The table shows some information about the planets in our Solar System.

planet	diameter in km	average distance from Sun, in million km	time to orbit Sun, in Earth units
Mercury	4800	57	88 days
Venus	12 200	108	224 days
Earth	12 800	149	1 year
Mars	7000	228	1.9 years
Jupiter	143 000	780	11.9 years
Saturn	120 000	1425	29.5 years
Uranus	49 000	2900	84 years
Neptune	50 000	4500	165 years

AO2 **(a)** Which planet has the largest diameter? [1]

AO2 **(b) (i)** An asteroid takes 4.7 Earth years to orbit the Sun. Between which two planets does it orbit? [1]

AO1 **(ii)** About 65 million years ago, a very large asteroid collided with Earth. Write down **three** effects of an asteroid colliding with Earth. [3]

AO3 **(c)** Astronauts have visited the Moon and unmanned spacecraft have explored the surface of Mars. Suggest why Mars may be the first planet to be visited by humans. [2]

AO2 **(d)** Our Sun is a medium-sized star. What will happen to our Sun at the end of its life when it has used up all of the hydrogen in its core? Arrange these four stages in the correct order:

black dwarf planetary nebula
red giant white dwarf [3]

[Total: 10]

AO1 **5** There have been many models of the Universe. One of the earliest was described by Ptolemy. A later model came from Galileo.

(a) What is the main difference between the models of Ptolemy and Galileo? [1]

(b) Most scientists today believe that the Universe started from a single point about 15 billion years ago and then exploded. What name is given to this theory? [1]

[Total: 2]

AO1 recall the science AO2 apply your knowledge AO3 evaluate and analyse the evidence

Worked Example – Foundation Tier

The Earth is getting warmer. The greenhouse effect contributes to global warming.

a What gases contribute to the greenhouse effect? [2]

The greenhouse effect is caused by the greenhouse gases in the atmosphere. These are carbon dioxide and sulfur dioxide. Also there is a hole in the ozone layer and this makes it hotter.

b (i) The picture shows smoke from chimneys drifting over a town.

Why does the smoke cause a rise in atmospheric temperature? [1]

The smoke is a good insulator and stops energy being released into the atmosphere.

(ii) The picture shows an ash cloud from a volcano.

Explain what happens to the temperature on Earth as a result of the ash cloud. [2]

The smoke reflects radiation back to Earth so the Earth gets colder.

How to raise your grade!

Take note of these comments – they will help you to raise your grade.

The answer started well by referring to greenhouse gases in the atmosphere and mentioning carbon dioxide; then followed two common misconceptions. The main problem from sulfur dioxide is acid rain and the depletion of the ozone layer leads to increased levels of ultraviolet radiation and does not contribute to global warming.
1/2

There is confusion here. Smoke from chimneys reflects radiation from the town back to Earth, which is why the Earth warms up.
0/1

The answer does mention reflecting radiation (which is correct) and the Earth cooling (also correct) BUT the reasoning is not correct. It is radiation from the Sun that is reflected back into space.
1/2

This student has scored 2 marks out of a possible 5. This is below the standard of Grade C. With a little more care the student could have achieved a Grade C.

P2 Exam-style questions

Higher Tier

1 The chart shows typical electricity generation efficiencies from different sources.

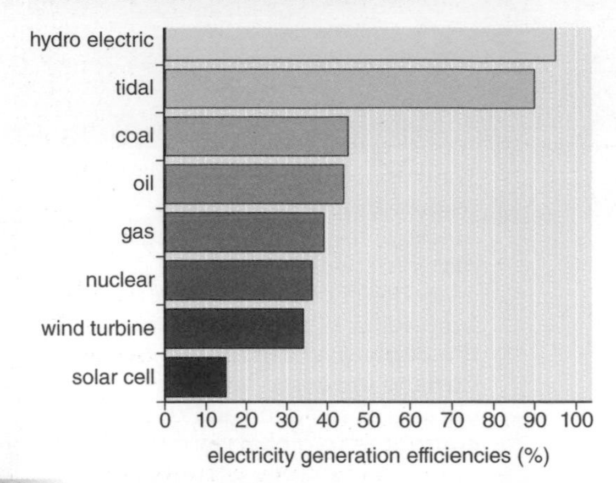

AO2 **(a)** What is the efficiency of a coal-fired power station? [1]

AO2 **(b)** The electrical output from a coal-fired power station in Yorkshire is 1960 million joules per second. Calculate the energy lost each second at the power station. [2]

AO1 **(c)** It has been estimated that the light emitted from a domestic bulb represents only 1% of the energy in the coal burned in the power station to produce that electrical power. Explain how electricity companies try to reduce energy loss as electricity is distributed around the country. [2]

[Total: 5]

2 A source of beta radiation is used to control the thickness of paper in a rolling mill.

Nasira is designing a similar piece of equipment to control the thickness of steel sheets up to 1 cm thick.

AO3 **(a)** Explain why she should choose a source that emits gamma radiation. [2]

AO1 **(b)** Once the equipment is no longer needed, there will be some radioactive waste. Suggest what should be done with the radioactive waste. [3]

[Total: 5]

3 The nearest star to our Sun is 4.22 light years away.

AO1 **(a)** What is meant by a light year? [1]

AO1
AO2 **(b)** A star system at the edge of the Milky Way is 20 000 light years away. Scientists have sent a coded radio message towards that star system. If there is a life form there and they reply to the message immediately, how long will it be before a reply might be received? [1]

[Total: 2]

4 The graph shows how carbon dioxide emissions from different regions of the world have changed since 1990 and how they are expected to change by 2030.

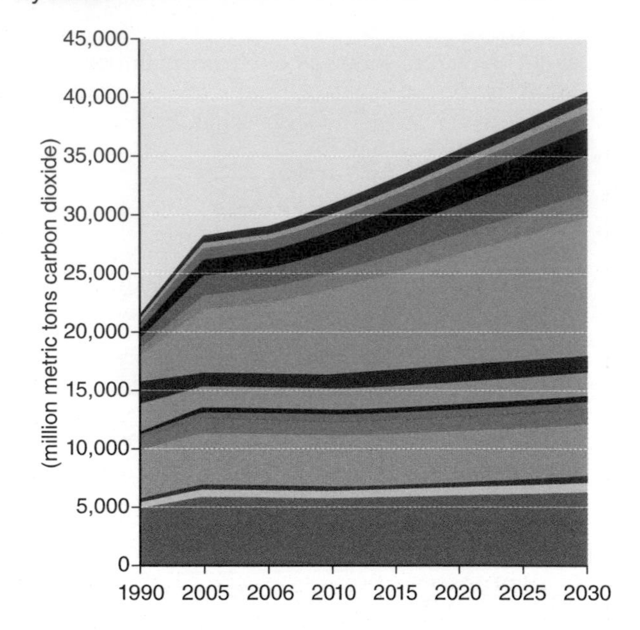

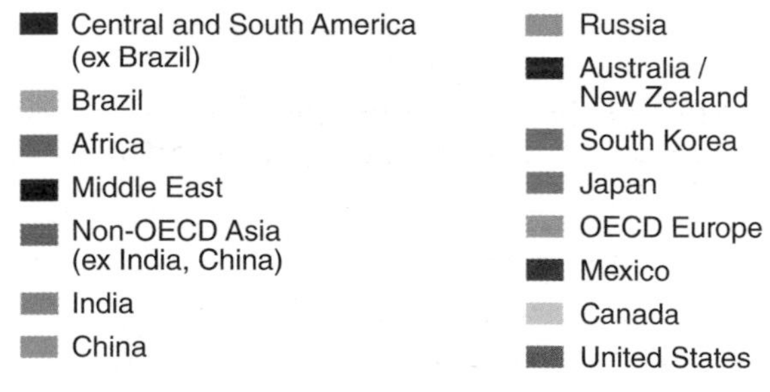

AO2 **(a)** Which region shows the largest expected increase in emissions by 2030? [1]

AO2 **(b)** Which region shows a reduction in emissions between 1990 and 2030? [1]

AO2 **(c)** What are the expected carbon dioxide emissions in 2020? [1]

AO1 **(d)** Why are governments trying to reach agreement on reducing the levels of carbon dioxide emissions? [2]

[Total: 5]

AO1 recall the science | AO2 apply your knowledge | AO3 evaluate and analyse the evidence

Worked Example – Higher Tier

A coastal footpath is lit by a 'streetlight' the electricity for which is obtained from batteries. The batteries are charged by electricity obtained from solar cells.

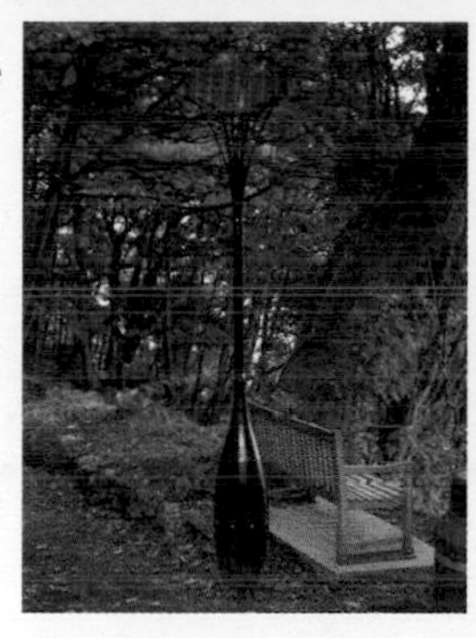

How to raise your grade!

Take note of these comments – they will help you to raise your grade.

(a) Write down four advantages of using solar cells to produce electricity. [4]

- *They don't cause pollution.*
- *They need no maintenance.*
- *They don't need electric cables.*
- *They work on DC.*

It is sometimes a good idea to use bullet points when asked to provide a specific number of answers.

The first three answers are worthy of credit. The last statement is also correct but is irrelevant to the question.
3/4

(b) Describe how light produces electricity in a photocell. [3]

Photocells contain two pieces of silicon that contain impurities. One type is known as n-type and the other as p-type. Together they make a p-n junction. When light photons are absorbed by the photocell, a current flows.

The first part of the answer is providing information about the structure of the photocell and although correct will not score marks. This is a good answer that can be improved by stating that the electrons are knocked loose from the silicon and that it is a flow of electrons that causes a current.
2/3

(c) The light shown in the photograph is on a tree-lined path. How does the presence of the trees affect the working of the solar cells? [2]

The trees block the light from the solar cells, which reduces the amount of current produced. The current produced is inversely proportional to the surface area.

The question asks how the trees affect the working of the solar cells so the answer goes into the detail of the relationship between the current and surface area exposed to the light but it should be directly proportional.
1/2

This student has scored 6 marks out of a possible 9. This is below the standard of Grade A. With a little more care the student could have achieved a Grade A.

When the evidence doesn't add up.

Sometimes people use what sound like scientific words and ideas to sell you things or persuade you to think in a certain way. Some of these claims are valid, and some are not. The activities on these pages are based on the work of Dr Ben Goldacre and will help you to question some of the scientific claims you meet. Read more about the work of Ben in his *Bad Science* book or at badscience.net.

How much to look younger?

There are many ways to make yourself look younger if you're an adult. These include the style and colour of your hair, the texture of your skin, your body shape and the clothes you wear. Manufacturers and retailers know this and recognise where there's money to be made.

Which of these do you think is more effective?

Are there other ways for adults to make themselves look younger?

How are these age-defying products promoted?

YOUNG SKIN FROM OLD?

Skin changes in appearance as people get older. These photographs show how older skin looks different to young skin.

> Examine the photographs. What are the differences?

> How might an anti-ageing skin cream work on the old skin? What would it need to do?

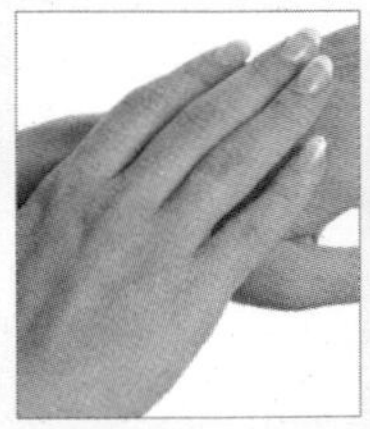

Young skin

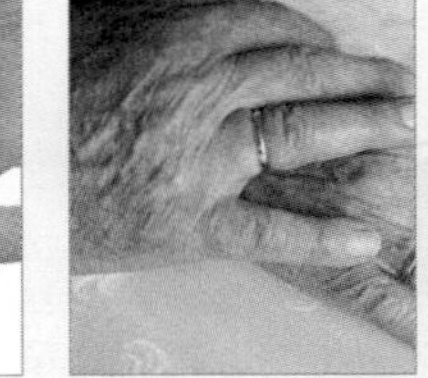

Old skin

THE SCIENCE BEHIND THE CLAIM

As you get older you may not like the appearance of wrinkles and crows feet. You can spend quite a lot of money on anti-ageing skin creams. Creams are advertised with appealing images and lavish claims, but do they really work?

One immediate gain from a cream is rehydration. Dried out skin doesn't look good so we can make it look better by moisturising it. This is easy and the active ingredients are really cheap. However something more is needed to make someone genuinely look younger.

These are three types of active ingredient commonly used:

> Alphahydroxy acids, such as vitamins A and C, are used to exfoliate the skin. Some of these work at high doses, but they are also irritants, so they can only be sold at low doses.

> Vegetable proteins, which are long chain molecules. As the cream dries on the skin the chain molecules tighten, applying tension and temporarily tightening it.

> Hydrogen peroxide, which is corrosive and will lightly burn the skin.

Why might someone who uses a cream with these ingredients think that it is working? Will the effects last?

We are learning to:

> find out how anti-ageing skin creams work

> examine claims that are made for them

NEW AND IMPROVED! ADVERTISER'S CLAIMS

Many anti-ageing skin creams are sold on the basis that there is a scientific reason that they work. Some claims are justified but others are pretty dubious, even if they look persuasive at first glance. You should think critically about what you are told by the advertisers.

> Claims sound more convincing if they are based on tests and if scientists have been involved. Powerful scientific words include *conclusive tests*, *laboratory*, *cleanse*, *purify* and *health*.

> They may claim to make you feel better, look younger, have more energy and be healthier. Some of the claims may be difficult to prove; they should have been tested, the full results published and independently checked in a scientific way. There are very few cases of anti-ageing skin creams being proved to get rid of wrinkles. Why do you think this is?

> Watch out for claims such as 'eight out of ten users said that…' if it's not clear what kind of people and how many were asked. What do you think ten company employees might say about their product and would this be representative of all their consumers?

THE PSYCHOLOGY OF COSMETICS

You can buy very cheap creams in the shops. You can even make your own skin cream using simple ingredients. If you did, it would be pretty good at moisturising, so your skin would feel soft and maybe a little smoother. It wouldn't, however, make you look younger for long.

Why do you think anti-ageing skin creams are sometimes quite expensive?

Do you think people who buy anti-ageing skin creams

a) genuinely believe that they make them look younger?

b) hope that they might but don't really believe it?

c) do it because it makes them feel good?

When the evidence doesn't add up.

Sometimes people use what sound like scientific words and ideas to sell you things or persuade you to think in a certain way. Some of these claims are valid, and some are not. The activities on these pages are based on the work of Dr Ben Goldacre and will help you to question some of the scientific claims you meet. Read more about the work of Ben in his *Bad Science* book or at badscience.net.

What we'll look like in the future

Here are some questions to get you thinking about the future of the human race.

> How do you think humans have evolved over the last few millions of years?

> What changes have taken place in the way we look and move?

> How might evolution affect us over the next few million years?

News stories that sound like they are about science can come from a number of different places. Sometimes they are fair reports of real scientific research, but sometimes they are just good stories.

The story opposite featured in a number of news reports, including *The Times*, where this version was printed, BBC, *Daily Telegraph* and *The Sun*. Your task is to work out if this story is good science or bad science.

You are going to investigate the predictions reported in this article. First you need to identify the predictions. This is the first one – that humans will evolve into two separate species. To decide whether you think this is good or bad science think about what you have learnt about evolution. Do you think this likely to happen?

From **THE TIMES**
October 17, 2006

The future ascent (and descent) of man

Within 100,000 years the divide between rich and poor could lead to two human sub-species

By Mark Henderson, Science Editor

The mating preferences of the rich, highly educated and well-nourished could ultimately drive their separation into a genetically distinct group that no longer interbreeds with less fortunate human beings, according to Oliver Curry.

Dr Curry, a research associate in the Centre for Philosophy of Natural and Social Science of the London School of Economics, speculated that privileged humans might over tens of thousands of years evolve into a "gracile" subspecies, tall, thin, symmetrical, intelligent and creative. The rest would be shorter and stockier, with asymmetric features and lower intelligence, he said.

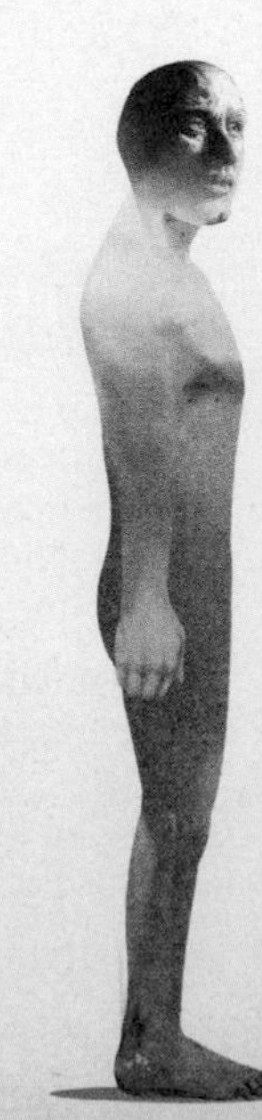

We are learning to:

- > analyse a piece of text about a scientific topic to identify the key features
- > set these features against a background of accepted scientific evidence
- > consider why some stories about science get written

People today are taller and live longer than people a few hundred years ago. Why do you think this has happened? Do you think that trend will continue?

Dr Curry is a research associate, but is this story based on scientific research? Why do you think Bravo asked Dr Curry to write this piece? Do you know if Bravo usually take an interest in stories about science?

Dr Curry's vision echoes that of H. G. Wells in *The Time Machine*. He envisaged a race of frail, privileged beings, the Eloi, living in a ruined city and coexisting uneasily with ape-like Morlocks who toil underground and are descended from the downtrodden workers of today.

Dr Curry also said that today's concept of race would be gone by the year 3000, relationships between people with different skin colours producing a "coffee-colour" across all populations.

In Brazil, the black African, white European and native American populations have been having children together for hundreds of years but there is still a lot of diversity in physical appearance. Can you think of any other examples that you know of that either back up the idea about the whole population being "coffee-coloured" or make you think it might not happen that way?

With improvements in nutrition and medicine, people would routinely grow to 6ft 6in and live to the age of 120, he said.

Genetic modification, cosmetic surgery and sexual selection — whereby mate preferences drive evolution — meant that people would tend to be better-looking than today.

Otherwise, humans will look much as they do now, with one exception: Dr Curry also suggested that increased reliance on processed food would make chewing less important, possibly resulting in less developed jaws and shorter chins. Ten thousand years from today this effect could be compounded as human faces grow more juvenile in appearance.

This effect — neotony — is known from domestic animals: dogs resemble young versions of wild relatives such as wolves.

Dr Curry raised the worrying possibility that reliance on technology could erode social skills and even health. As deaths from genetic diseases such as cancer are prevented, the genes themselves might become more common, no longer being "weeded out" of the gene pool. Increased use of medicine as a means of treating disease could lead to the deterioration of the body's immune system.

How have we ended up with a huge variety of breeds of domestic dog? Could the same happen with humans?

Dr Curry's predictions were commissioned by the television channel Bravo to celebrate its 21st anniversary on air.

"The Bravo Evolution Report suggests that the future of man will be a story of the good, the bad and the ugly," he said. "While science and technology have the potential to create an ideal habitat for humanity over the next millennium, there is the possibility of a genetic hangover due to an over-reliance on technology reducing our natural capacity to resist disease or get along with each other.

"After that, things could get ugly, with the possible emergence of genetic 'haves' and 'have-nots'."

Do you think that the "genetic 'haves' and 'have-nots'" is a likely future for our race? Is a world with two species of humans likely? What evidence would you use to support your claim?

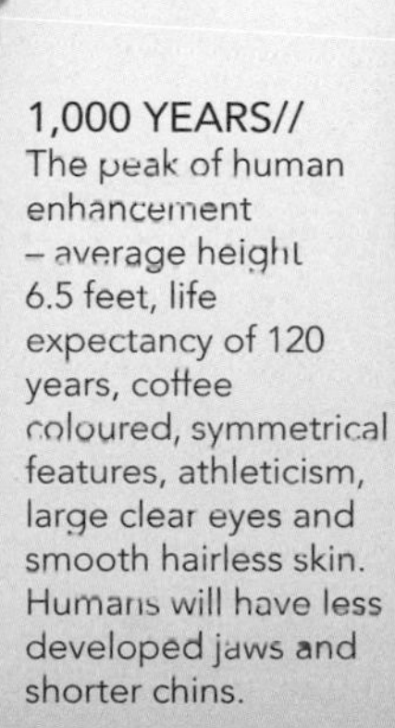

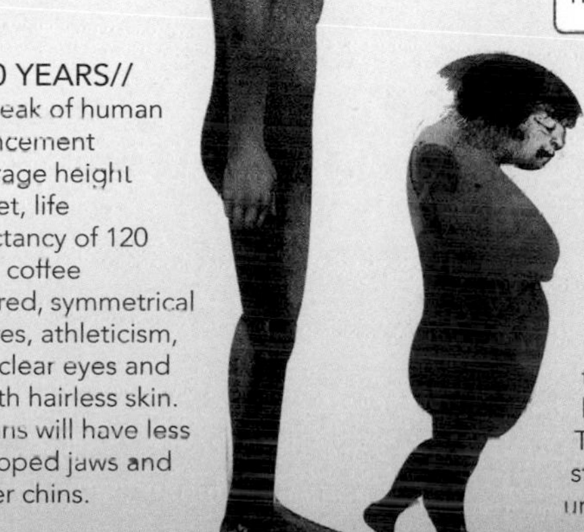

1,000 YEARS// The peak of human enhancement – average height 6.5 feet, life expectancy of 120 years, coffee coloured, symmetrical features, athleticism, large clear eyes and smooth hairless skin. Humans will have less developed jaws and shorter chins.

100,000 YEARS// Mankind will be divided into two distinct sub-species – the genetic 'haves' and the genetic 'have nots.' The 'haves' will be tall, thin, symmetrical, clean, healthy, intelligent and creative. The 'have nots' will be short, stocky, asymmetrical, grubby, unhealthy and less intelligent.

When the evidence doesn't add up.

Sometimes people use what sound like scientific words and ideas to sell you things or persuade you to think in a certain way. Some of these claims are valid, and some are not. The activities on these pages are based on the work of Dr Ben Goldacre and will help you to question some of the scientific claims you meet. Read more about the work of Ben in his *Bad Science* book or at badscience.net.

Keeping your brain fit

In science you learn about ideas that scientists have developed by collecting evidence from experiments; you are also learning to collect and evaluate evidence yourself. You can use this outside of the laboratory to weigh up information you come across everyday. Let's look at this example about how to prepare for exams.

When it comes to exam season you will get lots of different tips from teachers, other students and companies that all claim to help you do better in exams but who is right? If you had an important exam coming up, what would be a good way of making sure that your brain was going to function well?

GOOD ADVICE?

Here are three pieces of advice offered to students before exams. For each one:

> suggest why it might be true;

> suggest why you might be dubious about it;

> decide whether you think it's good advice and explain why.

Before sitting an aural exam (a listening test) spin round three times clockwise and three times anticlockwise. This stimulates the semicircular canals which are located in the inner ear, thus stimulating the cochlea.

A drink with caffeine in it is a good idea as it acts as a stimulant and will cause your brain to work quicker.

Before doing an exam in the morning make sure you have a good breakfast. Something like porridge is good as the energy is released slowly during the morning, so you don't get tired towards the end.

We are learning to:

> apply ideas about cells and body systems to evaluate advice on how to focus on mental activities;

> examine scientific claims for accuracy;

> function effectively in a group, developing ideas collaboratively.

THE SCIENCE BEHIND THE CLAIM

Let's look in more detail at some activities that some schools have used to try to improve students' concentration and learning. Your task is to work out which bits of science are good science and which are bad. To help you decide whether you think your activity is good or bad try discussing these questions:

> What *advice* is being given?

> What *claim* is being made?

> What *scientific ideas* are being used to justify that claim?

> What *scientific ideas* do you have that may tell you something about this topic?

> Is the advice *sound*?

This is the advice.

This is the claim.

Interlock the fingers of both your hands, holding your elbows out at the sides. This completes a circuit and allows positive energy to flow. Positive energy creates positive thoughts, stimulating the brain, stilling anxieties and clearing the way for a free flow of logical thought.

These are the scientific ideas used to back up this claim. Forming the arms in a loop creates no circuit that any kind of energy 'flows around' and 'positive energy' is a meaningless term.

So can you think of any reason why this might work? You know that regular exercise is good and could help to refocus on ideas and mental activities.

Is the advice sound? Well, it won't do you any harm and may even improve concentration, but not for the reasons claimed.

Now you have a go. Are these good or bad science?

Water is a vital ingredient of blood and blood is essential to transport oxygen to the brain. For the brain to work well you have to ensure your blood is hydrated. This needs water, little and often. The best way of rehydrating the blood taking oxygen to the brain is to hold water in the mouth for up to half a minute, thus allowing direct absorption.

Your carotid arteries are vital to supplying your brain with richly oxygenated blood. Ensure their peak performance by pressing your brain buttons. These are just below the collar bone, one on either side. Make 'C' shapes with forefinger and thumb to place over the brain buttons and gently massage.

WOULD YOU PAY MONEY FOR THIS?

Many products or services are sold on the basis that there is a scientific reason that they work. Some claims are justified but some are pretty dubious, even if they look persuasive at first glance. You should think critically about what you are told.

> Claims might sound more convincing if they use technical scientific terms, but sometimes they are used incorrectly, to make something sound scientific when it's not.

> Powerful scientific words include 'conclusive tests', 'energy', 'cleanse', 'purify' and 'health'.

> They may claim to make you feel better, look better, have more energy and be healthier.

> The claims may be true but they should have been tested, the full results published and independently checked.

Carrying out controlled assessments in GCSE science

Introduction

As part of your GCSE Science course, you will have to carry out a controlled assessment. This will be divided into three parts.

> Research and collecting secondary data

> Planning and collecting primary data

> Analysis and evaluation

The tasks will be set by OCR, the awarding body, and marked by the teachers in school. The marking will be checked to make sure standards are the same in every school.

Some of the work you do must be supervised and you will have to work on your own under examination conditions.

Some experimental work may be performed in groups and results shared.

Some research work may be done as part of a homework exercise.

As well as your scientific skills, the quality of written communication will also be assessed as part of the controlled assessment.

Controlled assessment is worth 25% of the marks for your GCSE. It's worth doing it well.

Part 1: Research and collecting secondary data

You will have to plan and carry out research. The task will be given to you in the form of a handout. You will be allowed to research in class and/or as a homework activity.

Secondary data needs to be appropriate and you will have to select the information from a variety of sources to answer the questions you have been set.

Secondary data can be collected by a variety of different methods:

> survey

> questionnaire

> interview

> textbook

> newspapers and magazines

> internet search

Always make sure that you reference the material you use from any secondary source.

The work you do for your GCSE examination will need to be hand written or typed and printed out. You will take the work to a supervised lesson where you will use the information you have collected to answer specific questions in an answer booklet. Your research will be retained. It will be needed when you complete the analysis and evaluation part of your controlled assessment.

Your research may also be needed when you are planning your experiment.

Assessment tip

Make sure that the data you look at is relevant and appropriate. The data is more likely to be reliable if the same results and conclusions are obtained by a number of different researchers.

Assessment tip

As part of your research, you may need to design your own survey or questionnaire as well as collecting the data.

Choose the search terms you use on the internet carefully.

Assessment tip

When referencing, make sure you record:

> book title, author and page

> newspaper or magazine title, date, author and page

> full website address, author (if possible).

Definition

Secondary data are measurements/observations made by anyone other than you.

Part 2: Planning and collecting primary data

As part of your GCSE Science course, you will develop practical skills which will help you to plan and collect primary data from a science experiment. Your experimental work will be divided into several parts:

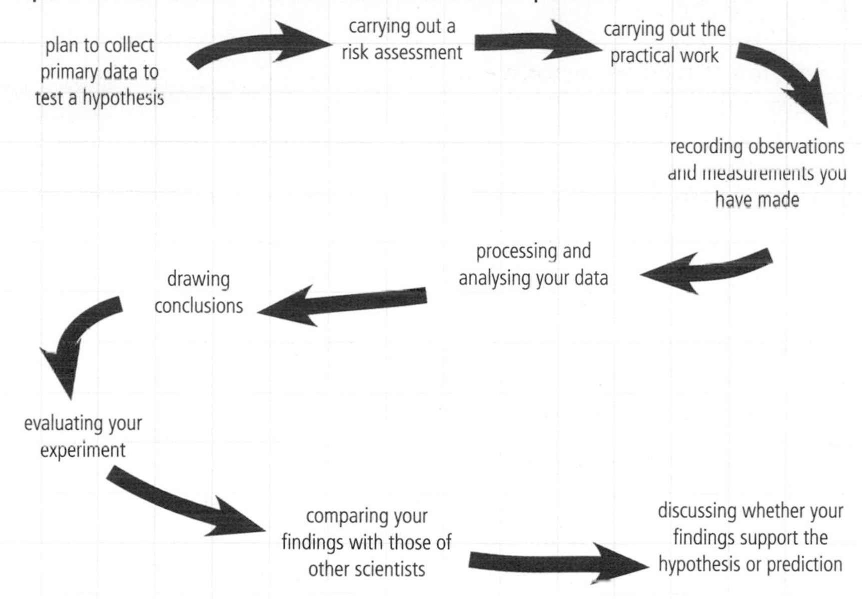

A scientific investigation usually begins with a scientist testing an idea, answering a question, or trying to solve a problem.

You first have to plan how you will carry out the investigation.

Your planning will involve testing a **hypothesis**. For example, you might observe during a fermentation with yeast investigation that beer or wine is produced faster at a higher temperature.

So your hypothesis might be 'as the temperature increases, the rate of fermentation increases'.

In Science, you will be given a hypothesis to test. In Additional Science, or if you're doing Separate Sciences, you will have to produce a hypothesis.

To formulate a hypothesis you may have to research some of the background science.

First of all, use your lesson notes and your textbook. The topic you've been given to investigate will relate to the science you've learned in class.

Also make use of the Internet, but make sure that your Internet search is closely focused on the topic you're investigating.

- ✔ The search terms you use on the Internet are very important. 'Investigating fermentation' is a better search term than just 'fermentation', as it's more likely to provide links to websites that are more relevant to your investigation.
- ✔ The information on websites also varies in its reliability. Free encyclopaedias often contain information that hasn't been written by experts. Some question and answer websites might appear to give you the exact answer to your question, but be aware that they may sometimes be incorrect.
- ✔ Most GCSE Science websites are more reliable, but if in doubt, use other information sources to verify the information.

If you do have to produce a hypothesis, you can use your lesson notes, the research you have already done and textbooks. The topic you've been given to investigate will relate to the science you've learned in class.

Assessment tip

You will be given a hypothesis to investigate as part of your controlled assessment.

Definition

A **hypothesis** is a possible explanation that someone suggests to explain some scientific observations.

Assessment tip

If you're formulating a hypothesis, it's important that it's testable. In other words, you must be able to test the hypothesis in the school lab.

Assessment tip

In the planning stage, scientific research is important if you are going to obtain higher marks.

Example 1

Investigation: Plan and research an investigation into the activity of enzymes

Your hypothesis might be 'When I increase the temperature, the rate of reaction increases'.

You may be able to add more detail, 'This is because as I increase the temperature, the frequency of collisions between the enzyme and the reactant increases'.

Assessment tip

Technology, such as data-logging and other measuring and monitoring techniques, for example heart sensors, may help you to carry out your experiment.

Definition

The **resolution** of the equipment refers to the smallest change in a value that can be detected using a particular technique.

Assessment tip

Carrying out a preliminary investigation, along with the necessary research, may help you to select the appropriate technique to use.

Choosing a method and suitable apparatus

As part of your planning, you must choose a suitable way of carrying out the investigation.

You will have to choose suitable techniques, equipment and technology, if this is appropriate. How do you make this choice?

You will have already carried out the techniques you need to use during the course of practical work in class (although you may need to modify these to fit in with the context of your investigation). For most of the experimental work you do, there will be a choice of techniques available. You must select the technique:

- ✔ that is most appropriate to the context of your investigation, and
- ✔ that will enable you to collect valid data, for example if you are measuring the effects of light intensity on photosynthesis, you may decide to use an LED (light-emitting diode) at different distances from the plant, rather than a light bulb. The light bulb produces more heat, and temperature is another independent variable in photosynthesis.

Your choice of equipment, too, will be influenced by measurements you need to make. For example:

- ✔ you might use a one-mark or graduated pipette to measure out the volume of liquid for a titration, but
- ✔ you may use a measuring cylinder or beaker when adding a volume of acid to a reaction mixture, so that the volume of acid is in excess to that required to dissolve, for example, the calcium carbonate.

In science, the measurements you make as part of your investigation should be as precise as you can, or need to, make them. To achieve this, you should use:

- ✔ the most appropriate measuring instrument
- ✔ the measuring instrument with the most appropriate size of divisions.

The smaller the divisions you work with, the more precise your measurements. For example:

- ✔ in an investigation on how your heart rate is affected by exercise, you might decide to investigate this after a 100 m run. You might measure out the 100 m distance using a trundle wheel, which is sufficiently precise for your investigation
- ✔ in an investigation on how light intensity is affected by distance, you would make your measurements of distance using a metre rule with millimetre divisions; clearly a trundle wheel would be too imprecise

✔ in an investigation on plant growth, in which you measure the thickness of a plant stem, you would use a micrometer or Vernier callipers. In this instance, a metre rule would be too imprecise.

 Variables

In your investigation, you will work with independent and dependent variables.

The factors you choose, or are given, to investigate the effect of are called **independent variables**.

What you choose to measure, as affected by the independent variable, is called the **dependent variable**.

 Independent variables

In your practical work, you will be provided with an independent variable to test, or will have to choose one – or more – of these to test. Some examples are given in the table.

Investigation	Possible independent variables to test
activity of yeast	> temperature > sugar concentration
rate of a chemical reaction	> temperature > concentration of reactants
stopping distance of a moving object	> speed of the object > the surface on which it's moving

Independent variables can be **discrete** or **continuous**.

> When you are testing the effect of different disinfectants on bacteria you are looking at discrete variables.

> When you are testing the effect of a range of concentrations of the same disinfectant on the growth of bacteria you are looking at continuous variables.

Range

When working with an independent variable, you need to choose an appropriate **range** over which to investigate the variable.

You need to decide:

✔ which treatments you will test, and/or

✔ the upper and lower limits of the independent variables to investigate, if the variable is continuous.

Once you have defined the range to be tested, you also need to decide the appropriate intervals at which you will make measurements.

Definition

Variables that fall into a range of separate types are called **discrete variables**.

Definition

Variables that have a continuous range are called **continuous variables**.

Definition

The **range** defines the extent of the independent variables being tested.

The range you would test depends on:

- ✔ the nature of the test
- ✔ the context in which it is given
- ✔ practical considerations, and
- ✔ common sense.

Example 2

1 Investigation: Investigating the factors that affect how quickly household limescale removers work in removing limescale from an appliance

You may have to decide on which acids to use from a range you're provided with. You would choose a weak acid, or weak acids, to test, rather than a strong acid, such as sulfuric acid. This is because of safety reasons, but also because the acid might damage the appliance you were trying to clean. You would then have to select a range of concentrations of your chosen weak acid to test.

2 Investigation: How speed affects the stopping distance of a trolley in the lab

The range of speeds you would choose would clearly depend on the speeds you could produce in the lab.

Assessment tip

Again, it's often best to carry out a trial run or preliminary investigation, or carry out research, to determine the range to be investigated.

Concentration

You might be trying to find out the best, or optimum, concentration of a disinfectant to prevent the growth of bacteria.

The 'best' concentration would be the lowest in a range that prevented the growth of the bacteria. Concentrations higher than this would be just wasting disinfectant.

If, in a preliminary test, no bacteria were killed by the concentration you used, you would have to increase it (or test another disinfectant). However, if there was no growth of bacteria in your preliminary test, you would have to lower the concentration range. A starting point might be to look at concentrations around those recommended by the manufacturer.

Dependent variables

The dependent variable may be clear from the problem you're investigating, for example the stopping distance of moving objects. But you may have to make a choice.

Assessment tip

The value of the ***depend***ent variable is likely to ***depend*** on the value of the independent variable. This is a good way of remembering the definition of a dependent variable.

Example 3

1 Investigation: Measuring the rate of photosynthesis in a plant

There are several ways in which you could measure the rate of photosynthesis in a plant. These include:

- counting the number of bubbles of oxygen produced in a minute by a water plant such as *Elodea* or *Cabomba*
- measuring the volume of oxygen produced over several days by a water plant such as *Elodea* or *Cabomba*
- monitoring the concentration of carbon dioxide in a polythene bag enclosing a potted plant using a carbon dioxide sensor
- measuring the colour change of hydrogencarbonate indicator containing algae embedded in gel.

2 Investigation: Measuring the rate of a chemical reaction

You could measure the rate of a chemical reaction in the following ways:

> the rate of formation of a product

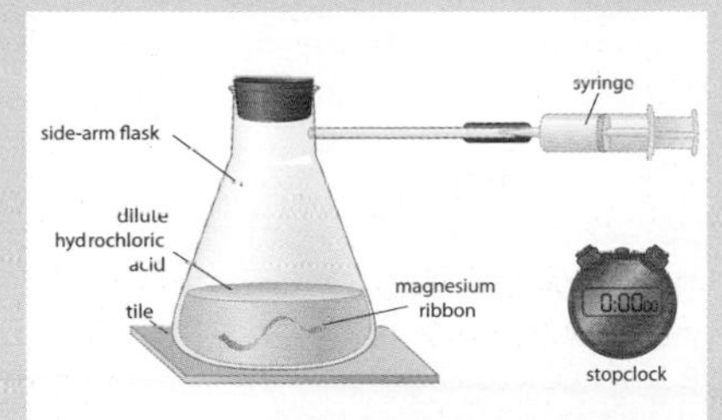

> the rate at which the reactant disappears

> a colour change

> a pH change.

Control variables

The validity of your measurements depend on you measuring what you're supposed to be measuring.

Some of these variables may be difficult to control. For example, in an ecology investigation in the field, factors such as varying weather conditions are impossible to control.

Experimental controls

Experimental controls are often very important, particularly in biological investigations where you're testing the effect of a treatment.

Definition

Other variables that you're not investigating may also have an influence on your measurements. In most investigations, it's important that you investigate just one variable at a time. So other variables, apart from the one you're testing at the time, must be controlled, and kept constant, and not allowed to vary. These are called **control variables**.

Definition

An **experimental control** is used to find out whether the effect you obtain is from the treatment, or whether you get the same result in the absence of the treatment.

Example 4

Investigation: The effect of disinfectants on the growth of bacteria

If the bacteria don't grow, it could be because they have been killed by the disinfectant. But the bacteria in your investigation may have died for some other reason. Another factor may be involved. To test whether any effects were down to the disinfectant, you need to set up the same practical, but this time using distilled water in place of the disinfectant. The distilled water is your control. If the bacteria are inhibited by the disinfectant, but grow normally in the dish containing distilled water, it's reasonable to assume that the disinfectant inhibited their growth.

Assessing and managing risk

Before you begin any practical work, you must assess and minimise the possible risks involved.

Before you carry out an investigation, you must identify the possible hazards. These can be grouped into biological hazards, chemical hazards and physical hazards.

Biological hazards include:

- microorganisms
- body fluids
- animals and plants.

Chemical hazards can be grouped into:

- irritant and harmful substances
- toxic
- oxidising agents
- corrosive
- harmful to the environment.

Physical hazards include:

- equipment
- objects
- radiation.

Scientists use an international series of symbols so that investigators can identify hazards.

Hazards pose risks to the person carrying out the investigation.

A risk posed by concentrated sulfuric acid, for example, will be lower if you're adding one drop of it to a reaction mixture to make an ester, than if you're mixing a large volume of it with water.

When you use hazardous materials, chemicals or equipment in the laboratory, you must use them in such a way as to keep the risks to absolute minimum. For example, one way is to wear eye protection when using hydrochloric acid.

Definition

A **hazard** is something that has the potential to cause harm. Even substances, organisms and equipment that we think of being harmless, used in the wrong way, may be hazardous.

Hazard symbols are used so that hazards can be identified

Definition

The **risk** is the likelihood of a hazard to cause harm in the circumstances it's being used in.

Risk assessment

Before you begin an investigation, you must carry out a risk assessment. Your risk assessment must include:

- ✔ all relevant hazards (use the correct terms to describe each hazard, and make sure you include them all, even if you think they will pose minimal risk)
- ✔ risks associated with these hazards
- ✔ ways in which the risks can be minimised
- ✔ results of research into emergency procedures that you may have to take if something goes wrong.

You should also consider what to do at the end of the practical. For example, used agar plates should be left for a technician to sterilise; solutions of heavy metals should be collected in a bottle and disposed of safely.

Assessment tip

Higher tier

To make sure that your risk assessment is full and appropriate:

> remember that for a risk assessment for a chemical reaction, the risk assessment should be carried out for the products and the reactants

> when using chemicals, make sure the hazard and ways of minimising risk match the concentration of the chemical you're using; many acids, for instance, while being corrosive in higher concentrations, are harmful or irritant at low concentrations.

Collecting primary data

- ✔ You should make sure that observations, if appropriate are recorded in detail. For example, it's worth recording the appearance of your potato chips in your osmosis practical, in addition to the measurements you make.
- ✔ Measurements should be recorded in tables. Have one ready so that you can record your readings as you carry out the practical work.
- ✔ Think about the dependent variable and define this carefully in your column headings.

- ✔ You should make sure that the table headings describe properly the type of measurements you've made, for example 'time taken for magnesium ribbon to dissolve'.
- ✔ It's also essential that you include units – your results are meaningless without these.
- ✔ The units should appear in the column head, and not be repeated in each row of the table.

Definition

When you carry out an investigation, the data you collect are called **primary data.** The term 'data' is normally used to include your observations as well as measurements you might make.

Repeatability and reproducibility of results

When making measurements, in most instances, it's essential that you carry out repeats.

These repeats are one way of checking your results.

Definition

One set of results from your investigation may not reflect what truly happens. Carrying out repeats enables you to identify any results that don't fit. These are called **outliers** or **anomalous results**.

Results will not be repeatable of course, if you allow the conditions the investigation is carried out in to change.

You need to make sure that you carry out sufficient repeats, but not too many. In a titration, for example, if you obtain two values that are within 0.1 cm^3 of each other, carrying out any more will not improve the reliability of your results.

This is particularly important when scientists are carrying out scientific research and make new discoveries.

Once you have planned your experiment and collected your primary data, your work will be retained. It will be needed when you complete the analysis and evaluation part of your controlled assessment.

Definition

If, when you carry out the same experiment several times, and get the same, or very similar results, we say the results are **repeatable**.

Definition

Taking more than one set of results will improve the **reliability** of your data.

Part 3: Analysis and evaluation

Calculating the mean

Using your repeat measurements you can calculate the arithmetical mean (or just 'mean') of these data. We often refer to the mean as the 'average.'

Temperature, °C	Number of yeast cells, mm^3			Mean number of yeast cells, mm^3
	Test 1	Test 2	Test 3	
10	1000	1040	1200	1080
20	2400	2200	2300	2300
30	4600	5000	4800	4800
40	4800	5000	5200	5000
50	200	1200	700	700

Definition

The **reproducibility** of data is the ability of the results of an investigation to be reproduced by someone else, who may be in a different lab, carrying out the same work.

Definition

The **mean** is calculated by adding together all the measurements, and dividing by the number of measurements.

You may also be required to use formulae when processing data. Sometimes, these will need rearranging to be able to make the calculation you need. Practise using and rearranging formulae as part of your preparation for assessment.

Significant figures

When calculating the mean, you should be aware of significant figures.

For example, for the set of data below:

18	13	17	15	14	16	15	14	13	18

The total for the data set is 153, and ten measurements have been made. The mean is 15, and not 15.3.

This is because each of the recorded values has two significant figures. The answer must therefore have two significant figures. An answer cannot have more significant figures than the number being multiplied or divided.

Definition

Significant figures are the number of digits in a number based on the precision of your measurements.

Using your data

When calculating means (and displaying data), you should be careful to look out for any data that don't fit in with the general pattern.

It might be the consequence of an error made in measurement. But sometimes outliers are genuine results. If you think an outlier has been introduced by careless practical work, you should ignore it when calculating the mean. But you should examine possible reasons carefully before just leaving it out.

Definition

An **outlier** (or **anomalous result**) is a reading that is very different from the rest.

Displaying your data

Displaying your data – usually the means – makes it easy to pick out and show any patterns. And it also helps you to pick out any anomalous data.

It is likely that you will have recorded your results in tables, and you could also use additional tables to summarise your results. The most usual way of displaying data is to use graphs. The table will help you decide which type to use.

Type of graph	When you would use the graph	Example
Bar charts or bar graph	Where one of the variables is categorical	'The diameters of the clear zones where the growth of bacteria was inhibited by different types of disinfectant'
Line graph	Where independent and dependent variables are both continuous	'The volume of carbon dioxide produced by a range of different concentrations of hydrochloric acid'
Scatter graph	To show an association between two (or more) variables	'The association between length and breadth of a number of privet leaves' In scatter graphs, the points are plotted, but not usually joined

If it's possible from the data, join the points of a line graph using a straight line, or in some instances, a curve. In this way graphs can also help us to process data.

Assessment tip

Remember when drawing graphs, plot the independent variable on the x-axis, and the dependent variable on the y-axis.

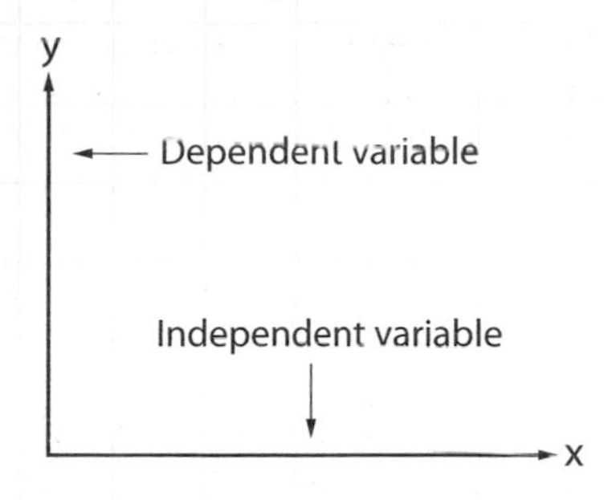

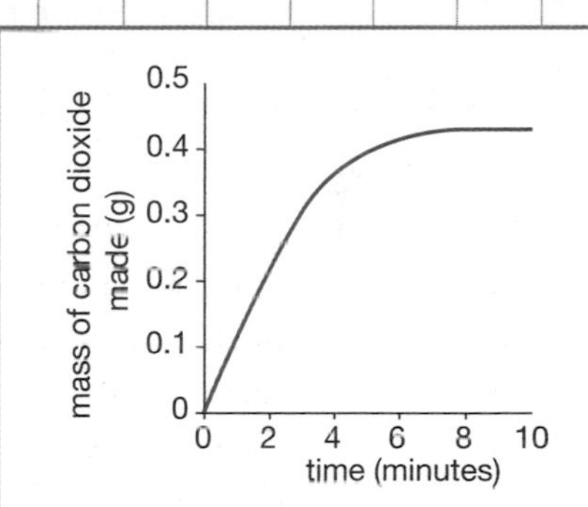

We can calculate the rate of production of carbon dioxide from the gradient of the graph

Conclusions from differences in data sets

When comparing two (or more) sets of data, we often compare the values of two sets of means.

Example 5

Investigation: Comparing the effectiveness of two disinfectants

Two groups of students compared the effectiveness of two disinfectants, labelled A and B. Their results are shown in the table.

Disinfectant	Diameter of zone of inhibition (clear zone), mm										Mean dia. mm
	1	2	3	4	5	6	7	8	9	10	
A	15	13	17	15	14	16	15	14	13	18	15
B	25	23	24	23	26	27	25	24	23	22	24

When the means are compared it appears that disinfectant B is more effective in inhibiting the growth of bacteria. But can we be sure? The differences might have resulted from the treatment of the bacteria using the two disinfectants. But the differences could have occurred purely by chance.

Scientists use statistics to find the probability of any differences having occurred by chance. The lower this probability is, which is found out by statistical calculations, the more likely it is that it was (in this case) the disinfectant that caused the differences observed.

Statistical analysis can help to increase the confidence you have in your conclusions.

Assessment tip

You have learnt about probability in your maths lessons.

Definition

If there is a relationship between dependent and independent variables that can be defined, we say there is a **correlation** between the variables.

Drawing conclusions

Observing trends in data or graphs will help you to draw conclusions. You may obtain a linear relationship between two sets of variables, or the relationship might be more complex.

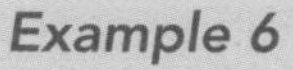

Example 6

Conclusion: The higher the concentration of acid, the shorter the time taken for the magnesium ribbon to dissolve.

Conclusion: The higher the concentration of acid, the faster the rate of reaction.

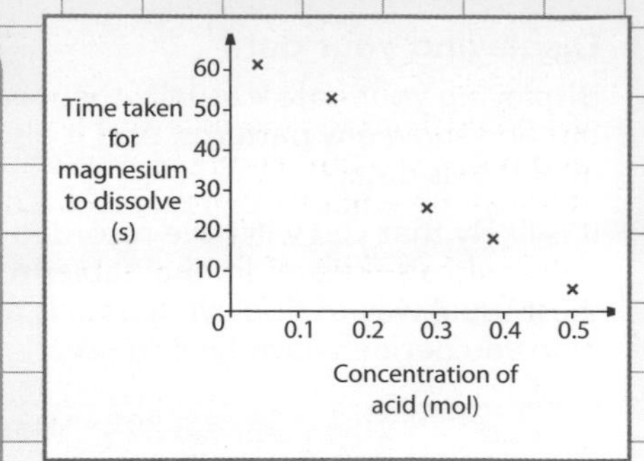

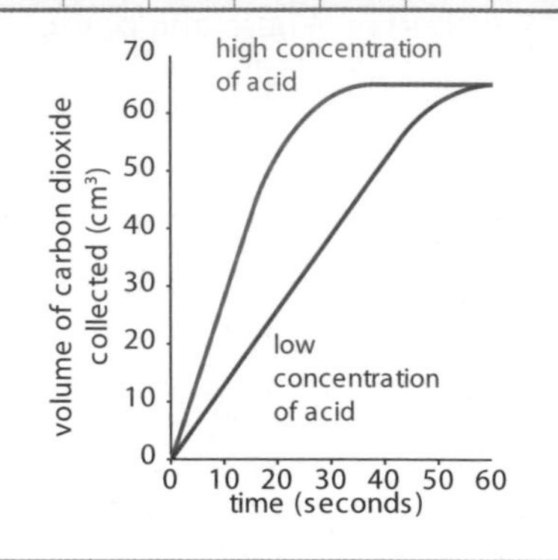

When drawing conclusions, you should try to relate your findings to the science involved.

> In the first investigation in Example 6, your discussion should focus on the greater possibility/increased frequency of collisions between reacting particles as the concentration of the acid is increased.

> In the second investigation in Example 6, there's a clear scientific mechanism to link the rate of reaction to the concentration of acid.

Example 7

Studies have shown that levels of vitamin D are very low in people with long-term inflammatory diseases. But there's no scientific evidence to suggest that these low levels are the cause of the diseases.

Evaluating your investigation

Your conclusion will be based on your findings, but must take into consideration any uncertainty in these introduced by any possible sources of error. You should discuss where these have come from in your evaluation.

The two types of errors are:

✔ random error

✔ systematic error.

This can occur when the instrument you're using to measure lacks sufficient sensitivity to indicate differences in readings. It can also occur when it's difficult to make a measurement. If two investigators measure the height of a plant, for example, they might choose different points on the compost, and the tip of the growing point to make their measurements.

They're either consistently too high or too low. One reason could be down to the way you are making a reading, for example taking a burette reading at the wrong point on the meniscus. Another could be the result of an instrument being incorrectly calibrated, or not being calibrated.

Definition

Error is a difference between a measurement you make, and its true value.

Definition

With **random error**, measurements vary in an unpredictable way.

Definition

With **systematic error**, readings vary in a controlled way.

Assessment tip

A pH meter must be calibrated before use using buffers of known pH.

Accuracy and precision

When evaluating your investigation, you should mention accuracy and precision. But if you use these terms, it's important that you understand what they mean, and that you use them correctly.

The terms accuracy and precision can be illustrated using shots at a target.

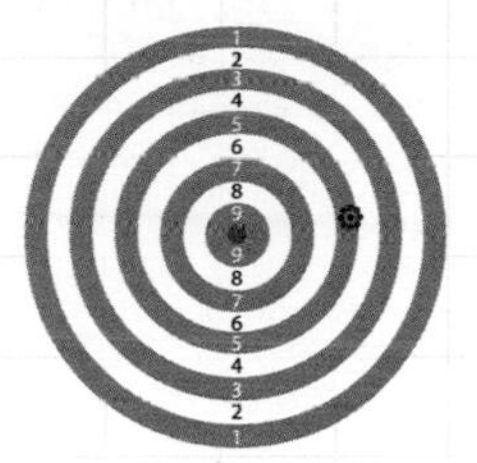

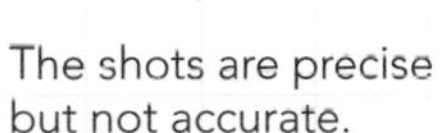

The shots are precise but not accurate.

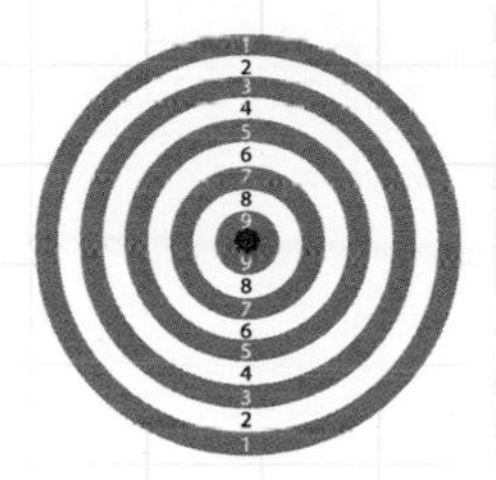

The shots are precise and accurate.

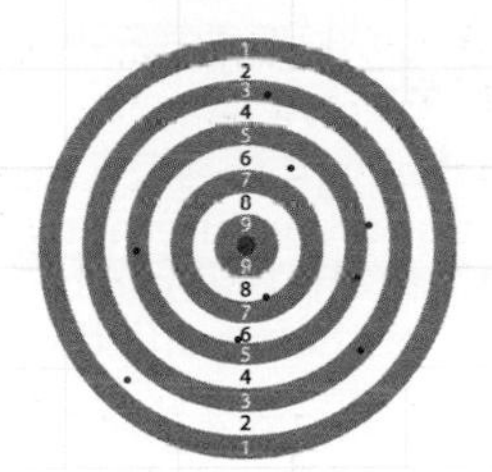

The shots are not precise and not accurate.

Definition

When making measurements:

> the **accuracy** of the measurement is how close it is to the true value

> **precision** is how closely a series of measurements agree with each other.

Improving your investigation

When evaluating your investigation, you should discuss how your investigation could be improved. This could be by improving:

- ✔ the reliability of your data. For example, you could make more repeats, or more frequent readings, or 'fine-tune' the range you chose to investigate, or refine your technique in some other way
- ✔ the accuracy and precision of your data, by using more precise measuring equipment.

Using secondary data

As part of controlled assessment, you will be expected to compare your data – primary data – with **secondary data** you have collected.

The secondary data you collected earlier and the primary data you collected from your experiment will now be returned to you. You will be provided with an answer booklet to complete. This will help you to organise the data you have collected and answer specific questions about the topic you have been investigating.

You should review secondary data and evaluate it. Scientific studies are sometimes influenced by the **bias** of the experimenter.

- ✔ One kind of bias is having a strong opinion related to the investigation, and perhaps selecting only the results that fit with a hypothesis or prediction.
- ✔ Or the bias could be unintentional. In fields of science that are not yet fully understood, experimenters may try to fit their findings to current knowledge and thinking.

There have been other instances where the 'findings' of experimenters have been influenced by organisations that supplied the funding for the research.

You must fully reference any secondary data you have used, using one of the accepted referencing methods.

Assessment tip

What you shouldn't discuss in your evaluation are problems introduced by using faulty equipment, or by you using the equipment inappropriately. These errors can, or could have been, eliminated, by:

> the checking of equipment, and

> practising techniques before your investigation, and taking care and patience when carrying out the practical.

 Do the data support the hypothesis?

You need to discuss, in detail, whether all, or which of your primary, and the secondary data you have collected, support your original hypothesis. They may, or may not.

You should communicate your points clearly, using the appropriate scientific terms, and checking carefully your use of spelling, punctuation and grammar. You will be assessed on this written communication as well as your science.

If your data do not completely match the hypothesis, it may be possible to modify the hypothesis or suggest an alternative one. You should suggest any further investigations that can be carried out to support your original hypothesis or the modified version.

It is important to remember, however, that if your investigation does support the hypothesis, it can improve the confidence you have in your conclusions and scientific explanations, but it can't prove your explanations are correct.

 Referencing methods

The two main conventions for writing a reference are the:

- ✔ Harvard system
- ✔ Vancouver system.

In your text, the Harvard system refers to the authors of the reference, for example 'Smith and Jones (1978)'.

The Vancouver system refers to the number of the numbered reference in your text, for example '... the reason for this hypothesis is unknown.[5]'.

Though the Harvard system is usually preferred by scientists, it is more straightforward for you to use the Vancouver system.

Harvard system

In your references list a book reference should be written:

> Author(s) (year of publication). *Title of Book*, publisher, publisher location.

The references are listed in alphabetical order according to the authors.

Vancouver system

In your references list a book reference should be written:

> 1 Author(s). *Title of Book.* Publisher, publisher location: year of publication.

The references are number in the order in which they are cited in the text.

Assessment tip

Remember to write out the URL of a website in full. You should also quote the date when you looked at the website.

How to be successful in your GCSE Science written examination

Introduction

OCR uses assessments to test how good your understanding of scientific ideas is, how well you can apply your understanding to new situations and how well you can analyse and interpret information you've been given. The assessments are opportunities to show how well you can do these.

To be successful in exams you need to:

- ✔ have a good knowledge and understanding of science
- ✔ be able to apply this knowledge and understanding to familiar and new situations, and
- ✔ be able to interpret and evaluate evidence that you've just been given.

You need to be able to do these things under exam conditions.

The language of the external assessment

When working through an assessment paper, make sure that you:

- ✔ re-read a question enough times until you understand exactly what the examiner is looking for
- ✔ make sure that you highlight key words in a question. In some instances, you will be given key words to include in your answer
- ✔ look at how many marks are allocated for each part of a question. In general, you need to write at least as many separate points in your answer as there are marks.

What verbs are used in the question?

A good technique is to see which verbs are used in the wording of the question and to use these to gauge the type of response you need to give. The table lists some of the common verbs found in questions, the types of responses expected and then gives an example.

Verb used in question	Response expected in answer	Example question
write down; state; give; identify	These are usually more straightforward types of question in which you're asked to give a definition, make a list of examples, or the best answer from a series of options	'Write down three types of microorganism that cause disease' 'State one difference and one similarity between radio waves and gamma rays'
calculate	Use maths to solve a numerical problem	'Calculate the relative formula mass for sodium hydrogen carbonate'

estimate	Use maths to solve a numerical problem, but you do not have to work out the exact answer	'Estimate the number of bacteria in the culture after five hours'
describe	Use words (or diagrams) to show the characteristics, properties or features of, or build an image of something	'Describe how energy is lost through the wall of a house from the inside to the outside'
suggest	Come up with an idea to explain information you're given	'Suggest one advantage of producing bio-diesel rather than bio-ethanol'
demonstrate; show how	Use words to make something evident using reasoning	'Show how enzyme activity changes with temperature'
compare	Look for similarities and differences	'Compare the structure of arteries and veins'
explain	To offer a reason for, or make understandable, information you're given	'Seat belts are important safety features in cars. Explain how they work'
evaluate	To examine and make a judgement about an investigation or information you're given	'Evaluate if fuel C is a sensible choice to heat a greenhouse'

What is the style of the question?

Try to get used to answering questions that have been written in lots of different styles before you sit the exam. Work through past papers, or specimen papers, to get a feel for these. The types of questions in your assessment fit the three assessment objectives shown in the table.

Assessment objective	Your answer should show that you can...
AO1 Recall the science	Recall, select and communicate your knowledge and understanding of science
AO2 Apply your knowledge	Apply skills, knowledge and understanding of science in practical and other contexts
AO3 Evaluate and analyse the evidence	Analyse and evaluate evidence, make reasoned judgements and draw conclusions based on evidence

Assessment tip

Of course you must revise the subject material adequately. But it's as important that you are familiar with the different question styles used in the exam paper, as well as the question content.

How to answer questions on: AO1 Recall the science

These questions, or parts of questions, test your ability to recall your knowledge of a topic. There are several types of this style of question:

- ✔ Describe a process
- ✔ Explain a concept
- ✔ Complete sentences, tables or diagrams
- ✔ Tick the correct statements
- ✔ Use lines to link a term with its definition or correct statement

Example 8

a What is meant by the term *metabolic rate*?

Tick (✓) **one** box.

☐ the amount of energy a person uses each hour
☐ the amount of exercise a person does each day
☐ the amount of food a person eats each day

How to answer questions on: AO1 Recall the science in practical techniques

You may be asked to recall how to carry out certain practical techniques; either ones that you have carried out before, or techniques that scientists use.

To revise for these types of questions, make sure that you have learnt definitions and scientific terms. Produce a glossary of these, or key facts cards, to make them easier to remember. Make sure your key facts cards also cover important practical techniques, including equipment, where appropriate.

Example 9

1 Describe how to test the pH of a solution.

2 Describe how to find the resistance of a wire from a voltage-current graph.

Assessment tip

Don't forget that mind maps – either drawn by you or by using a computer program – are very helpful when revising key points.

How to answer questions on: AO2 Apply skills, knowledge and understanding

Some questions require you to apply basic knowledge and understanding in your answers.

You may be presented with a topic that's familiar to you, but you should also expect questions in your Science exam to be set in an unfamiliar context.

Questions may be presented as:

- ✔ experimental investigations
- ✔ data for you to interpret
- ✔ a short paragraph or article.

The information required for you to answer the question might be in the question itself, but for later stages of the question, you may be asked to draw on your knowledge and understanding of the subject material in the question.

Practice will help you to become familiar with contexts that examiners use and question styles. But you will not be able to predict many of the contexts used. This is deliberate; being able to apply your knowledge and understanding to different and unfamiliar situations is a skill the examiner tests.

Practise doing questions where you are tested on being able to apply your scientific knowledge and your ability to understand new situations that may not be familiar. In this way, when this type of question comes up in your exam, you will be able to tackle it successfully.

Assessment tip

Work through the Preparing for Assessment: Applying your knowledge tasks in this book as practice.

Example 10

1 This table shows the atomic numbers of some elements.

element	symbol	atomic number
mercury	Hg	80
thallium	Tl	81
lead	Pb	82
bismuth	Bi	83
polonium	Po	84
astatine	At	85

a Lead-210 decays by emitting an alpha particle.

Which element is formed when lead-210 decays?

Use the table to help you.

b Finish and balance the equation to show what happens when lead-210 decays.

$$^{210}_{82}\text{Pb} \rightarrow \underline{\hspace{3cm}} + \underline{\hspace{3cm}}$$

How to answer questions on: AO2 Apply skills, knowledge and understanding in practical investigations

Some opportunities to demonstrate your application of skills, knowledge and understanding will be based on practical investigations. You may have carried out some of these investigations, but others will be new to you, and based on data obtained by scientists. You will be expected to describe patterns in data from graphs you are given or that you will have to draw from given data.

Again, you will have to apply your scientific knowledge and understanding to answer the question.

Example 11

1 Look at the graph on the right showing the resistance of the bacterium, *Streptococcus pneumoniae,* to three different types of antibiotic.

a Which antibiotic does there seem to be least resistance to, even when it has been used before?

b Can you explain why this might be the case?

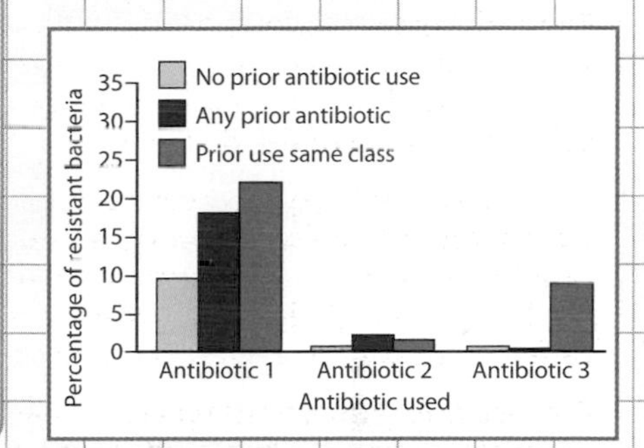

How to answer questions on: AO3 Analysing and evaluating evidence

For these types of questions, you will analyse and evaluate scientific evidence or data given to you in the question. It's likely that you won't be familiar with the material.

When describing patterns and trends in the data, make sure you:

✔ explain a pattern or trend in as much detail as you can

✔ mention anomalies where appropriate

You must also be able to evaluate the information you're given. This is one of the hardest skills. Think about the validity of the scientific data: did the technique(s) used in any practical investigation allow the collection of accurate and precise data?

Your critical evaluation of scientific data in class, along with the practical work and controlled assessment work, will help you to develop the evaluation skills required for these types of questions.

Assessment tip

Remember, when carrying out any calculations, you should include your working at each stage. You may get credit for getting the process correct, even if your final answer is wrong.

Example 12

1 The table shows the properties of some metals.

metal	melting point in °C	density in g/cm³	relative electrical conductivity	cost per tonne in £
aluminium	660	2.7	38	2491
copper	1083	8.9	60	9048
silver	962	10.5	63	1 125 276
zinc	420	7.1	17	2260

Pylon wires are made from metal.

Which metal would be most suitable for using to make pylon wires?

Use information about each of the metals in the table to explain your answer.

The quality of your written communication

Scientists need good communication skills to present and discuss their findings. You will be expected to demonstrate these skills in the exam. Questions will [end] with the sentence: The quality of your written communication will be assessed in your answer to this question.

- ✔ You must also try to make sure that your spelling, punctuation and grammar are accurate, so that it's clear what you mean in your answer. Again, examiners can't award marks for answers where the meaning isn't clear.
- ✔ Make sure your language is concise. When describing and explaining science, use correct scientific vocabulary.

Practise answering some longer 6 mark questions. These will examine your quality of written communication as well as your knowledge and understanding of science. Look at how marks are awarded in mark schemes provided by the awarding body. You'll find these in the specimen question papers, and past papers.

You will also need to remember the writing and communication skills you've developed in English lessons. For example, make sure that you understand how to construct a good sentence using connectives.

Assessment tip

You will be assessed on the way in which you communicate science ideas.

Assessment tip

When answering questions, you must make sure that your writing is legible. An examiner can't award marks for answers that he or she can't read.

Example 13

All long answer, 6 mark, questions will have the same general mark scheme which will be amplified and made specific to the question by examiners. Three levels of answer will gain credit.

Level 3

All information in answer is relevant, clear, organised and presented in a structured and coherent format. Specialist terms are used appropriately. Few, if any, errors in grammar, punctuation and spelling. (5–6 marks)

Level 2

For the most part the information is relevant and presented in a structured and coherent format. Specialist terms are used for the most part appropriately. There are occasional errors in grammar, punctuation and spelling. (3–4 marks)

Level 1

Answer gives two ways in which the rate of reaction can be increased. Answer may be simplistic. There may be limited use of specialist terms. Errors of grammar, punctuation and spelling prevent communication of the science. (1–2 marks)

Level 0

Insufficient or irrelevant science. Answer not worthy of credit. (0 marks)

Revising for your Science exam

You should revise in the way that suits you best. But it's important that you plan your revision carefully, and it's best to start well before the date of the exams. Take the time to prepare a revision timetable and try to stick to it. Use this during the lead up to the exams and between each exam.

When revising:

- ✔ find a quiet and comfortable space in the house where you won't be disturbed. It's best if it's well ventilated and has plenty of light
- ✔ take regular breaks. Some evidence suggests that revision is most effective when you revise in 30 to 40 minute slots. If you get bogged down at any point, take a break and go back to it later when you're feeling fresh. Try not to revise when you are feeling tired. If you do feel tired, take a break
- ✔ use your school notes, textbook and possibly a revision guide. But also make sure that you spend some time using past papers to familiarise yourself with the exam format
- ✔ produce summaries of each [topic] or [module]
- ✔ draw mind maps covering the key information on a [topic] or [module]

Assessment tip

Try to make your revision timetable as specific as possible – don't just say 'science on Monday, and Thursday', but list the [modules] that you'll cover on those days.

- ✔ set up revision cards containing condensed versions of your notes
- ✔ ask yourself questions, and try to predict questions, as you're revising topics or [modules]
- ✔ test yourself as you're going along. Try to draw key labelled diagrams, and try some questions under timed conditions
- ✔ prioritise your revision of topics. You might want to allocate more time to revising the topics you find most difficult.

Assessment tip

Start your revision well before the date of the exams, produce a revision timetable, and use the revision strategies that suit your style of learning. Above all, revision should be an active process.

How do I use my time effectively in the exam?

Timing is important when you sit an exam. Don't spend so long on some questions that you leave insufficient time to answer others. For example, in a 60-mark question paper, lasting one hour, you will have, on average, one minute per question.

If you're unsure about certain questions, complete the ones you're able to do first, then go back to the ones you're less sure of.

If you have time, go back and check your answers at the end of the exam.

What will my exam look like?

Your science exam consists of two papers.

Paper 1 contains three sections and lasts 1 hour 15 minutes.
There are 25 marks for each section.
You should spend about 25 minutes answering each section.
The questions in each section will test objectives AO1, AO2 and AO3 using structured questions.

Paper 2 contains four sections and lasts 1 hour 30 minutes.
The first three sections will be similar to the Paper 1 sections.
Section D contains a ten mark data response question which primarily assesses objective AO3. You will be required to analyse and evaluate evidence, make reasoned judgements and draw conclusions based on evidence.

You should spend about 15 minutes answering this section.

On exam day

A little bit of nervousness before your exam can be a good thing, but try not to let it affect your performance. When you turn over the exam paper keep calm. Look at the paper and get it clear in your head exactly what is required from each question. Read each question carefully. Don't rush.

If you read a question and think that you have not covered the topic, keep calm – it could be that the information needed to answer the question is in the question itself or the examiner may be asking you to apply your knowledge to a new situation.

Finally, good luck!

Mathematical skills

You will be allowed to use a calculator in all assessments.

These are the maths skills that you need, to complete all the assessments successfully.

You should understand:

- ✔ the relationship between units, for example, between a gram, kilogram and tonne
- ✔ compound measures such as speed
- ✔ when and how to use estimation
- ✔ the symbols $= \quad < \quad > \quad \sim$
- ✔ direct proportion and simple ratios
- ✔ the idea of probability.

You should be able to:

- ✔ give answers to an appropriate number of significant figures
- ✔ substitute values into formulae and equations using appropriate units
- ✔ select suitable scales for the axes of graphs
- ✔ plot and draw line graphs, bar charts, pie charts, scatter graphs and histograms
- ✔ extract and interpret information from charts, graphs and tables.

You should be able to calculate:

- ✔ using decimals, fractions, percentages and number powers, such as 10^3
- ✔ arithmetic means
- ✔ areas, perimeters and volumes of simple shapes

In addition, if you are a higher tier candidate, you should be able to:

- ✔ **change the subject of an equation**

and should be able to use:

- ✔ **numbers written in standard form**
- ✔ **calculations involving negative powers, such as 10^{-1}**
- ✔ **inverse proportion**
- ✔ **percentiles and deciles.**

Some key physics equations

With the written papers, there will be an equation sheet. Below are some of the key equations found on the sheet; it will help if you practise using them.

Equation	Meaning of symbol and its unit
specific heat capacity: $E = m \times c \times \theta$	E is energy transferred in joules m is mass in kilograms θ is temperature change in degrees Celsius c is specific heat capacity in J/kg °C
efficiency of a device: efficiency = $\frac{\text{useful energy out}}{\text{total energy in}}$ (× 100%) efficiency = $\frac{\text{useful power out}}{\text{total power in}}$ (× 100%)	← energy out and energy in are measured in the same units ← power out and power in are measured in the same units
amount of energy transfer: $E = P \times t$	E is energy transferred in kilowatt-hours P is power in kilowatts t is time in hours This equation may also be used when: E is energy transferred in joules P is power in watts t is time in seconds
cost of mains electricity used: *total cost* = E × *cost per kilowatt-hour*	*total cost* is in pence E is energy transferred in kilowatt-hours *cost per kilowatt-hour* is in pence per kilowatt-hour
kinetic energy KE = $\prod$ mv^2	KE is the energy in joules m is the mass in kilograms v is the speed in metres per second
wave equation: $v = f \times \lambda$	v is speed in metres per second f is frequency in hertz λ is wavelength in metres

Glossary

A

accommodation eyes ability to change focus

acetylcholine neuro-transmitter chemical that diffuses across synapse

acid Solution with a pH of less than 7

acid rain rain water which is made more acidic by pollutant gases

active immunity you have immunity if your immune system recognises a pathogen and fights it

adaptation features that organism have to help them survive in their environment

addition polymer a very long molecule resulting from polymerisation, e.g. polythene

aggregate gravel made of a range of particle sizes

alcohol substance made by the fermentation of yeast

algal bloom a thick mat of algae near the surface of water which stops sunlight getting through

alkali a soluble base. A substance which produces OH^- ions in water

alkanes a family of hydrocarbons found in crude oil with single covalent bonds, e.g. methane

alkenes a family of hydrocarbons with one double covalent bond between carbon atoms

allele inherited characteristics are carried as pairs of alleles on pairs of chromosomes. Different forms of a gene are different alleles

alloy mixture of two or more metals – used to make coins

alpha particles radioactive particles which are helium nuclei – helium atoms without the electrons (they have a positive charge)

alternating current an electric current that is not a one-way flow

amalgam an alloy which contains mercury

amphibians animals with a moist permeable skin

amplitude the maximum displacement of a point on a wave from its rest position

analogue signal a signal that shows a complete range of frequencies; sound is analogue

anion ion with a negative charge; they move to the anode during electrolysis

anode electrode with a positive charge

antibiotic therapeutic drug acting to kill bacteria which is taken into the body

antibody protein normally present in the body or produced in response to an antigen which it neutralizes, thus producing an immune response

antigen any substance that stimulates the production of antibodies – antigens on the surface of red blood cells determine blood group

antioxidant food additive to prevent the spoiling of food by oxidation

antiviral drug therapeutic drug acting to kill viruses

arteries blood vessels that carry blood away from the heart

asteroid composed of rock or metallic material orbiting the Sun in a region between Mars and Jupiter

atmosphere mixture of gases above the lithosphere, mainly nitrogen and oxygen

auxin a type of plant hormone

axon part of neurone that carries nerve impulse

B

backward reaction the reaction which goes from right to left in a reversible reaction

balanced symbol equation a symbolic representation showing the kind and amount of the starting materials and products of a reaction

basalt a rock forms when iron rich magma cools

base a substance that will react with acids

beta particles particles given off by some radioactive materials (they have a negative charge)

big bang the event believed by many scientists to have been the start of the universe

binocular vision part of vision seen by both eyes

binomial system the scientific way of naming an organism

biodegradable a biodegradable material can be broken down by micro-organisms

biomass waste wood and other natural materials which are burned in power stations

birds animals that have feathers

bitumen thick tar-like substance that does not boil in a fractionating column

black hole a region of space from which nothing, not even light, can escape

blind trial a drugs trial where volunteers do not know which treatment they are receiving

blood groups blood falls into one of four groups: A, B, AB or O

blood pressure force with which blood presses against the walls of the blood vessels

blood sugar level amount of glucose in the blood

body mass index (BMI) measure of someone's weight in relation to their height

boiling point temperature at which a liquid changes into a gas

brass an alloy which contains copper and zinc

breathable fabric Gore-tex® like material that allows water vapour to escape

bromine orange substance used to test alkenes. A liquid corrosive halogen

butane An alkane of 4 carbon atoms, part of LPG

C

captive breeding Breeding a species in zoos to maintain the wild population

carbon cycle a natural cycle through which carbon moves by respiration, photosynthesis and combustion in the form of carbon dioxide

carbon dioxide gas present in the atmosphere at a low percentage but important in respiration, photosynthesis and combustion

carbon footprint the total amount of greenhouse gases given off by a person in a given time

carbon monoxide poisonous gas made when fuels burn in a shortage of oxygen

carnivore an animal that eats other animals

catalyst substance added to a chemical reaction to alter the speed of the reaction

catalytic converters boxes fitted to vehicle exhausts which reduce the level of nitrogen oxides and unburnt hydrocarbons in the exhaust fumes

cathode electrode with a negative charge

cation ion with a positive charge

cell membrane layer around a cell which helps to control substances entering and leaving the cell

cement the substance made when limestone and clay are heated together

central nervous system (CNS) collectively the brain and spinal cord

centripetal force force acting on a body, travelling in a circle, which acts toward the centre of the circle and keeps the body moving in a circle

CFCs gases which used to be used in refrigerators and which harm the ozone layer

cholesterol fatty substance which can block blood vessels

colloid a liquid with small particles dispersed throughout it, forming neither solution nor sediment

combustion process where fuels react with oxygen to produce heat

comets lumps of rock and ice found in space – some orbit the Sun

complete combustion when fuels burn in excess of oxygen to produce carbon dioxide and water only

composite material a material which consists of identifiably different substances

concrete a form of artificial stone

conductors materials which transfer thermal energy easily; electrical conductors allow electricity to flow through them

conservation a way of protecting a species or environment

constellation a group of stars that has been given a name by astronomers because they seem to be close together when seen from Earth; members of a constellation may actually be many light years away from each other

consumer Organism in a food chain that gets its energy from eating food

contagious a disease that spreads directly from person to person

convection current when particles in a liquid or gas gain thermal energy from a warmer region and move into a cooler region, taking this energy with them

core the centre part of the Earth, made of iron

corrode to lose strength due to chemical attack

covalent bond bond between atoms where an electron pair is shared

cracking the process of making small hydrocarbon molecules from larger hydrocarbon molecules using a catalyst

critical angle angle at which a light ray incident on the inner surface of a transparent glass block just escapes from the glass

cross links links between two adjacent polymer chains that stop the movement of the molecules, which make the plastic more rigid

crude oil black material mined from the Earth from which petrol and many other products are made

crust surface layer of the Earth made of tectonic plates

cyclic fluctuation the rise and fall of a population

D

decay to rot

decolourise turn from a coloured solution to a colourless solution

decomposer organisms that break down dead animals and plants

decomposes chemically broken down

deforestation removal of large area of trees

degassing gases coming out of volcanoes

dehydration result of body losing too much water

denatured an enzyme is denatured if its shape changes so that the substrate cannot fit into the active site

denitrifying-bacteria bacteria that convert nitrates into nitrogen gas

depressant a drug that slows down the working of the brain

diastolic pressure the lowest point that your blood pressure reaches as the heart relaxes between beats

di-bromo compound colourless compound resulting from an alkene and bromine solution

diet what a person eats

diffraction a change in the directions and intensities of a group of waves after passing by an obstacle or through an opening whose size is approximately the same as the wavelength of the waves

digital signal a signal that has only two values – on and off

direct current an electric current that flows in one direction only

dispersion particles spreading out in a colloid

displayed formula when the formula of a chemical is written showing all the atoms and all the bonds

disposal getting rid of unwanted substances such as plastics

dominant allele/characteristic an allele that will produce the characteristic if present

double covalent bond covalent bond where each atom shares two electrons with the other atom

dynamo a device that converts energy in movement into energy in electricity

E

EAR for protein Estimated average daily requirement of protein in diet

ecological niche the role of an organism within an ecosystem

efficiency ratio of useful energy output to total energy input; can be expressed as a percentage

egestion the way an animal gets rid of undigested food waste called faeces

electrolysis when an electric current is passed through a solution which conducts electricity

electrolyte the liquid in which electrolysis takes place

electromagnetic spectrum electromagnetic waves ordered according to wavelength and frequency – ranging from radio waves to gamma rays

elliptical orbit a path that follows an ellipse – which looks a bit like a flattened circle

endangered a species where the numbers are so low they could soon become extinct

endoscope device using optical fibres which allows doctors to look inside the human body

energy the ability to 'do work' – the human body needs energy to function

epicentre the point on the Earth's surface directly above the focus of an earthquake

essential elements the three elements, nitrogen, phosphorus and potassium that are essential for the growth of plants

ethene a plant hormone that speeds up fruit ripening

eutrophication when waterways become too rich with nutrients (from fertilisers) which allows algae to grow wildly and use up all the oxygen

evolution the gradual change in organisms over millions of years caused by random mutations and natural selection

excretion the process of getting rid of waste from the body

exponential growth the ever increasing growth of the human population

extinct when all members of a species have died out

F

fault a crack in the Earth's crust resulting from the displacement of one side with respect to the other

finite resource resources such as oil that will eventually run out

first class protein proteins from meat and fish which contain all essential amino acids

fish animals with scales and gills

food part of our diet that provides energy

forward reaction the reaction which goes from left to right in a reversible reaction

fossil fuels fuels such as coal, oil and gas

frequency the number of waves passing a set point per second

G

galaxy a group made of billions of stars

gametes the male and female sex cells (sperm and eggs)

gamma rays ionising electromagnetic waves that are radioactive and dangerous to human health – but useful in killing cancer cells

gene section of DNA that codes for a particular characteristic

genotype the genetic makeup of an organism

geotropism a plant's growth response to gravity

germinate the growth of a seed into a plant

gibberellic acid a plant hormone that speeds up germination

global warming the increase in the Earth's temperature due to increases in carbon dioxide levels

granite an igneous rock

greenhouse gas any of the gases whose absorption of infrared radiation from the Earth's surface is responsible for the greenhouse effect, e.g. carbon dioxide, methane, water vapour

Haber process industrial process for making ammonia

habitat the place where an organism lives

haemoglobin chemical found in red blood cells which carries oxygen

hallucinogen A drug, like LSD, that gives the user hallucinations

heat stroke result of body being too hot; skin is cold, pulse is weak

herbivore animals that eat plants

hertz units for measuring wave frequency

heterozygous a person who has two different alleles for an inherited characteristic, e.g. someone with blond hair may also carry an allele for red hair

homozygous a person who has two alleles that are the same for an inherited feature, e.g. a blue-eyed person will have two blue alleles for eye colour

hormones chemicals that act on target organs in the body (hormones are made by the body in special glands)

hybrid the infertile offspring produced when two animals of different species breed

hydrated iron (III) oxide the chemical name for rust

hypothalamus small gland in brain, detects temperature of blood

hypothermia a condition caused by the body getting too cold, which can lead to death if untreated

I

igneous rock rock which has formed when liquid rock has solidified

indicator species organisms used to measure the level of pollution in water or the air

infrared waves non-ionising waves that produce heat – used in toasters and electric fires

insulation a substance that reduces the movement of energy; heat insulation in the loft of a house slows down the movement of warmth to the cooler outside

insulator a material that transfers thermal energy only very slowly

interfere waves interfere with each other when two waves of different frequencies occupy the same space; interference occurs in light and sound and can produce changes in intensity of the waves

invertebrate animal without a backbone

ionisation the formation of ions (charged particles)

ionosphere a region of the earth's atmosphere where ionization caused by incoming solar radiation affects the transmission of radio waves; it extends from 70 kilometres (43 miles) to 400 kilometres (250 miles) above the surface

J

joule a unit of energy

K

kilowatt 1000 watts

kilowatt hour unit of electrical energy equal to 3 600 000J

kinetic energy the energy that moving objects have

kwashiorkor An illness caused by protein deficiency due to lack of food. Sufferers often have swollen bellies caused by retention of fluid in the abdomen.

L

laser a special kind of light beam that can carry a lot of energy and can be focussed very accurately; lasers are often used to judge the speed of moving objects or the distance to them

latent heat the energy needed to change the state of a substance

legume a plant that has bacteria inside root nodules that provide the plant with nitrates

light emitting diode (LED) a very small light in electric circuits that uses very little energy

light-year a unit of distance equal to the distance light travels through space in one year

limestone a sedimentary rock, made of calcium carbonate

lithosphere the cold rigid outer part of the Earth which includes the crust and the upper part of the mantle

M

magma molten rock found below the Earth's surface

malleable bendable

mammal animals that have fur and produce milk for their young

mantle semi-liquid layer of the Earth beneath the crust

marble a metamorphic rock, made of calcium carbonate

melanin the group of naturally occurring dark pigments, especially the pigment found in skin, hair, fur, and feathers

melting point temperature at which a solid changes into a liquid

metamorphic rock rock which has been changed after it has formed

meteors bright flashes in the sky caused by rocks burning in Earth's atmosphere

microorganism very small organism (living thing) which can only be viewed through a microscope – also known as a microbe

microwaves non-ionising waves used in satellite and mobile phone networks – also in microwave ovens

monocular vision part of vision seen by only one eye

monohybrid cross a cross between two organisms that differ by a single characteristic. Used to follow the inheritance of a single pair of genes

Morse code a code consisting of dots and dashes that code for each letter of the alphabet

motor neurone nerve cell carrying information from the central nervous system to muscles

multiplexing combination of multiple signals into one signal transmitted over a shared medium

mutation where the DNA within cells have been altered (this happens in cancer)

mutualism A relationship in which both organisms benefit

N

NASA National Aeronautics and Space Administration – the organisation in the United States responsible for the space programme

National Grid network that carries electricity from power stations across the country (it uses cables, transformers and pylons)

natural selection process by which 'good' characteristics that can be passed on in genes become more common in a population over many generations ('good' characteristics mean that the organism has an advantage which makes it more likely to survive)

near-Earth object asteroid, comet or large meteoroid whose orbit crosses Earth's orbit

neutralisation reaction between H^+ ions and OH^- ions (acid and base react to make a salt and water)

nitinol a smart alloy which contains nickel and titanium

nitrifying-bacteria bacteria that convert ammonia into nitrates

nitrogen-fixing bacteria bacteria that convert nitrogen into ammonia or nitrates

nitrogenous fertiliser a fertiliser which contains a nitrogen compound

non-renewable energy energy which is used up at a faster rate than it can be replaced e.g. fossil fuels

nucleus central part of an atom that contains protons and neutrons

obesity a medical condition where the amount of body fat is so great that it harms health

omnivore animal that eats plants and other animals

optical fibre a flexible optically transparent fibre, usually made of glass or plastic, through which light passes by successive internal reflections

optimum conditions the conditions under which a reaction works most effectively

oscilloscope a device that displays a line on a screen showing regular changes (oscillations) in something. an oscilloscope is often used to look at sound waves collected by a microphone

oxidation a chemical reaction in which a substance gains oxygen and/or loses electrons

ozone layer layer of the Earth's atmosphere that protects us from ultraviolet rays

P

p wave longitudinal seismic wave capable of travelling through solid and liquid parts of the Earth

painkiller a drug that stops nerve impulses so pain is not felt

parasite organism which lives on (or inside) the body of another organism

pathogen harmful organism which invades the body and causes disease

payback time the time it takes for the original cost outlay to be recovered in savings

percentage yield the percentage of actual product made in a chemical reaction compared to the amount which ideally could be made

performance enhancer a drug used to improve performance in a sporting event

peripheral nervous system network of nerves leading to and from the brain and spinal cord

pH meter a device which measures the pH of a substance accurately

pH scale scale in which acids have a pH below 7, alkalis a pH of above 7 and a neutral substance a pH of 7

phase fraction of a complete wave that one wave disturbance is different to another

phenotype the characteristic that is shown/expressed

photocell a device which converts light into electricity

photon a photon is a unit or particle of electromagnetic energy; photons travel at the speed of light but have no mass

photosynthesis process carried out where carbon dioxide and water, in the presence of sunlight, produce glucose and oxygen

phototropism a plant's growth response to light

placebo a dummy pill

plant hormones hormones that control various plant processes such as growth and germination

plaque build up of cholesterol in a blood vessel (which may block it)

plutonium a radioactive metal often formed as a bi-product from a nuclear power station – sometimes used as a nuclear fuel

p–n junction the boundary between two special types of silicon in a photocell and other electronic components

pollination the process of transferring pollen from one plant to another

population the number of one species in a habitat

power the rate that a system transfers energy, power is usually measured in watts (W)

predator animal which preys on (and eats) another animal

preservative a substance which is added to food to help it keep longer

prey animals which are eaten by a predator

probe unmanned space vehicle designed to travel beyond Earth's orbit

producers organisms in a food chain that make food using sunlight

R

radiation thermal energy transfer which occurs when something hotter than its surroundings radiates heat from its surface

radio waves non-ionising waves used to broadcast radio and TV programmes

radioactive material which gives out radiation

radiotherapy using ionizing radiation to kill cancer cells in the body

receiver device which receives waves, e.g. a mobile phone

recessive allele/characteristic two recessive alleles needed to produce the characteristic

recycle to reuse materials

red shift when lines in a spectrum are redder than expected – if an object has a red shift it is moving away from us

reduction a chemical reaction in which a substance loses oxygen and/or gains electrons

reflex a muscular action that we take without thinking about

refraction when a light ray travelling through air enters a glass block and changes direction

reinforced concrete concrete with steel rods or mesh running through it

renewable energy energy that can be replenished at the same rate that it's used up e.g. biofuels

reptile cold blooded vertebrate having an external covering of scales or horny plates

respiration process occurring in living things where oxygen is used to release the energy in foods

rhyolite a rock which forms when silica rich magma cools

rust the substance made when iron corrodes, hydrated iron (III) oxide

S

s wave transverse seismic wave capable of travelling through solid but not liquid parts of the Earth

salt the substance formed when any acid reacts with a base

satellite a body orbiting around a larger body; communications satellites orbit the Earth to relay television and telephone signals

second class protein proteins from plants which only contain some essential amino acids

sedimentary rock rock which has formed when fragments of older rock or living things have stuck together, or by precipitation

seismic wave vibration transmitted through the Earth

seismometer a device used to detect movements in the Earth's crust

sensor device that detects a change in the environment

sensory neurone nerve cell carrying information from receptors to central nervous system

sex chromosomes a pair of chromosomes that determine gender, XX in female, XY in male

shock wave seismic wave that travels out from the epicentre of an earthquake

smart alloy an alloy which will return to a previous shape

solar cells devices which convert the Sun's energy into electricity

Solar System the collection of planets and other objects orbiting around the Sun

solder an alloy which contains lead and tin

species basic category of biological classification, composed of individuals that resemble one another, can breed among themselves, but cannot breed with members of another species

specific heat capacity the amount of energy needed to raise the temperature of 1 kg of a substance by 1 degree Celsius

specific latent heat the amount of energy needed to change the state of a substance without changing its temperature; for example the energy needed to change ice at 0°C to water at the same temperature

star bright object in the sky which is lit by energy from nuclear reactions

steel an alloy which contains iron

stimulant a drug that speeds up the working of the brain

stratosphere a layer in the atmosphere starting at 15 km above sea level and extending to 50 km above sea level; the ozone layer is found in the stratosphere

stroke sudden change in blood flow to the brain – can be fatal

subduction where one plate sinks below another

subsidence settling of the ground caused by mining

sustainable development managing a resource so that it does not run out

sustainable resource a resource that will not run out because it is being produced as fast as it is being used

synapse gap between two neurons

systolic pressure the highest point that your blood pressure reaches as the heart beats to pump blood through your body

T

tectonic plate a large section of the lithosphere which can move across the surface of the Earth

temperature a measure of the degree of hotness of a body on an arbitrary scale

thermal decomposition a reaction in which, when heated, one substance is chemically changed into at least two new substances

thermogram a picture showing differences in surface temperature of a body

thrombosis blood clot in a blood vessel causing it to be blocked

total internal reflection the reflection of light inside an optically denser material at its boundary with an optically less dense material (usually air)

toxin poisonous substance (pathogens make toxins which make us feel ill)

tracer a radioactive, or radiation-emitting, substance used in a nuclear medicine scan or other research where movement of a particular chemical is to be followed

transformer device by which alternating current of one voltage is changed to another voltage

transmitter a device which gives out some form of energy or signal, usually used to mean a radio transmitter which broadcasts radio signals

transverse in transverse waves, the vibration is at right angles to the direction in which the wave travels

trophic level the stages in a food chain

tsunami huge waves caused by earthquakes – can be very destructive

turbine device for generating electricity – the turbine moves through a magnetic field and electricity is generated

U

ultraviolet radiation electromagnetic waves given out by the Sun which damage human skin

units of alcohol measurement of alcoholic content of a drink

universe the whole of space

uranium a radioactive metal used in nuclear power stations

V

variation the differences between individuals (because we all have slight variations in our genes)

vasoconstriction in cold conditions the diameter of small blood vessels near the surface of the body decreases – this reduces the flow of blood

vasodilation in hot conditions the diameter of small blood vessels near the surface of the body increases – this increases the flow of blood

vector an animal that carries a pathogen without suffering from it

vegan a type of diet/a person who does not eat animals or animal products

vegetarian a type of diet/a person who does not eat meat or fish

vertebrate animal with a backbone

W

watt a unit of power, 1 watt equals 1 joule of energy being transferred per second

wavelength distance between two wave peaks

withdrawal symptoms reactions when a person stops taking a drug

Index

A

accommodation (eyes) 23
acetylcholine 25
acid rain 63, 81, 105, 108, 155
acids 164–7, 170–1
active immunity 20
adaptations 57, 72–5, 84
  natural selection 76–9
addiction, drug 28
addition polymerisation 111
addition reactions 113
additives, food 122–3
adipose tissue 15
aerials 206–7, 212
air
  composition of 106–7
  as convector 191
  as insulator 190, 191, 192
air pollution 80–1, 106–9, 238–41
alcohol 29, 30, 31
  in perfumes 125
alkalis 164–7, 170–1
alkanes 101, 112–13
alkenes 101, 111, 112–13
alleles 42, 43
alloys 152–3, 156, 157
alpha radiation 249, 251
alternating current 234–5, 237
aluminium 147, 155, 156, 157, 158–9
amalgam 152, 153
amino acids 15, 65
ammonia 65, 107, 160–3, 171
amplitude 194, 195
analogue signals 209, 210–11
angle of incidence 196, 201
angle of reflection 196
angle of refraction 201
animals
  adaptations 72–5
  in the carbon cycle 63
  classification 54–7
  competition 67
  in the food chain 58–61
  interdependence 66–9
  in the nitrogen cycle 64–5
  sustainability 84–7
  testing on 21, 126, 127
anti-ageing skin creams 272–3
antibiotics 20–1, 79
antibodies 20, 21, 26–7
antigens 20, 21
antioxidants 122
antiviral drugs 20
arteries 10, 12, 13
arthropods 54
artificial classification 55
aspirin 28
assessment, preparing for
  biology 26–7, 44–5, 70–1, 88–9
  chemistry 114–15, 132–3, 158–9, 176–7
  physics 202–3, 220–1, 246–7, 264–5
assessments, controlled 278–90
asteroids 256–7, 259
astronauts 255
atmosphere
  evolution 107
  gases in 106–7, 238–41
  microwaves 207
  ozone layer 80–1, 218, 219
  pollution 80–1, 106–9, 238–41
  radio waves 214, 215
atmospheric duct 207
atoms
  combustion 105
  conduction 191
  hydrocarbons 111–13
  nuclear radiation 249
  polymers 111–13, 119
  purifying copper 151
attraction, forces of 119, 125, 127
auxin 37, 38, 39
axons 24–5

B

bacteria
  adaptations 21, 75, 79
  causing infection 18, 20
  classification 57
  in the nitrogen cycle 65
  resistant forms 21, 79
baking powder 122, 123
balanced diet 14, 16
bases 164–7
bears 72, 74, 75
behavioural adaptations 75
beta radiation 249, 251
Big Bang 260–3
binding mediums 128, 129
binocular vision 22
biodegradable polymers 119
biomass 59, 244
bitumen 99
black dwarf 262
black holes 253, 262, 263
black robins 85
bleach 175
blind drug trials 21

blood cells 13, 20, 21, 26–7
blood clots 13
blood pressure 10–13
blood sugar levels 35
blood vessels 10, 12, 13, 33
BMI (body mass index) 17
body temperature 20, 32–3, 34
bowerbirds 67
brain fitness 276–7
brass 152, 153
bread 122
breathalysers 30, 31
brick 147
brine 174–5
bromine 113
bronze 153
buffalos 68
building materials 146–9
Bunsen burner 104
butane 112
butene 112

C

cacti 73
cakes, baking 122, 123
calcium carbonate 63, 148
calcium oxide 148
camouflage 57, 78
cancers
- diet and lifestyle 19, 30, 31
- radiation 248, 250
- radiotherapy 250
- screening 250
- sun 218, 219

carbohydrates 14, 15, 16
- cooking 120, 121

carbon cycle 62–3, 107
carbon dioxide
- in the air 63, 106–7, 238–41
- in cooking 122, 123
- from fuel combustion 102–5, 108
- as a greenhouse gas 109, 238–41

carbon footprint 81
carbon fuels *see* fossil fuels
carbon monoxide 13, 30, 104–5, 108–9
carbon sink 63
carbon-to-carbon double bonds 111–13
carbonates 63
carnivores 59
cars
- manufacturing 154–9
- pollution from 103, 106, 109
- recycling 156, 157

catalysts 160–1
catalytic converters 108–9
cats 57
cavity wall insulation 191, 192, 193
CDs (compact discs) 199
cells
- chromosomes 41, 42
- effect of radiation 248, 249, 250
- effect of smoking 30
- egg and sperm cells 42, 43
- neurones 24–5
- red blood cells 13
- white blood cells 20, 21, 26–7

cement 148–9
central nervous system 22–5
CFC gas 80, 219
checklists
- biology 46–7, 90–1
- chemistry 134–5, 178–9
- physics 222–3, 266–7

chemical plants 162–3
chlorine 174–5
cholesterol 12, 13, 16
chromosomes 40–3
cirrhosis 31
classification 54–7
cleaner species 68, 69
climate change 80–1, 238–41, 256
coal
- combustion 102–3, 195
- domestic fires 192, 193
- power stations 236–7, 244

cold environments, adapting to 75
colloids 129
colour blindness 23
combustion 102–5, 108–9
comets 253, 258–9
communication
- electrical signals 199
- light signals 198–9
- microwaves 206–7
- radio waves 195, 196, 199, 212–15
- with the Universe 254
- wireless 212–5

competition, plants and animals 66–7, 72
conclusions from data sets 287–8, 290
concrete 146, 147, 148–9
conduction 190, 191, 192, 193
conservation of animals 84–5
consumers (food chain) 58
continental drift 145
continental plates 143
control variables 283–4
convection 191, 193
convection currents (in Earth) 143
cooking 120–3, 205
Copernicus 260, 261
copper 150–1, 152, 153, 155

copper oxide 150
coral 63
core (Earth) 142, 143, 217
corneas 23
coronary artery disease 13
coronary heart disease 12–13
corrosion 154–5
cosmetic testing 126–7
cosmetics 124–7, 272–3
covalent bonds 119
cracking 100, 101
critical angle 201
crude oil 98–101
crust, Earth's 142, 143, 144
currents, electrical 234–5, 237, 244
cyclic fluctuation 69
cystic fibrosis 43

D

DAB (Digital Audio Broadcasting) 213
Darwin, Charles 39, 76–7
data (scientific)
  analysis 286–8
  collection 278–90
  displaying 287–8
data transmission 208–11
decay 62, 63, 64–5
decomposers 64–5
decomposition 63, 64–5
deep vein thrombosis (DVT) 13
defence mechanisms 74
deforestation 87, 107, 240
degassing 107
dehydration 33
denaturing 121
denitrifying bacteria 65
dependent variables 282–3
depressants 28, 29
designer polymers 116–19
diabetes 34–5
diet 12–13, 14–17
diffraction 197, 207, 214, 215
Digital Audio Broadcasting (DAB) 213
digital signals 199, 208–11, 213
disease
  causes 18–19
  fighting 20–1
  heart disease 12–13
  immunisation 20–1, 26–7
  incidence of 20
  smoking 30
  spreading 19, 20
  see also cancers
DNA 55, 77, 249
dogs 22, 56
dolphins 57
dominant characteristics 42, 43
double bonds 111–13
double glazing 191, 193
drink-driving 31
drugs 20–1, 28–31
dry environments, adapting to 73
dust, in the atmosphere 240
dynamos 234–5

E

EAR (estimated average requirement) 17
Earth
  asteroid collisions 256–7, 259
  earthquakes 142, 143, 195, 216–17
  global warming 238–41
  near-Earth objects 258, 259
  origins 256, 257
  in the Solar System 252, 253, 255
  structure 142–5, 217
earthquakes 142, 143, 195, 216–17
eclipse of the Sun 230
ecological niches 67
effectors 24–5
egg cells 42, 43
electrical appliances 242–3
electrical signals 199
electricity
  from biomass 244
  cost 242–3
  distribution 236, 244–5
  generating 234–7
  from the Sun 230–1
  from power stations 236–7, 244
  from the wind 232, 233
electricity meters 242–3
electrolysis 150–1, 174–5
electrolyte 151
electromagnetic spectrum 194, 196, 205
  see also waves
electrons 175, 191, 231, 249
elephants, evolution of 76
emulsifiers 122, 123
emulsion paint 129
emulsions 123, 129
endoscopes 201
energy
  to change temperature 188, 189
  in the food chain 61
  kinetic energy 127, 187, 191, 205
  nuclear 248–9
  power 242–5
  from power stations 236–7

renewable 230–3, 244
saving 192, 193
from the Sun 230–3
and wavelength 204, 205
from the wind 232, 233
energy efficiency 192–3, 237
energy flow 61, 187
energy loss 192–3
energy transfer 188–93
by cooking 205
in food chain 61
by heating 188–93
environmental issues *see* extraction; global warming; pollution; population; sustainability
equations, physics 299
erosion, acid rain 63, 108
esters 125
ethane 112, 113
ethene 39, 111, 112, 116
eutrophication 169
evaluation of investigations 288–9
evaporation 127, 129
evolution 57, 76–9, 274–5
evolutionary tree 55
exam-style questions
biology 48–51, 92–5
chemistry 136–9, 180–3
physics 224–7, 268–71
examination techniques 291–7
exercise 10–11
experiments, carrying out 279–86
extinction 84–7
extraction
copper 150, 151
crude oil 100
salt 173
eyes, structure 22–3
eyesight 22–3, 74

F

famine 17, 176
farm crops 168–9, 176–7
fats 12, 13, 14, 15
emulsifiers 122, 123
fertilisers 160, 168–71, 176–7
fever 20
fibreglass 192
fires, as heaters 192
fishing quotas 87
fitness 10–12
flu 18
focussing (eyes) 23
food
additives 122–3
balanced diet 14, 16
chemistry of 15
competition for 66, 67
cooking 120–3, 205
labelling 17
food chains 58, 61, 68
food pyramids 59, 61
food webs 60, 70–1
fossil fuels
combustion 102–5
crude oil 98–101
global warming 238
pollution from 80–1, 238
power stations 236–7, 244
fractional distillation 98–9
frequency 195–7, 210, 213, 235
fruit growing 36, 38, 39
fuels *see* fossil fuels; renewable energy
fungal infections 18, 20
fungi 64, 65

G

galaxies 252, 253, 261
Galileo 260, 261
gametes 41, 42
gamma rays 196, 249, 251
gases
in the atmosphere 106–7
combustion 102–5, 108–9
greenhouse gases 81, 102–5, 109, 238–41
heating 191
see also individual gases
generators 234–5, 237
genes 40–3, 55, 77
genetic disorders 40, 41, 43
genotype 43
geotropism 37
germination 38
ghosting 212, 213
giraffes 79
glass 147, 233
global warming 80–1, 238–41
glucose 35
GORE-TEX® 115, 117
granite 146, 147
gravity 253, 255, 257, 261, 263
greenhouse gases 81, 102–5, 109, 238–41
greenhouses 233

H

Haber process 160–1, 162, 163
habitats 56, 72–3, 74, 75, 84–5
haemoglobin 13
haemophilia 41
hallucinogens 28, 29
hamsters 56

healthcare
- use of polymers 118
- use of radiation 251
- use of smart alloys 153

heart attack 12, 13
heart disease 12–13, 30
heat 186–93
- body temperature 20, 32–3
- from the Sun 232–3

heat stroke 33
helium 263
herbivores 59
heterozygous 43
homeostasis 32–3
homozygous 43
hormones 34–5
- plants 36–9

hot environments, adapting to 73
houses, heating 190–3, 202–3, 231
human evolution 274–5
Human Genome Project 40
hunting, adaptations 57, 74
hybrids 57
hydrocarbons
- alkanes and alkenes 101, 111–13
- combustion 102–5
- crude oil 98–101
- polymers 113
- unsaturated and saturated 111, 113

hydrochloric acid 166
hydrogen
- in ammonia 160–2
- electrolysis of brine 174–5
- in stars 262–3

hydrogen ions 165
hydrophilic 123
hydrophobic 117, 123
hydroxide ions 165
hypothalamus 33
hypothermia 33
hypothesis 279, 290

I

igneous rock 144, 145, 147
immune system 20–1, 26–7
immunisation 20–1, 26–7
independent variables 281–2
indicator species 82–3
indicators 166
infections 18–21
infrared radiation
- data transmission 196, 208–9
- global warming 239
- heat transmission 190, 191, 204–5, 232, 233

inheritance 40–3, 77
inherited characteristics 40–2, 43
insulation 190, 191, 192
insulin 34–5
interdependence 66–9
interference 206–7, 210, 211, 213
intermolecular forces 99, 119
ionising radiation 248, 249, 250
ionosphere 214
ions 151, 165, 175, 191, 249
iron 154, 155, 156

J

jellyfish 238
Jupiter 257

K

kilowatt hours (kWh) 242–3
kilowatts (kW) 242–3
kinetic energy 127, 187, 191, 205
kingdoms 54
koala bears 72

L

Lamarck, Jean Baptiste de 79
lasers 198–9
latent heat 188, 189
lava 144, 145
LED (light emitting diode) 209
legumes 65, 69
lemmings 69
lenses 23
leuco dye 131
leukaemia 248
light
- lasers 198–9
- and mirrors 196, 197
- optical fibres 200–1
- and plant growth 36–7, 39
- rays 196, 197, 200–1
- red shift 261
- signals 199
- speed 194
- from the Sun 230–1

light emitting diode (LED) 209
light years 255
lightning 64, 65
limestone 63, 146, 147, 148
liquid crystals 131
lithosphere 143
liver, effects of alcohol 31
loft insulation 192, 193
long-sightedness 23
long-wave signals 213, 214

M

magma 143, 144, 145
magnets (in generators) 234–5, 237
malaria 19
mantle 142, 143, 144
marble 146, 147, 148
Mars 254, 255
mates, competition for 66, 67
mathematical skills 299
medicine
    fighting infection 20–1
    use of polymers 118
    use of radiation 251
    use of smart alloys 153
melanin 218
memory cells 21
metal carbonates 166, 167
metal hydroxides 165, 166–7
metal oxides 165, 166–7
metals
    alloys 152–3
    car manufacturing 154–7
    extraction 150
    rusting and corrosion 154–5
    see also individual metals
metamorphic rock 147
meteorites 252, 253, 256
meteors 253
methane 105, 112, 238–41
microorganisms 18–20
microporous membranes 117
microwaves
    communication 199, 206–7, 212, 214
    cooking 204–5
    properties 196, 204, 207
mining, salt 173
mirrors 196, 197
mobile phones 206–7
molecules
    alkanes and alkenes 112–13
    combustion 105
    cooking 121
    cracking 100, 101
    emulsifiers 123
    forces of attraction 119, 125, 127
    fractional distillation 99
    intermolecular forces 99, 119
    polymers 101, 110–13, 117, 119
    volatile liquids 125
molten rock 144
monocular vision 22
monohybrid cross 43
monomers 110–11
Moon 255, 256, 257
Morse code 199
mortar 149
mosquitoes 19
moths 78
motor neurones 24–5
MRSA 21
mules 57
multiplexing 211
mutations 41
mutualism 68, 69

N

National Grid 236, 244–5
natural classification 55
natural selection 76–9
nature v nurture 41
near-Earth objects 258, 259
nebula 262, 263
negative feedback systems 33, 35
nerve impulses 24–5, 29
nerves 23, 24–5, 29
nervous system 22–5, 29
neurones 24–5
neutralisation 164, 165, 166
neutron stars 263
Newton, Isaac 261
niches, ecological 67
nicotine 13, 29–30
nitrates 64, 65, 69
nitric acid 167
nitrifying bacteria 65
nitrogen
    in the air 106–7, 108–9
    in ammonia 160–2
    in fertilisers 168–71
nitrogen cycle 64–5
nitrogen-fixing bacteria 65, 69
nitrogen oxides 108–9
nitrogenous fertilisers 170–1
non-renewable fuels 98–101, 248
NPK fertilisers 168
nuclear fusion 263
nuclear power 248–51
nuclear radiation 248–51
nuclear waste 248–9
nylon 116–17

obesity 14, 15, 17
oceanic plates 143
oil, emulsions 123
oil, crude 98–101
oil paints 128, 129
omnivores 59
optic nerve 23
optical fibres 200–1, 210

orbits 253, 258, 259
ores 150
oscilloscopes 235
oxidation 151, 154, 155, 175
oxpeckers 68
oxygen
  in the air 106–7
  in the blood 10, 13
  combustion 102–5
  rusting and corrosion 154–5
ozone layer 80–1, 218, 219

P

painkillers 28, 29
paints 128–31
pandas 84–5
paraffin 100
parasites 19, 69
particles, force of attraction 119, 125, 127
passive immunity 21
passive solar heating 232, 233
pathogens 18–20, 21
payback time 193
penguins 75, 78
performance enhancers 28, 29
perfumes 124–5
peripheral nervous system 24–5
periscopes 197
petrol 99, 100, 101, 102–3
pH scale 164, 165, 166
phenotype 43
phosphorescent paints 130–1
phosphorous 168
photocells 230–1
photons 231
photosynthesis 61, 63, 106–7
phototropism 36, 37, 39
pigments 128–31
placebo drugs 21
planetary nebula 262
planets 252, 253, 254, 255, 257
plants
  in the carbon cycle 63
  competition for light 66
  controlling growth 36–9
  in the food chain 58–61
  in the nitrogen cycle 64–5
  response to light and gravity 36–7
  use of fertilisers 168–9
plaques, in arteries 12
plastics *see* polymers
plate tectonics 142, 143, 145
platypus 55
Pluto 253
plutonium 248
polar bears 75
polio 26–7
pollution
  in the air 80–1, 102–5, 106–9, 238–41
  from chemical plants 163
  from fertilisers 169
  from fuel combustion 103, 105, 109
  global warming 80–1, 238–41
  measuring 82–3
  oil spills 100
  ozone layer 80–1, 218, 219
  from paints 129
  in water 82–3, 100, 169
poly(ethene) 111, 116
polymerisation 110–11
polymers
  designer 116–19
  disposing of 118–19
  making 110–13
  stretchy and rigid 119
poly(propene) 116
poly(styrene) 116
population
  cyclic fluctuation 66, 69
  increasing 80–1
power, electrical 242–5
power stations 236–7, 244
predator-prey relationship 68, 69, 74, 75, 78
primary data 132–3, 279–86
propane 112
propanol 113
propene 112, 116
proteins 14, 15, 16, 17, 64–5
  cooking 121
protostars 263
protozoa 18, 19
PTFE 117
Ptolemy 260, 261
pupils 23, 24
PVC 116
pyramids of biomass 59
pyramids of numbers 59

Q

quarries 148
quasars 260

R

radiation *see* infrared radiation; nuclear radiation
radio, digital 211, 213
radio stations 213
radio waves 195, 196, 199, 212–15
radioactive material 248–51
radioactive paint 131
radioactive waste 248–9, 251
radiotherapy 250

reaction times 24–5
receivers 206, 212, 213
receptors, in the body 23, 45
recessive characteristics 42, 43
recycling
  cars 156, 157
  copper 150, 151
  in nature 62–5
red blood cells 13
red giants 262
red shift 261
reduction 151, 175
reflection 196, 212, 214
reflex actions 24–5
reflex arc 24
refraction 196, 201, 207, 213, 217
reinforced concrete 148, 149
relay neurones 24–5
remote controls 208, 209
renewable energy 230–3, 244
reproduction
  competition in 66, 67
  genetics 40–3
reptiles 75
researching data 44–5, 88–9, 176–7
resources, shortages of 80, 87
respiration 106–7
retina 22–3
reversible reactions 160–1
revision techniques 297–8
ripening fruit 36, 39
risk assessment 284–5
river pollution 82
rock salt 172, 173
rocks
  building materials 146–8
  formation 144, 145, 147
rooting powder 38
rusting 154–5

S

salmon 74
salt 13, 172–5
salts 166–7, 171
Sankey diagrams 192
satellites 206, 214, 215
saturated fats 12, 13
saturated hydrocarbons 111, 113
scientific claims 273, 277
secondary data 44–5, 88–9, 176–7, 278, 289
sedimentary rock 147
seismic waves 143, 195, 216–17
seismometer 216, 217
senses 22–3
sensory neurones 25
sex chromosomes 41, 42, 43
sharks 69
short-sightedness 23
signals
  analogue 209, 210–11
  digital 199, 208–11, 213
  interference 207
  microwaves 206–7
  optical fibres 200–1, 210
  radio waves 195, 196, 199, 212–5
  transmitting signals 197–9, 212, 213, 214, 215
  wireless 212–5
silicon 231
skin 44–5
skin cancer 218, 219
skin creams 272–3
sky wave propagation 214
smart alloys 153
smells 124–5
smog 106
smoke, in the atmosphere 240
smoke alarms 251
smoking 13, 29, 30, 31
snowy owls 69
sodium chloride 13, 166, 172–5
sodium hydrogencarbonate 123
sodium hydroxide 166–7, 174–5
solar cells 230
solar energy 190, 230–3
solar panels 190, 231, 232–3
Solar System 252–9, 260, 261, 262
solder 152, 153
solubility 126
solvents 126, 128, 129
soot 104, 105
space exploration 252, 254, 255
spacecraft 254, 255
species 56–7
specific heat capacity 189
specific latent heat 189
speed of light and sound 194
speed of waves 194–5
sperm cells 42, 43
spinal reflex 24–5
squirrels 67
stars 253, 261, 262–3
steel 153
  building material 146, 147, 149
  car manufacturing 154, 156, 157
step-down transformers 236, 244–5
step-up transformers 236, 244–5
steroids 29
stimulants 28, 29
stimuli
  human 24–5
  plants 36–7

stratosphere 218
strokes 11, 12, 13
subduction 143
subsidence 173
sulfur dioxide 81, 105, 108
sulfuric acid 171
Sun
  eclipse 230
  energy from 190, 230–3
  sunbathing 218–9
  Solar System 252–9, 260, 261, 262
sun index 218–19
sunbathing 218–19
sunburn 218
sunscreens 218–19
superbugs 21
supernova 263
surface area to volume ratio 73, 88–9
survival, adaptations for 72–5, 78
survival of the fittest 76–9
sustainability 84–7
sustainable development 87
sweating 33
synapses 25, 29

T

tectonic plates 142, 143, 145
telescopes 197
television signals 211
temperature 186–9
  body temperature 20, 32 3, 34
  changes of state 188
  global warming 238–41
  measuring 187
thermal decomposition 148
thermochromic paints 130–1
thermogram 187
thrombosis 13
tobacco smoke 30
total internal reflection 200–1, 214
transformers 236, 244–5
transmitters 206
transmitting signals 197–9, 206–7, 209–11, 212–15
trophic levels 58–9, 61
tsunamis 216, 256, 257

U

ultraviolet radiation 218–19
units of alcohol 31
universal indicator 166
Universe
  Big Bang 260–3
  early models 260, 261
  expanding 260, 261
  Solar System 252–5
  star formation 262–3
  threats to Earth 256–9
unsaturated hydrocarbons 111, 113
uranium 248

V

vaccination 20–2, 26–7
vacuum, as insulator 190, 191
variables in experimental work 281 4
variation 41–3
vasoconstriction 33
vasodilation 33
vectors of disease 19
viruses 18, 26–7
volatile liquids 125
volcanoes 142, 143, 144, 145
voltage 235, 236, 244–5
vultures 72

W

water
  by-product of combustion 104–5
  emulsifiers 123
  heated by solar panels 232
  pollution 82–83, 100, 169
  shortages 80
  waves 195, 197
water vapour 106, 238, 241
watts (W) 242–3
wavelength 194, 195, 196, 197
  infrared radiation 205, 233
  microwaves 205, 207
  radio waves 213–14
  relationship with energy 204
  ultraviolet radiation 219, 233
waves
  diffraction 197, 207, 214, 215
  properties 194–7
  reflection 196, 212, 214
  refraction 196, 201, 207, 213, 217
  see also data transmission; infrared radiation; microwaves; radio waves; seismic waves; ultrasound radiation
weedkillers 38
weight 14, 15, 17
whales 56, 57, 84, 86–7
white blood cells 20, 26–7
wind farms 232, 233
wireless signals 212–15

X

X and Y chromosomes 41, 43

 Internet research

The Internet is a great resource to use when you are working through your GCSE Science course.

Below are some tips to make the most of it.

1 Make sure that you get information at the right level for you by typing in the following words and phrases after your search: 'GCSE', 'KS4', 'KS3', 'for kids', 'easy', or 'simple'.

2 Use OR, AND, NOT and NEAR to narrow down your search.

> Use the word OR between two words to search for one or the other word.

> Use the word AND between two words to search for both words.

> Use the word NOT, for example, 'York NOT New York' to make sure that you do not get unwanted results (hits).

> Use the word NEAR, for example, 'London NEAR Art' to bring up pages where the two words appear very close to each other.

3 Be careful when you search for phrases. If you search for a whole phrase, for example, A Room with a View, you may get a lot of search results matching some or all of the words. If you put the phrase in quote marks, 'A Room with a View' it will only bring search results that have that whole phrase and so bring you more pages about the book or film and less about flats to rent!

4 For keyword searches, use several words and try to be specific. A search for 'asthma' will bring up thousands of results. But, a search for 'causes of asthma' or 'treatment of asthma' will bring more specific and fewer returns. Similarly, if you are looking for information on cats, for example, be as specific as you can by using the breed name.

5 Most search engines list their hits in a ranked order so that results that contain all your listed words (and so most closely match your request) will appear first. This means the first few pages of results will always be the most relevant.

6 Avoid using lots of smaller words such as A or THE unless it is particularly relevant to your search. Choose your words carefully and leave out any unnecessary extras.

7 If your request is country-specific, you can narrow your search by adding the country. For example, if you want to visit some historic houses and you live in the UK, search 'historic houses UK' otherwise it will search the world. With some search engines you can click on a 'web' or 'pages from the UK only' option.

8 Use a plus sign (+) before a word to force it into the search. That way only hits with that word will come up.